T0293783

Handbook of Nanoplasmonics

Handbook of Nanoplasmonics

Editor: Jonah Holmes

New York

Published by NY Research Press
118-35 Queens Blvd., Suite 400,
Forest Hills, NY 11375, USA
www.nyresearchpress.com

Handbook of Nanoplasmonics
Edited by Jonah Holmes

International Standard Book Number: 978-1-64725-385-1 (Hardback)

Cataloging-in-publication Data

Handbook of nanoplasmonics / edited by Jonah Holmes.
p. cm.
Includes bibliographical references and index.
ISBN 978-1-64725-385-1
1. Plasmons (Physics). 2. Nanostructured materials--Optical properties.
3. Nanoelectronics--Materials. 4. Nanoparticles. I. Holmes, Jonah.
QC176.8.P55 H36 2023
530.44--dc23

Contents

Preface

The study of optical phenomena in the nanoscale vicinity of metal surfaces is referred to as nanoplasmonics. Researches in nanoplasmonics focus nanoscale light below the diffraction limit by converting free photons into localized charge-density oscillations or surface plasmons on noble-metal nanostructures. These serve as nanoscale analogs of radio antennas and are typically designed by using antenna theory concepts. The phenomenon when plasmonic resonance leads to optical field enhancement at the metal interface is called as plasmonic field enhancement. It has several applications in areas such as surface-enhanced spectroscopy, sensing, non-linear optics, and light-activated cancer treatments. It is also used for enhancement of light absorption in photovoltaics and photocatalysis. Photonic-plasmonic microcavity for ultrasensitive protein detection is another practical application of plasmonic field enhancement. This book elucidates the concepts and innovative models around prospective developments with respect to nanoplasmonics. It is appropriate for students seeking detailed information in this area of study as well as for experts.

The information contained in this book is the result of intensive hard work done by researchers in this field. All due efforts have been made to make this book serve as a complete guiding source for students and researchers. The topics in this book have been comprehensively explained to help readers understand the growing trends in the field.

I would like to thank the entire group of writers who made sincere efforts in this book and my family who supported me in my efforts of working on this book. I take this opportunity to thank all those who have been a guiding force throughout my life.

Editor

1

Nanoscale Plasmon Sources: Physical Principles and Novel Structures

Hamed Ghodsi and Hassan Kaatuzian

Abstract

Started by M. Stockman with his proposed idea of a nanoscale quantum generator of plasmons that he called surface plasmon amplification by stimulated emission of radiation (SPASER) in 2002, during the last two decades various devices have been proposed, fabricated, and tested for SPASERs or plasmonic nanolasers which have almost the same meaning. Despite all these efforts, there are still serious barriers in front of these devices to be an ideal nanoscale coherent source of surface plasmons. The main challenges are the difficulty of fabrication, over-heating, low output powers, high loss rates, lack of integration capability with commercial fabrication processes, inefficient performance in room temperature, and so on. In this chapter, governing principles of nanolaser operation are discussed. Important parameters, limitations, and design challenges are explained, and some of the proposed or fabricated structures are presented and their merits and demerits are expressed. Eventually, several novel structures resulting from our works are introduced, and their performances are compared to the state-of-the-art structures.

Keywords: nanoscale plasmon source, SPASER, plasmonic nanolaser, nanoplasmonics, stimulated emission

1. Introduction

Theoretical postulation [1] and realization of laser [2] in the twentieth century changed both science and technology forever. Potential applications of lasers later have enormously expanded by the invention of the semiconductor diode laser in 1962 which brought them into the commercial market and in almost every device we know [3]. In the 1990s by the introduction of vertical-cavity surface-emitting laser (VCSEL) diodes, semiconductor lasers have pushed to their size limits [4]. The size of a dielectric cavity laser cannot be smaller than $\lambda/2$ in each dimension, and this limitation is known as the diffraction limit. With this in mind, modern VCSEL sizes are limited to a few microns [5]. On the other hand from the beginning of the millennia due to the rapid development of fabrication methods and tools [6], submicron manipulation of light using plasmonic devices has got lots of attention [7–10]. Plasmonic structures using metal/insulator interfaces broke the size limitation of the photonic devices and paved the way for integrating electronics, photonics, and optoelectronics on a single monolithic chip [11].

In order to bring benefits of the plasmonics into the field of laser research and fabrication, M. Stockman proposed the idea of a nanoscale quantum generator called surface plasmon amplification by stimulated emission of radiation (SPASER) in 2002 [12]. This device utilizes a plasmonic feedback mechanism in a gain medium for exciting stimulated emission in local plasmon modes of metallic nanoparticles. In the next two decades, various mechanisms and devices have been proposed and fabricated for the realization of a nanoscale coherent plasmon source or plasmonic nanolaser [13]. These devices can be categorized in nanoparticles [12], waveguide-based nanolasers [15], nanowires [16], nanoresonators [17], nanopatches [18], nanodisks [19], plasmonic crystals [20], and so on. Although these devices have shown significant potentials, there are still serious problems with the nanoscale coherent sources of surface plasmons. For instance, the difficulty of fabrication, over-heating, low output powers, high loss rates, lack of integration capability with commercial fabrication processes like CMOS, inefficient perfor-mance in room temperature, and so on can be noted [12, 13].

In this chapter and in Section 2, we start with basic principles of nanoplasmonics like the definition of surface plasmon polariton (SPP) modes in classical and quan-tum mechanical pictures, different sources of plasmon loss, and specific properties of plasmons focusing on special characteristics of plasmons in metallic nanoresonators. Then, the interaction of plasmons with carriers in a cavity will be briefly discussed according to plasmonic cavity quantum electrodynamics. In Section 3, three different methods for analyzing plasmonic nanolasers are discussed, and in Section 4 several previously introduced nanolaser structures are briefly reviewed. In Section 5, the proposed nanolaser structures by the authors are intro-duced, and this chapter will be concluded in Section 6.

2. Nanoplasmonics and quantum treatment of plasmons

The modern era of plasmonic began with the investigation of wood anomalies in the early twentieth century [21]. Later in 1957 Ritchie published a paper on plasma loss due to the electrons at the interface of a thin metal film [11]. In the next few years, theoretical works on collective oscillations of electrons at the surface of metals led to the introduction of plasmons as the quasiparticle corresponding to these oscillations [11]. However, applications of plasmons as a tool for nanoscale manipulation of light has gained significant attention with the paper by H. Atwater in 2007 named “Promise of plasmonics” [21]. In the past two decades, plasmonics has been developed both in theoretical and experimental aspects, and many differ-ent devices like switches [22], detectors [23], routers [24], amplifiers [25], and sources [17] have been introduced.

2.1 Basic principles

In order to find an appropriate model for surface plasma waves at the surface of a metal, we should deal with a charge density wave in an infinite electron gas which is often modeled by hydrodynamic equations [11]. An electromagnetic wave propagating in a material polarizes it and results in a mechanical excitation in electric charges and their movement. Therefore, oscillations in the electric field and mechanical oscillations are coupled. This coupled oscillation is called polariton. In case of metals, the electromagnetic field causes a longitudinal wave of charge density, and the coupled oscillations are known as plasmon polariton waves [11].

According to **Figure 1** at the interface of metal with a dielectric interaction of an electromagnetic field with the surface electrons, a specific type of plasmon polariton waves called surface plasmon polaritons or SPP modes results. Although

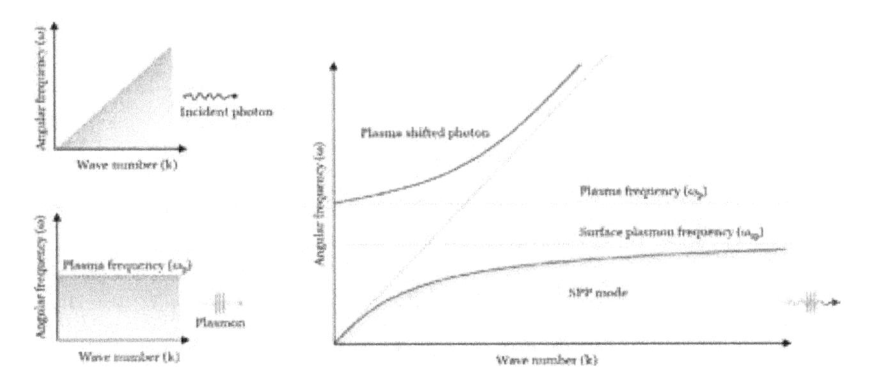

Figure 1.
Interaction of an electromagnetic field with surface plasma waves and excitation of surface plasmon polariton waves.

there are different types of plasmons like bulk plasmons and local surface plasmons (LSPs), SPP and LSP modes have significant roles in many plasmonic devices.

There are several models for plasmons, and we are going to briefly overview them here. The most well-known and simple model is Drude's model which describes the metal as a free electron gas system and models the system using the classical spring-mass model with the external force exerted from the incident field "**E**" equals to "-q**E**" acting on the system. We are not going to derive the equations here and only use the final result as shown in Eq. (1) which can be derived as mentioned in many related references like [22]:

$$\varepsilon_r(\omega) = 1 - \frac{\omega_p^2}{\omega^2 + i\gamma(\omega)\omega}, \qquad \omega_p^2 = \frac{ne^2}{\varepsilon_0 m}, \qquad \gamma = \frac{1}{\tau} \tag{1}$$

where "n" is number of electrons in the unit volume of the metal, "e" is the electron charge, "m" is the electron mass, "ε_0" is permittivity of vacuum, "ω_p" is the plasma frequency of the metal, "γ" is the total loss, and "τ" is the effective lifetime of the electrons associated with all of the decay processes.

According to Eq. (1), permittivity of a metal $\varepsilon_r(\omega)$ can be used in solving Helmholtz equations and finding the behavior of electromagnetic waves propagating at the metal/dielectric interface which are also known as SPP waves. However, Drude's model suffers from several shortcomings which leads to considerable errors especially near the plasma frequency of the metal. This is because in Drude's model, the effect of electrons in other energy bands (not just free electrons) is not included, and nonlocal effects are also not included [26]. To overcome these problems, Drude-Lorentz's model is introduced for the first problem which can be written in general multi-oscillator form as Eq. (2) and Landau damping correction according to Eq. (3) for the second problem. We are not going to further discuss these models either, and you can find details in [11, 26].

$$\epsilon_{Drude-Lorentz}(\omega) = 1 - \frac{f_0\omega_{p,0}^2}{\omega^2 + i\gamma_0\omega} + \sum_{j=1}^{j_{\max}} \frac{f_j\omega_{p,j}^2}{\omega_j^2 - \omega^2 - i\gamma_j\omega} \tag{2}$$

where the first sentence corresponds to Drude's model; "f_j" is the power of the j'th oscillator; and "$\omega_{p,j}$," "ω_j," and "γ_j" are plasma frequency, resonant frequency, and loss coefficient of the j'th oscillator, respectively.

$$\epsilon_r(\omega) = 1 - \frac{\omega_p^2}{\omega^2 + i\gamma(\omega)\omega - \beta k^2}, \quad \omega_p^2 = \frac{ne^2}{\epsilon_0 m}, \quad \gamma = \frac{1}{\tau} \tag{3}$$

where "β" is the Landau nonlocal parameter which becomes important for large values of wavenumber.

More precise treatment of surface plasmons can be done using the hydrodynamic model which includes solving Bloch equations, i.e., continuity, and Bernoulli and Poisson's equations simultaneously. According to Eqs. (4–6), one can describe collective oscillations of electrons in an arbitrary system using electron density (n) and hydrodynamic velocity ($v(r,t) = -\nabla\psi(r,t)$) [23].

$$\frac{d}{dt}n(r,t) = \nabla n(r,t) = \nabla.[n(r,t)\nabla\psi(r,t)] \tag{4}$$

$$\begin{aligned} \frac{d}{dt}\psi(r,t) &= \frac{1}{2}|\nabla\psi(r,t)|^2 + \frac{\delta G[n]}{\delta n} + \phi(r,t) \\ G[n] &= \frac{3}{10}(3\pi^2)^{\frac{2}{3}}[n(r,t)]^{\frac{5}{3}} \end{aligned} \tag{5}$$

$$\nabla^2\phi(r,t) = 4\pi n(r,t) \tag{6}$$

In the general form, Bloch equations are nonlinear and quite difficult to solve. However, using the perturbation theory, one can find linearized equations of Eq. (7) which helped Ritchie and his team to find plasmon dispersion equation in Eq. (8) for the first time [23].

$$\begin{aligned} &n(r,t) = n_0(r) + n_1(r,t) + \ldots \\ &\psi(r,t) = 0 + \psi_1(r,t) + \ldots \\ &\begin{cases} \frac{d}{dt}n_1(r,t) = \nabla.[n_0(r)\nabla\psi_1(r,t)] \\ \frac{d}{dt}\psi_1(r,t) = [\beta(r)]^2\frac{n_1(r,t)}{n_0(r)} + \phi(r,t) \\ \nabla^2\phi_1(r,t) = 4\pi n_1(r,t) \end{cases} \end{aligned} \tag{7}$$

$$\begin{aligned} &\omega^2 = \frac{1}{2}\left[\omega_p^2 + \beta^2k^2 + \beta k\sqrt{2\omega_p^2 + \beta^2k^2}\right] \\ &\frac{\beta k}{\omega_p} \ll 1 \rightarrow \omega = \frac{\omega_p}{\sqrt{2}} + \frac{\beta k}{2} \end{aligned} \tag{8}$$

The most accurate model for dealing with surface plasmons in atomic scales is solving Schrodinger's equation and calculating dynamical structure factor in Eq. (9) which is related to the oscillations of particle density in a many-particle system [23]:

$$S(r,r';\omega) = \sum_n \delta\hat{\rho}_{0n}(r_1)\delta\hat{\rho}_{n0}(r_2)\delta(\omega - E_n + E_0) \tag{9}$$

where the first two terms are elements of the operator "$\rho(r)$-$n_0(r)$" relating the ground state "ψ_0" with energy "E_0" and "δ" is the Dirac function, "$n_0(r)$" represents ground state density of particles, and "$\rho(r)$" is the particle density operator.

Using this model one can precisely calculate electron density profile in a many-electron system like a metal. However, solving the required equations is not easy, and most often approximations like random phase approximation or time-dependent density functional theory is used [23].

2.2 Specific properties of surface plasmons

Various applications of plasmonic technology in development of nanoscale devices and systems are all based on the same fundamental properties of plasmons. These specific properties include field confinement, enhancement of local density of optical states, and ultrawide bandwidth and fast response [11].

Confinement of electromagnetic fields in scales much smaller than the wavelength is the most crucial property of surface plasmon modes and can be defined in both parallel and orthogonal planes. Due to the high rate of loss, propagation length of the surface plasma waves in any direction is inversely related to the imaginary part of the wavenumber "$1/\mathrm{Im}(k_{sp})$." This length for good plasmonic metals like gold and silver is limited to a few microns and is considered as the upper limit of confinement [11]. The lower limit of confinement is exerted by Fourier transform properties with considering a monochromatic field with frequency "ω" and wavenumber "$k = \omega/c$" in the vacuum with far from any surface. It can be concluded that in the "x" direction "$\Delta x \Delta \alpha \geq 2\pi$" in which α is the x component of the wavenumber. Therefore, the lower limit of field confinement is "$2\pi/\alpha_{max} = \lambda$" which is also known as the diffraction limit. However, for surface plasmons, the wavenumber according to the dispersion relation (see **Figure 1**) can be much higher than "ω/c" which implies that surface plasmon modes can be confined in extremely tiny dimensions (much smaller than the wavelength) [11].

Enhancement of local density of optical states (LDOS) for surface plasmons can be investigated both near the metal surface and in a metallic nanoresonator. In a metallic nanoresonator, this effect which is also known as Purcell effect or enhancement of spontaneous emission is the vital property of plasmonic nanolasers. Purcell factor (F_p) is defined by the ratio of decay rate due to the spontaneous emission in a cavity over the decay rate in the free space. It can be calculated by Fermi's golden rule in a two-level atomic system and expressed by Eq. (10) [24].

$$\frac{\Gamma_{cav}}{n_1\Gamma_0} = \frac{3}{4\pi^2}\left(\frac{\lambda_{em}}{n_1}\right)^3 \frac{Q}{V_{eff}} \frac{\left|\hat{u}\cdot\vec{f}\left(\vec{r}\right)\right|^2}{1+4Q^2\left(\frac{\omega_{em}-\omega_c}{\omega_c}\right)^2} = F_p \frac{\left|\hat{u}\cdot\vec{f}\left(\vec{r}\right)\right|^2}{1+4Q^2\left(\frac{\omega_{em}-\omega_c}{\omega_c}\right)^2} \tag{10}$$

$$F_p = \frac{3}{4\pi^2}\left(\frac{\lambda_{em}}{n_1}\right)^3 \frac{Q}{V_{eff}}$$

In which "Γ_{cav}" is the decay rate in the cavity, "Γ_0" is the decay rate in the free space, "n_1" is the refractive index of the propagation medium, "λ_{em}" and "ω_{em}" are the emission wavelength of the medium, "ω_c" is the cavity resonance frequency, "Q" is the quality factor of the cavity, "V_{eff}" is the effective mode volume of the propagating mode in the cavity, and the dot product of the nominator corresponds to the mismatch between directions of transition dipole and the field.

In a dielectric microcavity despite the large quality factor, large mode volume results in infinitesimal Purcell factors, but nanoscale metallic resonators (the building block of a plasmonic nanolaser) provide a very small equivalent mode volume expressed by Eq. (11) which results in a large Purcell factor which is crucial for the nanolaser operation. Moreover, since the emission rate is proportional to the LDOS, the higher Purcell factor means the higher local density of optical states [24].

$$V_{eff} = \frac{\int \varepsilon\left(\vec{r}\right)\left|\vec{E}\left(\vec{r}\right)\right|^2 dr}{Max\left[\varepsilon\left(\vec{r}\right)\left|\vec{E}\left(\vec{r}\right)\right|^2\right]} \tag{11}$$

where "$\varepsilon(r)$" is the permittivity as a function of position inside the resonator volume in which the integral is calculated and "$E(r)$" is the electrical field related to the propagating mode. However, this condition for a plasmonic nanocavity may not be satisfied. Therefore, the electromagnetic energy density of a dispersive and dissipative medium should be used in Eq. (11). A dispersive lossless medium like the dielectric side of the interface (12) can provide a good estimation and a lossy medium like the metal side of the interface (13) should be used [25].

$$\langle u\rangle = \frac{1}{2}\varepsilon_0\left|E\left(\vec{r}\right)\right|^2\left(\varepsilon_r'(r,\omega) + \omega\frac{\partial\varepsilon_r'(r,\omega)}{\partial\omega}\right) \tag{12}$$

where "$\varepsilon'(r)$" is the real part of the permittivity.

$$\langle u\rangle = \frac{1}{4}\varepsilon_0\left|E\left(\vec{r}\right)\right|^2\left(\varepsilon_r'(r,\omega) + \frac{2\varepsilon_r''(r,\omega)\omega}{\gamma}\right) \tag{13}$$

where "$\varepsilon''(r)$" is the real part of the permittivity and "γ" is the loss rate introduced in Drude's model.

The third specific property of surface plasmons is their ultrawide bandwidths and fast response. In plasmonic nanoresonators due to considerable loss levels, quality factor is limited and in many cases between 10 and 100. Low-quality factor despite its negative effect on the Purcell factor provides an ultrawide bandwidth of several terahertz. This wide bandwidth resulting in fast time response has applications in generating femtosecond pulses in nanoscale dimensions and ultra-wideband nanoantennas [11].

2.3 Plasmon loss mechanisms

Inherent lossy nature of plasmon propagations requires more attention to the loss mechanisms both for using the loss as a beneficial application like biomarkers and biosensors [13] and minimizing its unwanted effects like plasmonic nanolasers.

Surface plasmons decay because of several elastic or inelastic loss channels, for instance, scattering because of other electrons, phonons, or crystal defects and so on. We will categorize them into three groups. The first is bulk decay rate or (γ_b) which can be expressed by Eq. (14) [26].

$$\gamma_b = \gamma_{e-e} + \gamma_{e-phonon} + \gamma_{e-defect} + \cdots \tag{14}$$

which the first term is due to electron-electron scattering as can be derived by Eq. (15) and the second term is due to electron-phonon scattering mechanism. Furthermore, the third term is the electron-defect decay rate. It should be noticed that for metals in the room temperature, electron-phonon decay rate is about 10^{14} Hz and increases with the temperature. But, the other two factors remain constant with the temperature and exist even in the absolute zero [26].

$$\gamma_{e-e} \approx 10^{15}.\left(\frac{\hbar\omega}{E_F}\right)^2 \text{ Hz} \tag{15}$$

As can be seen from Eq. (15), the electron-electron scattering has a direct relation with the frequency and in the visible frequencies is in the same order of magnitude as the first term [26].

In the plasmonic structures with dimensions about few nanometers, i.e., shorter than the mean free path of electrons, the second type of decay should be considered.

Surface scattering due to roughness of the surface (γ_{sc}) is inversely related to the mean free path due to surface roughness (a) and directly proportional to the Fermi velocity of electrons [26]. Fermi velocity is the maximum velocity an electron can achieve due to Fermi-Dirac distribution. For instant, for gold Fermi velocity is about 2.5×10^6 m/s.

The third type of losses in plasmonic devices is because of a process called Landau's damping [26]. Landau damping arises from the electrons oscillating with the velocity equal to the phase velocity of the surface plasmon mode. During acceleration, these electrons absorb energy from the plasmon mode. In other words, a quanta (plasmon) of the surface plasma wave is annihilated, and an electron-hole pair is generated in this procedure. Simply put, Landau damping is the plasmon-electron interaction mechanism. Landau damping rate (γ_L) which is significant for the large wavenumber values (see **Figure 2**) can be estimated by Eq. (16) [26].

$$\gamma_L \approx \omega_p \cdot \left(\frac{k_D}{k}\right)^3 . \exp\left(\frac{-k_D^2}{2k^2}\right), k_D = \frac{\omega_p}{v_F} \tag{16}$$

2.4 Surface plasmons in quantum mechanical picture

Plasmonic nanolasers are considered to be quantum nanogenerators of surface plasmons. Similar to lasers, these devices also operate based on both particle and wave properties of the electromagnetic waves. Therefore, finding a valid quantization approach for surface plasmon modes is necessary for explaining operation principles of the nanolasers.

The first attempts for finding a quantized description of plasmons are done by Bohm and Pines in the 1950s, and their works lead to the Pines model. In the Pines model, metal is considered to be a free electron gas material and electrons share long-range correlations in their positions in the form of collective oscillations in the whole system [27]. Pines model describes a quantized model of these collective

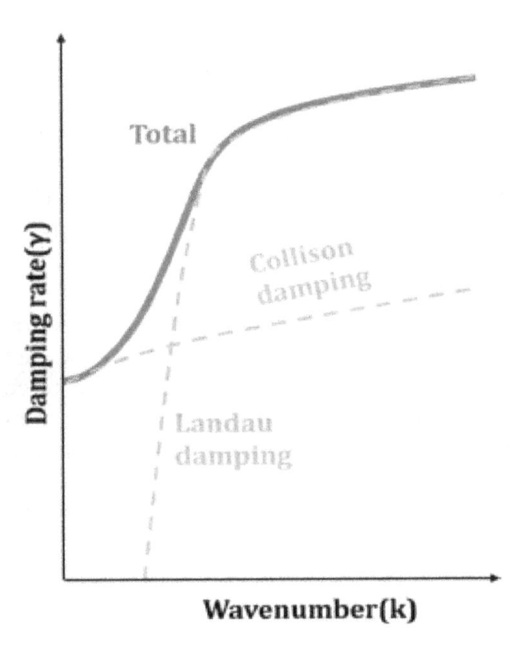

Figure 2.
Landau damping effect on the total damping rate [26].

oscillations which have both wave and particle properties, and the corresponding quanta (plasmon) is a boson [27].

Polariton is a joint state of light and matter introduced by Hopefield for providing a quantum model for the polarization field describing the response of matter to light [30]. Based on Hopfield's model, Ritchie and Elson proposed the first quantized description of surface plasma waves called Surface Plasmon Polariton or SPP. However, Hopefield's model did not consider the scattering and loss in the metal and effects of valance electrons and later Huttner and Barnett propose a model based on the Hopefield model including dispersion and loss, and recently a macroscopic quantization model based on Green's functions has also been published [27].

3. Physical models for analyzing plasmonic nanolasers

In order to analyze a plasmonic nanolaser, we need theoretical tools for describing the carrier-plasmon dynamics in the cavity. In this section, several models are discussed with different precision, but every method has its own limitations and should be used for a specific category of devices or a certain purpose.

3.1 Plasmon cavity quantum electrodynamics (PCQED)

Interaction of electron-hole pairs and plasmons in a nanocavity is the fundamental mechanism in any plasmonic nanolaser. As an example of this interaction, energy transfer diagram in a quantum well based nanolaser is illustrated in **Figure 3**.

This interaction should be treated similar to light-matter interaction in a laser cavity by the cavity quantum electrodynamics (CQED). However, in a plasmonic nanocavity due to Purcell enhancement of spontaneous emission and nanoscale dimensions and considerable loss and dispersion, there should be considerable differences that lead to a new cavity electrodynamic model for plasmonic cavities or PCQED [14].

The key difference between CQED and PCQED is in the method of controlling the interaction of electromagnetic fields with the medium. One of the most fundamental differences between them is the enhancement of spontaneous emission rate in a plasmonic cavity by the Purcell factor. The physical structure of the cavity affects the spectral characteristics of the plasmonic mode oscillations and results in a difference in the local density of optical states and the Purcell factor based on the designer's will. In other words, CQED controls interaction dynamics by their relationship with the quality factor of the resonator, while in PCQED dynamic of interactions is controlled by the Purcell factor. In dielectric microcavities, the quality factor is very high (even 10^{10}), while modal volume is limited to the refraction limit (few microns in each dimension), and Purcell enhancement does not occur. On the other hand, for plasmonic nanocavities because of intense mode confinement, equivalent mode volume is far smaller than the diffraction limit and results in considerable Purcell factor and density of state manipulation [14].

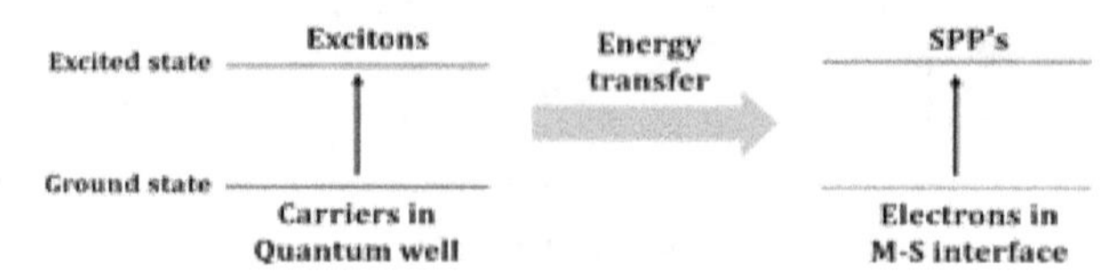

Figure 3.
Energy transfer diagram of a quantum-well based nanolaser.

Moreover, in PCQED loss and dispersion are critical factors and are necessary for correct modeling. Finding precise quantum mechanical models for these phenomena in plasmonic nanocavities still needs more research. However, we can use the photon/plasmon analogy and developed methods and tools of the photons like the density of state matrix and decay channels for estimating quantum mechanical behavior of plasmonic nanocavities [14].

3.2 Quantum mechanical atomic-scale model

Modeling phenomena like quantum fluctuations, spectral narrowing, coherency, threshold behavior, and precise dynamic analysis of plasmon nanolasers need an atomic-scale quantum mechanical model. However, in order to find closed-form equations, several simplifications are necessary, and thus this model just pro-vides a theoretical tool for investigating fundamental properties of plasmonic nanolasers.

To do so, consider an N-atom system in a low-quality factor nanocavity in which the decay rate of the cavity (κ) is the fastest decay rate of this system. This condi-tion is called the "bad cavity assumption" [28]. Therefore, resonator mode can be adiabatically eliminated, and the system state is totally determined by "N" active atoms. Considering two energy levels for each atom which are coupled to a cavity with resonance frequency (ω) and plasmon lifetime (1/2κ), one can describe the interactions between the atoms and field by Tavis-Cumming Hamiltonian of Eq. (17) [28].

$$H_{AF} = i\hbar g\left(a^{\dagger}J_{-} - aJ_{+}\right) \tag{17}$$

In which "g" is the coupling factor which is identical for all of the atoms, "a" and "$a^{\dagger}$" are annihilation and creation operators of plasmons, respectively, and "J_{α}" is the operator of collective atomic oscillations in the "α" direction and can be defined by Eq. (18) in which "σ_{jx}" and "σ_{jy}" are Pauli matrices [28].

$$J_{\alpha} = \sum_{j=1}^{N} \sigma_{j\alpha}, \qquad \alpha = \{x, y, z\}, \qquad \sigma_{j\pm} = \frac{\sigma_{jx} \pm \sigma_{jy}}{2} \tag{18}$$

Using atomic density operator "ρ" in a quantum system with state vector "ψ" and by considering the Hamiltonian of Eq. (17), Schrodinger's equation leads us to the dynamic equation of Eq. (19) in which "$\gamma_{\uparrow}$" is the pumping rate and "$\gamma_{\downarrow}$" is the spontaneous emission rate and "γ_{p}" is the dephasing rate of oscillating atoms [28]. The last term describes interaction of active atoms through the cavity mode [28].

$$\begin{aligned}\dot{\rho} = & -i\frac{1}{2}\omega[J_z, \rho] + \frac{\gamma_{\uparrow}}{2}\left. \sum_{j=1}^{N} 2\sigma_{j+}\rho\sigma_{j-} + \frac{1}{2}J_z\rho + \frac{1}{2}\rho J_z - N\rho \right) \\ & + \frac{\gamma_{\downarrow}}{2}\left(\sum_{j=1}^{N} 2\sigma_{j-}\rho\sigma_{j+} - \frac{1}{2}J_z\rho - \frac{1}{2}\rho J_z - N\rho\right) + \frac{\gamma_p}{2}\left(\sum_{j=1}^{N} 2\sigma_{jz}\rho\sigma_{jz} - N\rho\right) + \frac{g^2}{\kappa}(2J_{-}\rho)\end{aligned} \tag{19}$$

After several mathematic manipulations on Eq. (19) and by defining "$X_N(\xi, \xi^*, \eta) = tr\left(\rho e^{i\xi^* J_+} e^{i\eta J_z} e^{i\xi J_-}\right)$" and its Fourier transform "$\tilde{P}(v, v^*, m)$" as the atomic polarization operator, we can conclude Eq. (20) as a closed-form dynamic equation describing the system by collective atomic operators [28].

$$\begin{aligned}
&\frac{\partial \tilde{P}}{\partial t} = L\left(v, v^*, m, \frac{\partial}{\partial v}, \frac{\partial}{\partial v^*}, \frac{\partial}{\partial m}\right)\tilde{P} \\
&L = \frac{\gamma_\uparrow}{2}\left[\left(e^{-2\frac{\partial}{\partial m}} - 1\right)(N-m) + \frac{\partial^4}{\partial v^2 \partial v^{*2}} e^{2\frac{\partial}{\partial m}}(N+m) + 2N\frac{\partial^2}{\partial v \partial v^*}\right] + \\
&+\frac{\gamma_\uparrow}{2}\left(2e^{-2\frac{\partial}{\partial m}} - 1 + \frac{2\partial^2}{\partial v \partial v^*}\right)\left(\frac{\partial}{\partial v}v + \frac{\partial}{\partial v^*}v^*\right) + \\
&+\frac{\gamma_\downarrow}{2}\left[\left(e^{2\frac{\partial}{\partial m}} - 1\right)(N+m) + \frac{\partial}{\partial v}v + \frac{\partial}{\partial v^*}v^*\right] + \\
&+\gamma_p\left[\frac{\partial}{\partial v}v + \frac{\partial}{\partial v^*}v^* + \frac{\partial^2}{\partial v \partial v^*}e^{2\frac{\partial}{\partial m}}(N+m)\right] + i\omega\left[\frac{\partial}{\partial v}v + \frac{\partial}{\partial v^*}v^*\right] + \\
&+\frac{g^2}{\kappa}\left[2\left(1 - e^{-2\frac{\partial}{\partial m}}\right)vv^* - \left(\frac{\partial}{\partial v}vm + \frac{\partial}{\partial v^*}v^* m\right) + \frac{\partial^2}{\partial v^2}v^2 + \frac{\partial^2}{\partial v^{*2}}v^{*2}\right]
\end{aligned} \tag{20}$$

It should be noticed that Eq. (20) has not an analytical solution in this form and should be linearized or be solved numerically. In order to find a more familiar form of the dynamic rate equations, we should use linearization. By defining plasmon mode using dimensionless polarization "σ" and number of carriers by normalized population inversion "n," one can write Eq. (21) for a plasmonic nanolaser with N active atoms and the mentioned assumptions [28]:

$$\begin{aligned}
&\frac{d\left(\frac{\sigma}{n_s}\right)}{\Gamma dt} = -\left(1 - \wp\frac{n}{n_s}\right)\frac{\sigma}{n_s}, \sigma = tr\left[\rho J_- e^{i\omega t}\right]/N, n = tr[\rho J_z]/N \\
&\frac{d\left(\frac{n}{n_s}\right)}{\Gamma dt} = \frac{\frac{n}{s} - 1}{\Gamma T_1} - 4\wp\left|\frac{\sigma}{n_s}\right|^2 \\
&T_1 = \frac{1}{\gamma_\uparrow + \gamma_\downarrow}, n_s = \frac{\gamma_\uparrow - \gamma_\downarrow}{\gamma_\uparrow + \gamma_\downarrow}, \Gamma = \gamma_p + \frac{1}{2T_1} \\
&\wp = \wp_0 n_s, \quad \wp_0 = \frac{Ng^2}{\kappa\Gamma}
\end{aligned} \tag{21}$$

where for threshold parameter, "$\wp>1$" stimulated oscillations are dominant and for "$\wp<1$" nanolaser is working in the subthreshold region, and generated plasmons are not coherent. Using this method, quantum fluctuations of plasmon and carrier numbers in the cavity even for a few numbers of plasmons can be estimated. Eventually, you can find first- and second-order correlation functions of the generated plasmons over time for above threshold pumping and the resulting linewidth "D" in Eq. (22). Significant linewidth narrowing with respect to the natural broadening "Γ" for "$\wp>1$" implies proper laser operation, and time damping quantum fluctuations can be seen from the second-order correlation function [28].

$$\begin{aligned}
&g_>^{(1)}(\tau) = e^{-(i\omega+D)\tau}\left[1 - DT_1 + DT_1 e^{\frac{\tau}{2T_1}}\cos\left(\frac{\sqrt{2\Gamma T_1(\wp-1)}\tau}{T_1}\right)\right] \\
&g_>^{(2)} = 1 + 4DT_1 e^{\frac{-\tau}{2T_1}}\cos\left(\sqrt{2\Gamma T_1(\wp-1)}\frac{\tau}{T_1}\right) \\
&D = \gamma_p\frac{\Gamma T_1}{N}\frac{\wp_0(\wp_0+1)}{\wp-1}
\end{aligned} \tag{22}$$

3.3 Mean-field atomic-scale model (optical Bloch)

In this model, the classical wave equation is applied to the electric field, while plasmons are assumed quantized, and using Fermi's golden rule, a kinetic equation describing the behavior of plasmons is derived. This model proved to be consistent with the previous model above the threshold while having the advantage to be used for quantum dot gain mediums. Also, Einstein's spontaneous and stimulated emission coefficients can be calculated from this model. Therefore, it can prove the positive effect of increased spontaneous emission due to the Purcell effect on the stimulated emission of the nanolaser. However, similar to the model described in Section 3.2, it needs some simplifying assumptions and works in atomic scales [12].

In this model, the nanoscale system consists of a metal layer with permittivity "$\varepsilon(\omega)$" over a dielectric with permittivity "ε_h." The classic eigenvalue wave equation can be written for the plasmonic eigenmodes according to Eq. (23) [12].

$$\nabla.\left[\theta\left(\vec{r}\right) - s_n\right]\nabla\varphi\left(\vec{r}\right) = 0 \tag{23}$$

where "$\theta(r)$" is equal to "1" inside the dielectric and "0" inside the metal. Corresponding eigenvalues to the nth mode can be derived by Eq. (24) where "Ω_n" is the complex frequency of the nth eigenmode in which the real part is equal to the resonant frequency of nth mode "ω_n" and the imaginary part corresponds to the plasmon decay rate "γ_n" [12].

$$s(\Omega_n) = s_n, \qquad s(\omega) \equiv \left[1 - \frac{\varepsilon(\omega)}{\varepsilon_h}\right]^{-1} \tag{24}$$

Assuming "$\gamma_n << \omega_n$" we can write Eq. (25).

$$\gamma_n = \frac{\mathrm{Im}[s(\omega_n)]}{s'_n}, \qquad s'_n \equiv \left.\frac{d\,\mathrm{Re}\,[s(\omega_n)]}{d\omega}\right|_{\omega=\omega_n} \tag{25}$$

For the times shorter than the plasmon lifetime "$\tau_n = 1/\gamma_n$," corresponding Hamiltonian to the electric field of the quantized surface plasmons can be expressed by Eq. (26) in which "T" is the integration time and should satisfy "$\tau_n >> T >> 1/\omega_n$."

$$H = \frac{1}{4\pi T}\int_{-\infty}^{-\infty} \frac{d\left[\omega\varepsilon\left(\vec{r},\omega\right)\right]}{d\omega}\vec{E}\left(\vec{r},\omega\right)\vec{E}\left(\vec{r},-\omega\right)\frac{d\omega}{2\pi}d^3r \tag{26}$$

Using the extension of the electric field based on the quantized eigenmodes of the system, one can write Eq. (27), and Hamiltonian of Eq. (26) can be written as Eq. (28) which has the standard form of a quantum mechanical harmonic oscillator.

$$\vec{E}\left(\vec{r},t\right) \equiv -\nabla\phi\left(\vec{r},t\right), \qquad \phi\left(\vec{r},t\right) = \sum_n \sqrt{\frac{2\pi\hbar s_n}{\varepsilon_h s'_n}}\varphi_n\left(\vec{r}\right)e^{-\gamma_n t}\left[a_n e^{-i\omega_n t} + a_n^{\dagger}e^{i\omega_n t}\right] \tag{27}$$

$$H = \sum_n \hbar\omega_n\left(a_n^{\dagger}a_n + \frac{1}{2}\right) \tag{28}$$

where "a" and "$a\dagger$" are annihilation and creation operators of plasmons, respectively. Consider dipolar emitters (quantum dots) with carrier population densities of the ground state and excited state equal to "$\rho_1(r_a)$" and "$\rho_2(r_a)$," respectively, where "r_a" corresponds to the location of the a'th emitter with transition dipole moment equal to "d^a." Transition matrix element [12] for this transition "d_{10}" can be estimated by the Kane theory according to Eq. (29) [12] in which "e" is the electron charge, "f" is the power of the transition oscillator, "K" is the Kane constant, "m" is the electron mass, and "ω_n" is the plasmon frequency of the nth mode. "d_{10}" is proportional to the rate of spontaneous emission and Purcell effect.

$$d_{10} = e\sqrt{\frac{fK}{2m_0\omega_n^2}} \tag{29}$$

Interaction between the gain medium and the plasmon modes can be described by Hamiltonian of Eq. (30) which is exerted to the system. Accordingly, using the Fermi's golden rule, the kinetic equation of the system can be written for the number of plasmons in the nth mode by Eq. (31) in which "A_n" and "B_n" are the stimulated and spontaneous emission coefficients, respectively [12].

$$H' = \sum_a \vec{d}^{(a)} \nabla\phi\left(\vec{r}_a\right) \tag{30}$$

$$\dot{N}_n = A_n N_n - \gamma_n N_n + B_n \tag{31}$$

According to the mentioned model, Einstein emission coefficients can be derived by Eqs. (32) and (33).

$$A_n = \frac{4\pi}{3\hbar} \frac{s_n' s_n |d_{10}|^2 p_n q_n}{\varepsilon_h [\mathrm{Im}\, s(\omega_n)]^2} \gamma_n \tag{32}$$

$$B_n = \frac{4\pi}{3\hbar} \frac{s_n' s_n |d_{10}|^2 r_n q_n}{\varepsilon_h [\mathrm{Im} s(\omega_n)]^2} \gamma_n \tag{33}$$

where "p_n" and "r_n" are spatial overlap factors of the nth mode with the gain medium and "q_n" is the spectral overlap factor [12]. These parameters can be derived by Eqs. (34–35), respectively.

$$p_n = \int \left[\nabla\varphi_n\left(\vec{r}\right)\right]^2 \times \left[\rho_1\left(\vec{r}\right) - \rho_0\left(\vec{r}\right)\right] d^3r \tag{34}$$

$$r_n = \int \left[\nabla\varphi_n\left(\vec{r}\right)\right]^2 \times \rho_1\left(\vec{r}\right) d^3r \tag{35}$$

$$q_n = \int F(\omega) \left[1 + \frac{(\omega - \omega_n)^2}{\gamma_n^2}\right]^{-1} d\omega \tag{36}$$

where "$F(\omega)$" is the spectral characteristic of the transition dipole moments.

3.4 Semiclassical rate equations

The aforementioned methods will give us much useful information about the operating principles of the plasmonic nanolasers. However, a consistent model with macroscopic measurable parameters is also needed for larger-scale systems. To do so, a modified version of an initially proposed rate equation for the microcavity lasers in the 1990s can be used [29, 30]. This model as shown in Eq. (37) according to many recent pieces of research [15, 16, 29] can adequately explain the plasmon/exciton carrier dynamics of a plasmon nanolaser. Furthermore, the macroscopic parameters like output power and pumping current can be easily derived.

$$\begin{aligned} \frac{dn}{dt} &= R_p - An - \beta\Gamma As(n - n_0) - \frac{nv_s S_a}{V_a} \\ \frac{ds}{dt} &= \beta An + \beta\Gamma As(n - n_0) - \gamma s \end{aligned} \tag{37}$$

The first equation of Eq. (6) is expressing the rate of carrier changes, and the second one is describing the temporal behavior of the plasmon generation. Plasmon generation is determined by the spontaneous plasmons coupled in the lasing mode (the first term), stimulated emission (the second term), and plasmon loss rate (the last term) [29].

In these equations "n" is the excited state population of the carriers, "s" is the number of plasmons in the lasing mode, and "R_p" is the carrier generation rate. The coupling factor (β) is defined by the ratio of the spontaneous emission rate into the lasing mode and the spontaneous emission rate into all other modes. A possible calculation method for this parameter can be seen in Eq. (38) [15].

$$\beta = \frac{F_{cav}^{(1)}}{\sum_k F_{cav}^{(k)}} \tag{38}$$

where "$F_{cav}(k)$" is the Purcell factor of k'th mode. k = 1 corresponds to the lasing mode, and the summation is over all of the possible propagation modes in the cavity.

Mode overlap with the gain medium which is also known as Γ-factor is defined by the overlap between the spatial distributions of the gain medium and the lasing mode. In a homogenous medium, spontaneous emission rate "A" is equal to "$1/\tau_{spO}$" and "τ_{spO}" is the spontaneous emission lifetime of the material. However, in a nanocavity, Purcell effect [24] modifies the spontaneous emission rate via "$A = F_p A_O$," where "F_p" is the Purcell factor and "A_O" is the natural spontaneous emission rate in a homogenous medium. "n_O" is the excited state population of carriers in transparency, "v_s" is surface recombination velocity at the sidewalls of the resonator, and "S_a" and "V_a" are the area of sidewalls of the resonator and volume of the gain medium, respectively. Finally, "γ" is the total loss rate of plasmons in the cavity. In order to calculate it, the loss coefficient per unit length should be multiplied by the modal speed. Loss coefficient is calculated by "$\gamma_m + \gamma_g$" where "γ_m" and "γ_i" are resonator mirror loss and intrinsic cavity loss per unit length, respectively.

4. Different structures of metallic nanoscale plasmon sources

A plasmonic nanolaser needs a metallic nanocavity, gain medium, and a feedback mechanism. In the past two decades, several structures and materials have been introduced for this purpose. Some of these devices are presented in **Figure 4** [13]. These structures can be subwavelength in one dimension like plane nanolaser (see **Figure 5(a)**) [13], in two dimensions like nanowire-based plasmonic nanolaser (see **Figure 5(b)**) [16], and in three dimensions like nanocavity plasmon laser of [17] (see **Figure 5(c)**).

The gain medium of plasmonic nanolasers can be any material capable of radiative electron decay like any traditional laser. In the proposed structures, a variety of

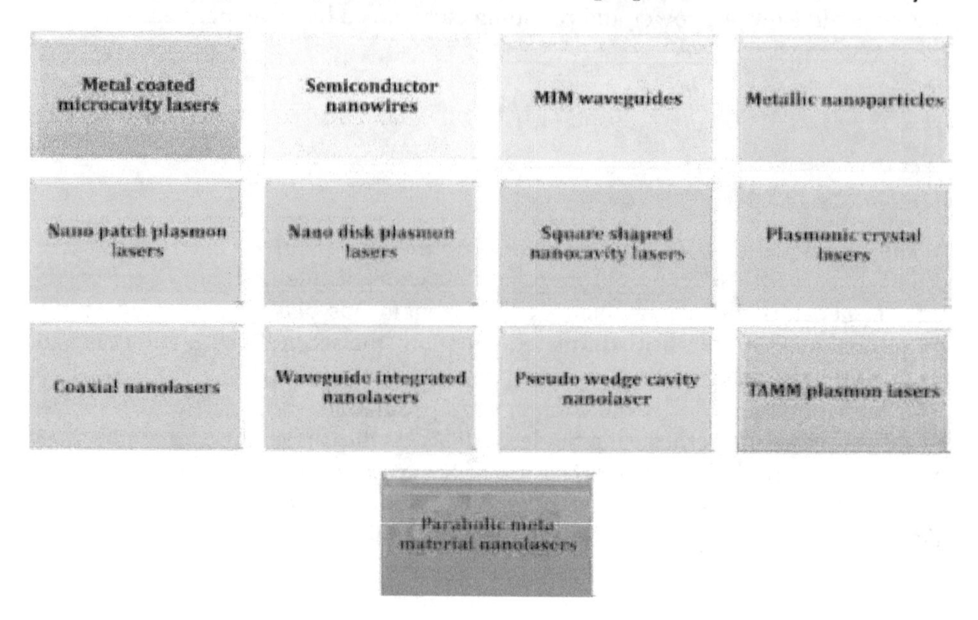

Figure 4.
Different structures of the plasmonic nanolasers [13].

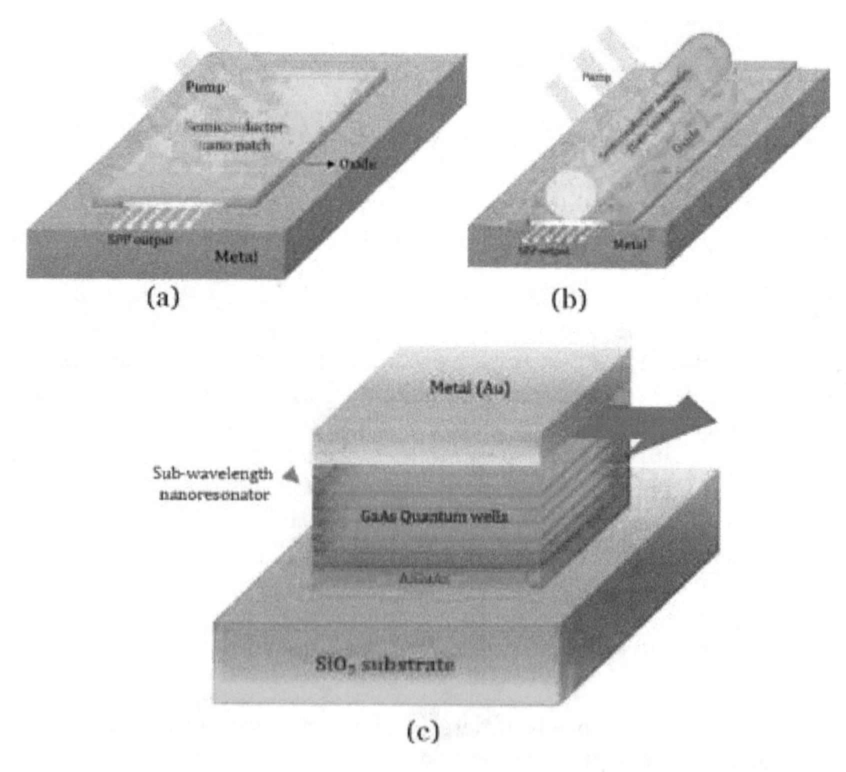

Figure 5.
Different structures of the plasmon nanolasers: (a) a plane plasmon nanolaser [13] (subwavelength in one dimension), (b) typical nanowire-based plasmon nanolaser [16] (subwavelength in two dimension), and (c) quantum well-based nanocavity plasmon laser of [15] (subwavelength in three dimension).

materials like die molecules, bulk semiconductors, semiconductor quantum wells, and quantum dots can be witnessed [13]. Many of the semiconductors were typical optoelectronic materials (III-V and II-IV alloys) like GaAs, AlGaAs, ZnS, InGaAs, InP, and so on [13].

5. Proposed nanoresonator structures

According to our most recent publications [17, 31, 32], we have proposed four nanolaser structures that are discussed in this section. All of these structures are electrically pumped in the room temperature, have subwavelength footprints, and have considerable performance characteristics. The first structure is a GaAs quantum dot-based nanocavity integrated into a plasmonic waveguide [31]. The second is a metal strip nanocavity structure which is based in tensile-strained germanium quantum wells [32]. The next one has a notched nanocavity and germanium quantum wells as the gain medium [17] and the last one is a corrugated metal–semiconductor–metal nanocavity structure utilizing two sets of germanium quantum dot arrays as the gain medium [32].

The first structure is a GaAs/AlGaAs QD nanocavity plasmon laser, which can be integrated into plasmonic waveguides for the realization of integrated plasmonic chips. This proposed nanolaser as sketched in **Figure 6** has several advantages over the previously introduced ones.

For instance, it has a high coupling efficiency to the waveguide plasmonic modes because of its thin structure and the monolithic metal layer. In addition, the pro-posed nanolaser structure benefits from a large beta factor that means lower threshold and also a high Purcell factor, which leads to higher gain and better laser performance. The MSM structure of this device also can provide an efficient heat transfer performance. Therefore, it predicted to efficiently operate without overheating and needs less chip area for fabrication of heatsink. Nevertheless, the threshold pumping current of the proposed device is considerably high, and this structure cannot provide output power in the mW range in the optimal pumping region. Design characteristics related to the first structure can be seen in **Table 1**.

The second device is a germanium/silicon-germanium ($Ge/Si_{0.11}Ge_{0.89}$) multiple quantum well plasmonic nanolaser as shown in **Figure 7**. This device utilizes a thin gold metal strip layer, sandwiched between Ge quantum wells in order to maximize both field confinement and exciton-plasmon interaction possibility, which means higher Purcell factor and better gain medium with mode overlap factor. Using two aluminum electrical contacts, one on top of the resonator and one beside it, an electrical pump current can be applied. Moreover, it can be coupled into

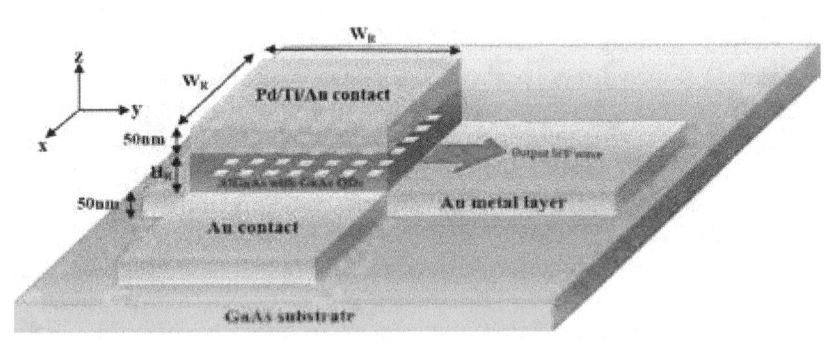

Figure 6.
3D schematic of the GaAs quantum dot-based nanoresonator.

	Symbol	Value
Cavity height	H_R	50 nm
Cavity size	W_R	260 nm
QD size	D_{QD}	5 nm
QD separation	D_{QD2QD}	10 nm
Distance of QDs from gold plate	H_{QD}	25 nm
Top/bottom metal thickness	H_{metal}	50 nm
Doping level (p-type)	N_A	10^{17} cm^{-3}
Doping level (n-type)	N_D	10^{19} cm^{-3}
Number of QDs	N_{QD}	256
QD volume	V_{QD}	1.25×10^{19} cm^{-3}

Table 1.
Design characteristics of the first structure.

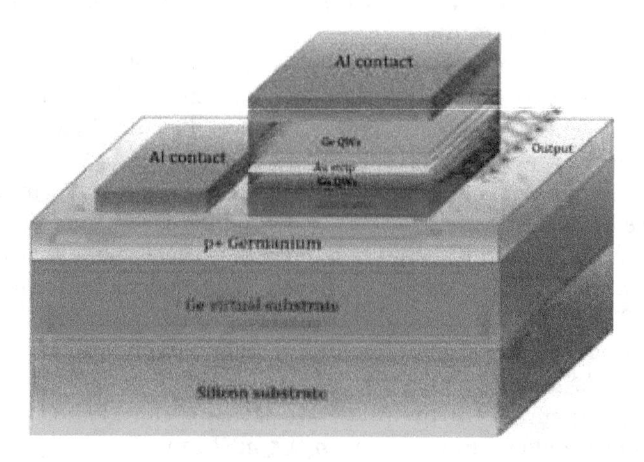

Figure 7.
3D schematic of the metal strip nanocavity structure with germanium quantum wells.

silicon-based waveguides similar to [15] or used in the far-field configuration in which plasmon modes will be converted into photons through the cavity interface. Our device benefits from a metal–semiconductor–metal–semiconductor (MSMS) structure, which can perform well in the 1550 nm regime by means of incorporating highly doped strained Ge quantum wells as the direct bandgap gain medium [33, 34]. Design characteristics of this structure can be found in **Table 2**.

It should be noticed that for transforming germanium into a direct bandgap material, strong tensile strain levels could be applied in the fabrication process. This will reduce the Γ-valley direct bandgap of the material below the L-valley indirect bandgap (0.664 eV) [33, 34]. This will result in an output wavelength about several micrometers in which efficient plasmonic nanocavities cannot be designed. Alternatively, much lower strain level can be utilized, which in combination with extreme level of donor doping for occupying the remaining indirect L-valley states below the Γ-valley results in a direct energy gap diagram [33, 34].

The third structure as shown in **Figure 8** has a cubic nanoresonator with two parabolic notches at both sides. This device provides a high-quality factor and Purcell factor because of the notches which can effectively decrease the output loss (amount of energy escaping the resonator) and improve energy confinement in the cavity. The gain medium of this structure consists of four Germanium quantum

Description	Symbol	Value	Unit
Resonator size	W_R	270	nm
Resonator height	H_R	130	nm
Bottom metal thickness	X_{Strip}	10	nm
Metal thickness	$X_{Contact}$	40	nm
Bottom buffer thickness	X_{Bottom}	15	nm
Top buffer thickness	X_{Top}	15	nm
Number of QWs	N_{QW}	4	—
QW thickness	X_{QW}	7	nm
Barrier wall thickness	$X_{Barrier}$	10	nm
Thickness of p-doped Ge buffer	X_{Buffer}	16	nm
Ge alloy percent	x	89	%
Doping concentration of the QWs and barriers	N_D	7.6×10^{19}	cm^{-3}
Doping concentration of the Ge buffer	N_A	1×10^{19}	cm^{-3}

Table 2.
Design characteristics of the first structure.

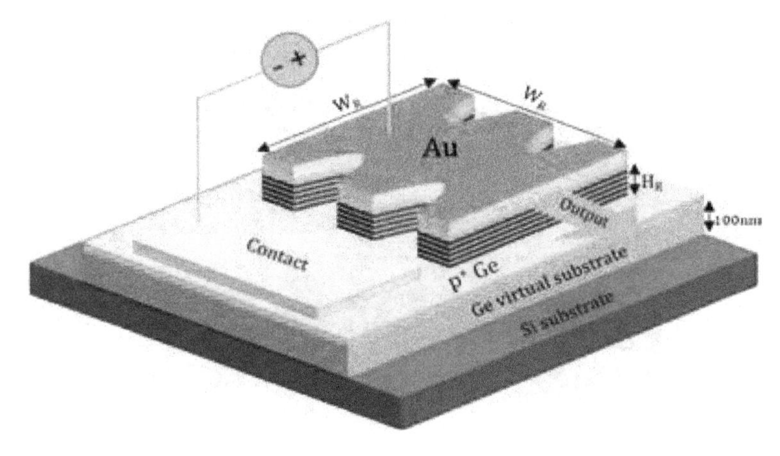

Figure 8.
3D schematic of the notched cavity nanolaser structure.

dots (tensile-strained direct energy bandgap). This structure can be easily inte-grated into different plasmonic and photonic waveguides with considerable cou-pling factors. Therefore, it is an appropriate choice for on-chip applications. Also, it can be simply used in the far-field lasing mode. An efficient integration approach can be found in [15]. This device also provides the output free-space wavelength of 1.55 μm, which means it is compatible with commercial photonic devices and systems. The design characteristics of the third structure can be witnessed in **Table 3**.

The last proposed structure is a corrugated metal-semiconductor-metal nanocavity device that can be seen in **Figure 9**. The specific design of the cavity leads to a significant plasmonic mode with gain medium interaction and also an increase in mode confinement and quality factor. Furthermore, this cavity design using the two-side contacts can provide efficient electrically pumping with a reasonable threshold current. The gain medium of our nanolaser consists of several germanium quantum dots provided on both sides for maximizing the output power.

Description	Symbol	Value	Unit
Resonator size	W_R	350	nm
Resonator height	H_R	98	nm
Top metal thickness	X_{Au}	40	nm
Bottom buffer thickness	X_{Bottom}	10	nm
Bottom buffer thickness	X_{Top}	10	nm
Number of QWs	N_{QW}	4	—
QW thickness	X_{QW}	7	nm
Barrier wall thickness	$X_{Barrier}$	10	nm
Thickness of p-doped Ge buffer	X_{Buffer}	20	nm
Ge alloy percent	x	85	%
Doping concentration of the QWs	N_D	4×10^{19}	cm^{-3}
Doping concentration of the Ge buffer	N_A	1×10^{19}	cm^{-3}

Table 3.
Design parameters of the third structure.

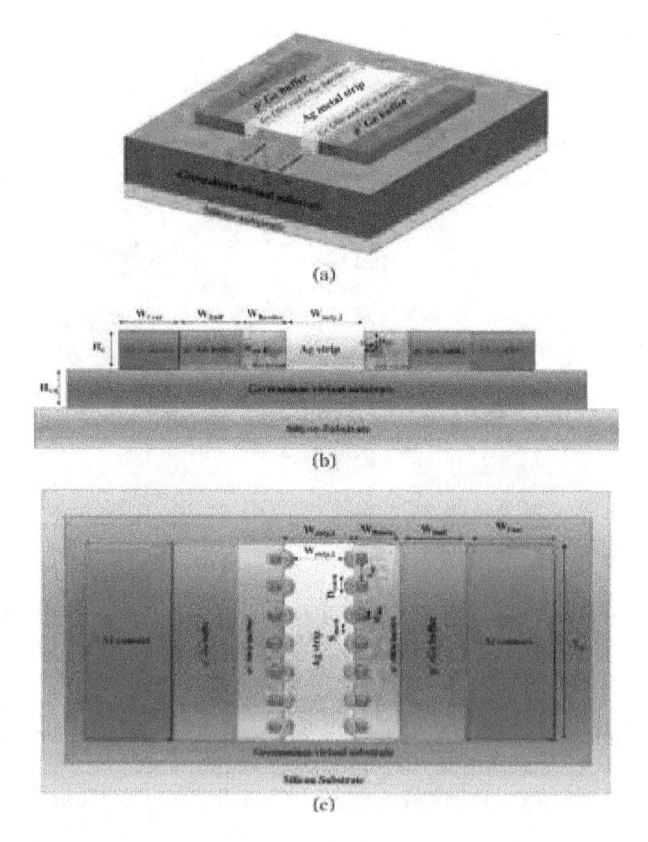

Figure 9.
Schematic illustration of the corrugated lateral MSMSM structure: (a) 3D schematic, (b) transverse cross section, and (c) top view.

The proposed structure for the nanolaser according to **Figure 9** consists of a corrugated metal nanostrip with two arrays of n + doped tensile-strained germanium quantum dots (QDs) at both sides. In addition, two high-doped p + germanium layers are used for better field confinement and providing higher carrier generation rates in the QDs. This structure also has two aluminum contacts for electrical pumping into the gain medium. In this structure, the electrical pump current flows perpendicular to the plasmonic mode propagation direction into the germanium quantum dots (QDs) in order to produce excitons. This electron – hole pairs

Description	Symbol	Value	Unit
Width of the thick area of the strip	$W_{Strip,1}$	60	nm
Width of the thin area of the strip	$W_{Strip,2}$	50	nm
Height of Ge virtual substrate	H_{VS}	200	nm
Resonator height	H_C	40	nm
Al contact width	W_{Cont}	40	nm
Ge buffer width	W_{Buff}	40	nm
Si-Ge barrier width	$W_{Barrier}$	30	nm
Quantum dot size	W_{QD}	5	nm
Quantum dot distance from the strip	D_{QD}	10	nm
Quantum dot distance from the top	H_{QD}	20	nm
Number of QDs	N_{QD}	34	—
Spacing between QDs	S_{QD}	20	nm
Spacing between two notches	S_{notch}	5	nm
Diameter of notches	D_{notch}	10	nm
Cavity length	L_C	270	nm
Ge alloy percent	x	85	%
Doping level of the QDs and barriers	N_D	4×10^{19}	cm^{-3}
Doping level of the Ge buffer	N_A	1×10^{19}	cm^{-3}

Table 4.
Design characteristics of the lateral MSMSM structure.

Parameter	1st structure	2nd structure	3rd structure	4th structure	Liu et al. [7]
Output wavelength (nm)	850	1550	1550	1550	850
Area (μm^2)	0.07	0.073	0.1125	0.076	0.06
Threshold current (mA)	4.7	29	21	1.9	1.87
Output power in mW at threshold	0.198	4.16	15.6	10.59	0.08
Output power in μW at 10 μA	0.44	2.8	3	50	0.25
Pump current in mA for 1 mW output power	20	~7	~3	0.2	N.A
Modulation bandwidth in GHz at threshold	3.37	5.7	2.98	28.5	—
Spectral bandwidth in THz at threshold	0.541	1.46	1.98	21.68	>0.08
Purcell factor (lasing mode)	66	291	700	2965	15
Quality factor (Q)	30	26	58	138	32

Table 5.
Performance characteristics of the proposed nanolasers.

recombine through the radiative recombination process and transfer their energy to the surface plasmon polaritons (SPPs) propagating at interfaces of the Ag metal strip and its side semiconductor layers. The design characteristics of this device are provided in **Table 4**.

The introduced nanolaser devices can be adequately analyzed using the aforementioned theoretical phenomena. By means of Finite difference Time Domain (FDTD) method for mode analysis and numerically solving nonlinear rate equations of (37) output performance of the proposed structures can be derived. Also, needed parameters are either extracted from experimental papers or found using numerical methods according to [17, 31, 32]. Resulting from our analysis, the performance of nanolaser structures of 5.1 can be concluded in **Table 5** which demonstrates a considerable performance with respect to a reference similar device [15].

6. Conclusion

In this chapter, we have briefly covered fundamental theories and models related to plasmonic nanolasers. To conclude, nanolasers are one of the most critical building blocks of the future integrated circuits containing both nanophotonics and electronic parts. After two decades of development, recent devices are more promising for the realization of a commercially available nanoscale plasmon source or plasmon nanolaser. Such a device will open a portal to the vast number of potential applications.

Author details

Hamed Ghodsi and Hassan Kaatuzian*
Photonics Research Laboratory (PRL), Electrical Engineering Department, Amirkabir University of Technology (Tehran Polytechnique), Tehran, Iran

*Address all correspondence to: hsnkato@aut.ac.ir

References

[1] Einstein A. Physikalische Zeitschrift. 1917;**18**:121-128

[2] Maiman T. Stimulated optical radiation in ruby. Nature. 1960; **187**(4736):493-494

[3] Hall R, Fenner G, Kingsley J, Soltys T, Carlson R. Coherent light emission from GaAs junctions. Physical Review Letters. 1962;**9**(9): 366-368. DOI: 10.1103/physrevlett. 9.366

[4] Koyama F et al. Room temperature cw operation of GaAs vertical cavity surface emitting laser. Transactions of IEICE. 1988;**E71**(11):1089-1090

[5] Ledentsov N, Ledentsov N, Agustin M, Kropp J, Shchukin V. Application of nanophotonics to the next generation of surface-emitting lasers. Nano. 2017;**6**(5):813-829

[6] Yatsui T. Nanophotonic Fabrication. Heidelberg: Springer; 2012

[7] Taheri AN, Kaatuzian H. Design and simulation of a nanoscale electro-plasmonic 1 × 2 switch based on asymmetric metal–insulator–metal stub filters. Applied Optics. 2014;**53**(28):6546-6553

[8] Rastegar Pashaki E, Kaatuzian H, Mallah Livani A, Ghodsi H. Design and investigation of a balanced silicon-based plasmonic internal-photoemission detector. Applied Physics B. Dec. 2018; **125**(1)

[9] Moazzam MK, Kaatuzian H. Design and investigation of a N-type metal/insulator/semiconductor/metal structure2port electro-plasmonic addressed routing switch. Applied Optics. 2015;**54**(20):6199-6207

[10] Livani AM, Kaatuzian H. Design and simulation of an electrically pumped Schottky - junction - based plasmonic amplifier. Applied Optics. 2015;**54**(9):2164-2173

[11] Enoch S, Bonod N. Plasmonics. Heidelberg: Springer; 2012

[12] Stockman M. The spaser as a nanoscale quantum generator and ultrafast amplifier. Journal of Optics. 2010;**12**(2):024004

[13] Ma R, Oulton R. Applications of nanolasers. Nature Nanotechnology. 2018;**14**(1):12-22

[14] Ginzburg P. Cavity quantum electrodynamics in application to plasmonics and metamaterials. Reviews in Physics. 2016;**1**:120-139

[15] Liu K, Li N, Sadana D, Sorger V. Integrated nanocavity plasmon light sources for on-chip optical interconnects. ACS Photonics. 2016; **3**(2):233-242

[16] Ho J, Tatebayashi J, Sergent S, Fong C, Ota Y, Iwamoto S, et al. A nanowire-based plasmonic quantum dot laser. Nano Letters. 2016;**16**(4): 2845-2850

[17] Ghodsi H, Kaatuzian H. High purcell factor achievement of notched cavity germanium multiple quantum well plasmon source. Plasmonics. 2019. Available from: https://link.springer.com/article/10.1007%2Fs11468-019-01012-w

[18] Yu K, Lakhani A, Wu MC. Subwavelength metal-optic semiconductor nanopatch lasers. Optics Express. 2010;**18**:8790-8799

[19] Kwon SH et al. Subwavelength plasmonic lasing from a semiconductor nanodisk with silver nanopan cavity. Nano Letters. 2010;**10**:3679-3683

[20] Lakhani AM, Kim MK, Lau EK, Wu MC. Plasmonic crystal defect

nanolaser. Optics Express. 2011;**19**: 18237-18245

[21] Atwater H. The promise of plasmonics. Scientific American. 2007; **296**(4):56-62

[22] Kaatuzian H, Taheri A. Applications of Nano-Scale Plasmonic Structures in Design of Stub Filters—A Step Towards Realization of Plasmonic Switches. 2017. [Online]. Available: http://www.intechopen.com/books/photonic-crystals

[23] Pitarke JM et al. Theory of surface plasmons and surface-plasmon polaritons. Reports on Progress in Physics. 2006;**70**(1):1-87

[24] Francs G et al. Plasmonic Purcell factor and coupling efficiency to surface plasmons. Implications for addressing and controlling optical nanosources. Journal of Optics. 2016; **18**(9):094005. DOI: 10.1088/2040-8978/18/9/094005

[25] Nunes FD, Vasconcelos TC, Bezerra M, Weiner J. Electromagnetic energy density in dispersive and dissipative media. Journal of the Optical Society of America B: Optical Physics. 2011;**28**:1544-1552

[26] Boriskina S et al. Losses in plasmonics: From mitigating energy dissipation to embracing loss-enabled functionalities. Advances in Optics and Photonics. 2017;**9**(4):775. DOI: 10.1364/aop.9.000775

[27] Tame M, McEnery K, Özdemir Ş, Lee J, Maier S, Kim M. Quantum plasmonics. Nature Physics. 2013; **9**(6): 329-340. DOI: 10.1038/nphys2615[Accessed: 09 June 2019]

[28] Parfenyev V, Vergeles S. Quantum theory of a spaser-based nanolaser. Optics Express. 2014;**22**(11):13671. DOI: 10.1364/oe.22.013671

[29] Ma R, Oulton R, Sorger V, Zhang X. Plasmon lasers: Coherent light source at molecular scales. Laser and Photonics Reviews. 2012;**7**(1):1-21

[30] Yokoyama H, Nishi K, Anan T, Nambu Y, Brorson S, Ippen E, et al. Controlling spontaneous emission and threshold-less laser oscillation with optical microcavities. Optical and Quantum Electronics. 1992;**24**(2):S245-S272

[31] Ghodsi H, Kaatuzian H. Design and analysis of an electrically pumped GaAs quantum dot plasmonic nanolaser. Optik. 2020;**203**:164027. DOI: 10.1016/j. ijleo.2019.164027

[32] Ghodsi H, Kaatuzian H, Pashaki E. Design and simulation of a Germanium multiple quantum well metal strip nanocavity plasmon laser. Optical and Quantum Electronics. 2020;**52**(1). DOI: 10.1007/s11082-019-2172-6

[33] Liu J, Sun X, Pan D, Wang X, Kimerling LC, Koch TL, et al. Tensile-strained, n-type Ge as a gain medium for monolithic laser integration on Si. Optics Express. 2007;**15**:11272-11277

[34] Chang G-E, Chen S-W, Cheng HH. Tensile-strained Ge/SiGe quantum-well photodetectors on silicon substrates with extended infrared response. Optics Express. 2016;**24**:17562-17571

2

Stimulated Scattering of Surface Plasmon Polaritons in a Plasmonic Waveguide with a Smectic: A Liquid Crystalline Core

Boris I. Lembrikov, David Ianetz and Yossef Ben Ezra

Abstract

We considered theoretically the nonlinear interaction of surface plasmon polaritons (SPPs) in a metal-insulator-metal (MIM) plasmonic waveguide with a smectic liquid crystalline core. The interaction is related to the specific cubic optical nonlinearity mechanism caused by smectic layer oscillations in the SPP electric field. The interfering SPPs create the localized dynamic grating of the smectic layer strain that results in the strong stimulated scattering of SPP modes in the MIM waveguide. We solved simultaneously the smectic layer equation of motion in the SPP electric field and the Maxwell equations for the interacting SPPs. We evaluated the SPP mode slowly varying amplitudes (SVAs), the smectic layer dynamic grating amplitude, and the hydrodynamic velocity of the flow in a smectic A liquid crystal (SmALC).

Keywords: surface plasmon polariton (SPP), smectic liquid crystals, stimulated light scattering (SLS), plasmonic waveguide

1. Introduction

Nonlinear optical phenomena based on the second - and third-order optical nonlinearity characterized by susceptibilities $\chi^{(2)}$ and $\chi^{(3)}$, respectively, are widely used in modern communication systems for the optical signal processing due to their ultrafast response time and a large number of different interactions [1–5]. The second-order susceptibility $\chi^{(2)}$ exists in non-centrosymmetric media, while the third-order susceptibility $\chi^{(3)}$ exists in any medium [6]. The second-order susceptibility $\chi^{(2)}$ may be used for the second harmonic generation (SHG), sum, and difference frequency generation; the ultrafast Kerr-type third-order susceptibility $\chi^{(3)}$ results in such effects as four-wave mixing (FWM), self-phase modulation (SPM), cross-phase modulation (XPM), third harmonic generation (THG), bistability, and different types of the stimulated light scattering (SLS) [1–6]. Optical-electrical-optical conversion processes can be replaced with the optical signal processing characterized by the femtosecond response time of nonlinearities in optical materials [2, 3]. All-optical signal processing, ultrafast switching, optical generation of ultrashort pulses, the control over the laser radiation frequency spectrum, wavelength exchange, coherent detection, multiplexing/demultiplexing, and tunable optical delays can be realized by using the nonlinear optical effects [1–4]. However, optical nonlinearities are weak and usually occur only with high-intensity laser beams [1, 6]. An effective nonlinear optical response can be substantially increased by using the plasmonic effects caused by the coherent oscillations of conduction electrons near the surface of noble metal structures [1]. In the case of the extended metal surfaces, the surface plasmon polaritons (SPPs) may occur [1, 7, 8]. SPPs are electromagnetic excitations propagating at the interface between a dielectric and a conductor, evanescently confined in the perpendicular direction [1, 7]. The SPP electromagnetic field decays exponentially on both sides of the interface which results in the subwavelength confinement near the metal surface [1]. The SPP propagation length is limited by the ohmic losses in metal [1, 7, 8].

Nonlinear optical effects can be enhanced by plasmonic excitations as follows: (i) the coupling of light to surface plasmons results in strong local electromagnetic fields; (ii) typically, plasmonic excitations are highly sensitive to dielectric properties of the metal and surrounding medium [1]. In nonlinear optical phenomena, such a sensitivity can be used for the light-induced nonlinear change in the dielectric properties of one of the materials which result in the varying of the plasmonic resonances and the signal beam propagation conditions [1]. Plasmonic excitations are characterized by timescale of several femtoseconds which permits the ultrafast optical signal processing [1]. The SPP field confinement and enhancement can be changed by modifying the structure of the metal or the dielectric near the interface [1]. For example, plasmonic waveguides can be created [1, 7–9]. Nanoplasmonic waveguides can confine and enhance electric fields near the nanometallic surfaces due to the propagating SPPs [9]. Nanoplasmonic waveguide consists of one or two metal films combined with one or two dielectric slabs [9]. Typically, two types of the plasmonic waveguides exist: (i) an insulator/metal/insulator (IMI) heterostructure where a thin metallic layer is placed between two infinitely thick dielectric claddings and (ii) a metal/insulator/metal (MIM) heterostructure where a thin dielectric layer is sandwiched between two metallic claddings [7]. The MIM waveguides for nonlinear optical applications require highly nonlinear dielectrics [9]. The nonlinear metamaterials can significantly increase the nonlinearity magnitude [10]. Investigation of nonlinear metamaterials is related in particular to nonlinear plasmonics and active media [10]. One of the metamaterial nonlinearity mechanisms is based on liquid crystals (LCs) [10]. Tunability and a strongly nonlinear response of metamaterials can be obtained by their integration with LCs offering a practical solution for controlling metamaterial devices [11].

The integration of LCs with plasmonic and metamaterials may be promising for applications in modern photonics due to the extremely large optical nonlinearity of LCs, strong localized electric fields of surface plasmon polaritons (SPPs), and high operation rates as compared to conventional electro-optic devices [12]. Practically all nonlinear optical processes such as wave mixing, self-focusing, self-guiding, optical bistabilities and instabilities, phase conjugation, SLS, optical limiting, interface switching, beam combining, and self-starting laser oscillations have been observed in LCs [13]. LC can be incorporated into nano- and microstructures such as a MIM plasmonic waveguide. Nematic LCs (NLCs) characterized by the orientation long-range order of the elongated molecules are mainly used in optical applications including plasmonics and nanophotonics [11–14]. For instance, light-induced control of fishnet metamaterials infiltrated with NLCs was demonstrated experimentally where a metal-dielectric (Au-MgF_2) sandwich nanostructure on a glass substrate with the inserted NLC was used [11]. However, the NLC applications are limited by their large losses and relatively slow response [14, 15]. The light scattering in smectic A LC (SmALC) waveguides had been studied theoretically and experimentally, and it was shown that the scattering losses in SmALC are much lower than in NLC due to a higher degree of the long-range order [15]. SmALC can be useful in nonlinear optical applications and low-loss active waveguide devices for integrated optics [14, 15].

SmALCs are characterized by a positional long-range order in the direction of the elongated molecular axis and demonstrate a layer structure with a layer thickness $d_{SmA} \approx 2nm$ [14]. Inside a smectic layer, the molecules form a two-dimensional liquid [14]. Actually, SmALC can be considered as a natural nanostructure. The structures of NLC with the elongated molecules directed mainly along the vector director $\vec{n}$ and the homeotropically oriented SmALC with the layer plane parallel to the claddings are shown in **Figure 1a** and **b**, respectively.

The nonlinear optical phenomena in SmALC such as a light self-focusing, self-trapping, SPM, SLS, and FWM based on the specific mechanism of the third-order optical nonlinearity related to the smectic layer normal displacement had been investigated theoretically [16–28]. In particular it has been shown that at the interface of a metal and SmALC, the counter-propagating SPPs created the dynamic grating of the smectic layer normal displacement $u(x,z,t)$, and the SLS of the interfering SPPs occurred [22, 23, 26]. We also investigated the behavior of SPP mode in a MIM waveguide with the SmALC core [24, 26]. In such a waveguide, SPP behaves as a strongly localized transverse magnetic (TM) mode which creates the localized smectic layer normal deformation and undergoes SPM [24, 26].

In this chapter we consider theoretically the interaction of the counter-propagating SPP modes in the MIM waveguide with the SmALC core. The interfering SPP TM modes with the close optical frequencies $\omega_{1,2}$ create a localized dynamic grating of the smectic layer normal displacement $u(x,z,t)$ with the frequency $\Delta\omega = \omega_1 - \omega_2 \ll \omega_1$ which results in the nonlinear polarization and stimulated scattering of SPPs. We solved simultaneously the equation of motion for smectic layers in the electric field of the interfering SPP modes and the Maxwell equations for the SPPs in the MIM waveguide taking into account the nonlinear polarization. We used the slowly varying amplitude (SVA) approximation for the SPPs [6]. We evaluated the magnitudes and phases of the coupled SPP SVAs. It is shown that the energy exchange between the coupled SPPs and XPM takes place. We also evaluated the SPP-induced smectic layer displacement and SmALC hydrodynamic velocity. We have shown that the high-frequency localized electric field can occur in the MIM waveguide with the SmALC core due to the flexoelectric effect [28].

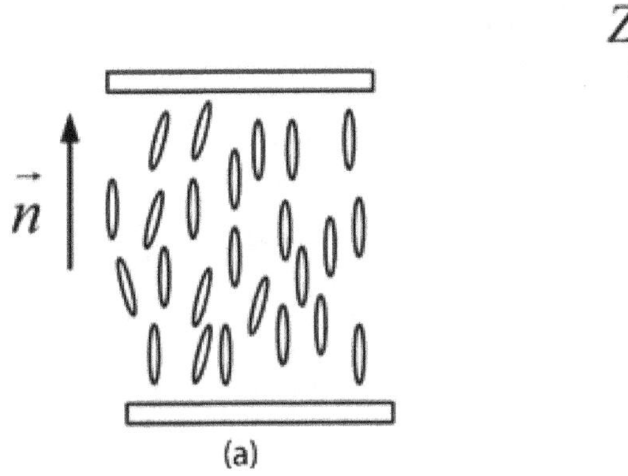

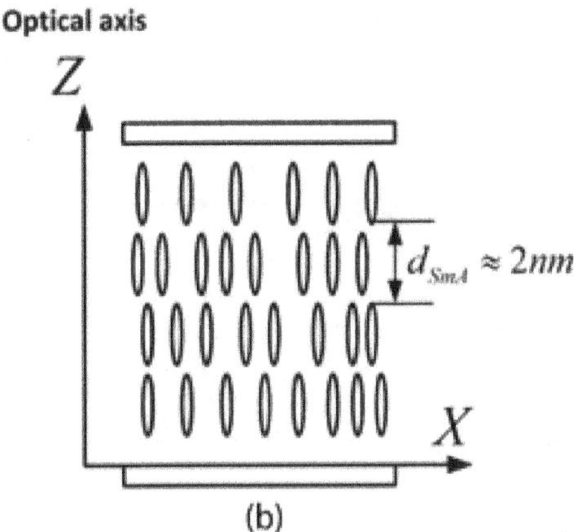

Figure 1.
The structure of molecular alignment of a nematic liquid crystal (NLC) (a) and the homeotropically oriented smectic A liquid crystal (SmALC). The molecules are perpendicular to the layer plane (b).

The chapter is constructed as follows. The hydrodynamics of SmALC in the external electric field is considered in Section 2. The SPP modes of the MIM waveguide are derived in Section 3. The SPP SVAs, the smectic layer dynamic grating amplitude, and the SmALC hydrodynamic velocity are evaluated in Section 4. The conclusions are presented in Section 5.

2. Hydrodynamics of SmALC in the external electric field

In this section we briefly discuss the SmALC hydrodynamics and derive the equation of motion for the smectic layer normal displacement $u(x,y,z,t)$ in the external electric field $\vec{E}(x,y,z,t)$. SmALC can be described by the one-dimensional periodic density wave due to its layered structure.

Smectic layer oscillations $u(x,y,z,t)$ in the external electric field $\vec{E}(x,y,z,t)$ are shown in **Figure 2**. Hydrodynamics of SmALC in general case is very complicated because SmALC is a strongly anisotropic viscous liquid including the layer oscillations, the mass density, and the elongated molecule orientation variations [29–31]. However, the elastic constant related to the SmALC bulk compression is much larger than the elastic constant $B \approx 10^6 - 10^7 \text{J m}^{-3}$ related to the smectic layer compression [29–31]. The layers can oscillate without the change of the mass density [29–31]. For this reason two uncoupled acoustic modes can propagate in SmALC: the ordinary longitudinal sound wave caused by the mass density variation and the second-sound (SS) wave caused by the layer oscillations [29–31]. SS wave is characterized by strongly anisotropic dispersion relation being neither purely transverse nor longitudinal. It propagates in the direction oblique to the layer plane

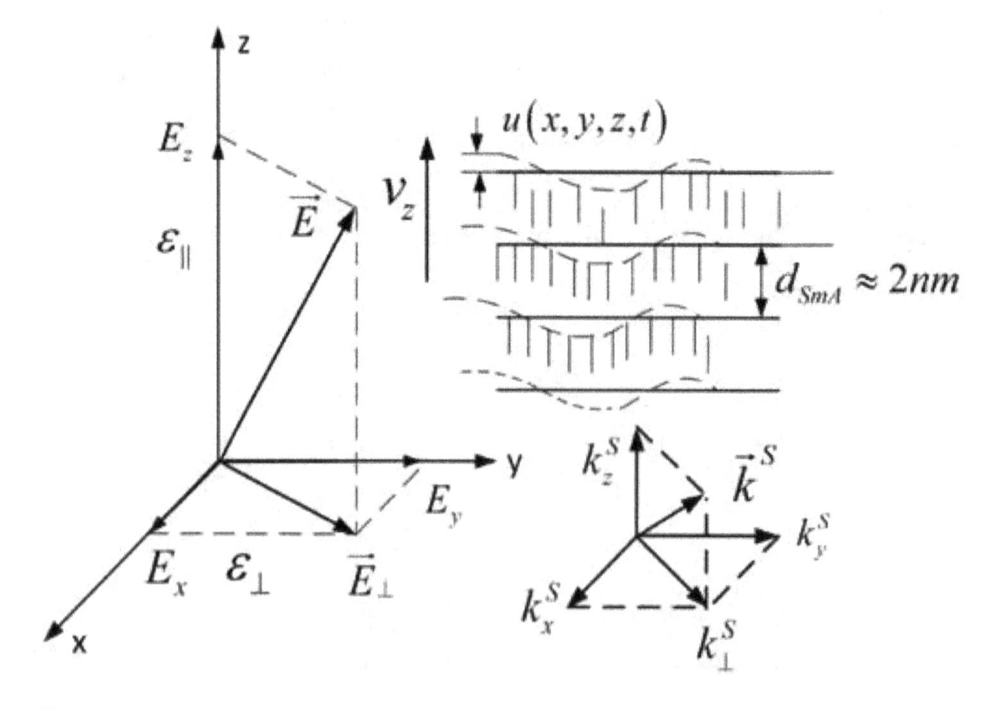

Figure 2.

The SmALC layer oscillations $u(x,y,z,t)$ in the external electric field $\vec{E}(x,y,z,t)$. $\vec{k}^S$ is the second-sound (SS) wave vector, v_z is the hydrodynamic velocity perpendicular to the layer plane; $\varepsilon_{||}$ and $\varepsilon_{\perp}$ are the diagonal components of the permittivity tensor parallel and perpendicular to the optical axis, respectively.

and vanishes for the wave vector $\vec{k}^S$ perpendicular or parallel to the layer plane [29]. SmALC is characterized by the complex order parameter, and SS represents the oscillations of the order parameter phase [29]. SS in SmALC has been observed experimentally by different methods [32–34]. The system of hydrodynamic equations for the incompressible SmALC under the constant temperature far from the phase transition has the form [29–31]

$$div\ \vec{v} = 0 \tag{1}$$

$$\rho\frac{\partial v_i}{\partial t} = -\frac{\partial \Pi}{\partial x_i} + \Lambda_i + \frac{\partial \sigma'_{ik}}{\partial x_k} \tag{2}$$

$$\Lambda_i = -\frac{\delta F}{\delta u_i} \tag{3}$$

$$\sigma'_{ik} = \alpha_0\delta_{ik}A_{ll} + \alpha_1\delta_{iz}A_{zz} + \alpha_4A_{ik} + \alpha_{56}(\delta_{iz}A_{zk} + \delta_{kz}A_{zi}) + \alpha_7\delta_{iz}\delta_{kz}A_{ll} \tag{4}$$

$$A_{ik} = \frac{1}{2}\left(\frac{\partial v_i}{\partial x_k} + \frac{\partial v_k}{\partial x_i}\right) \tag{5}$$

$$v_z = \frac{\partial u}{\partial t} \tag{6}$$

Here, $\vec{v}$ is the hydrodynamic velocity, $\rho \approx 10^3 \mathrm{kg\,m^{-3}}$ is the SmALC mass density, Π is the pressure, $\vec{\Lambda}$ is the generalized force density, σ'_{ik} is the viscous stress tensor, $\alpha_i \approx 10^{-1}\mathrm{kg(s\,m)^{-1}}$ are the viscosity Leslie coefficients, $\delta_{ik} = 1, i = k; \delta_{ik} = 0, i \neq k$, and F is the free energy density of SmALC. Typically, SmALC is supposed to be an incompressible liquid according to Equation (1) [29]. For this reason, we assume that the pressure $\Pi = 0$ and the SmALC free energy density F do not depend on the bulk compression [29– 31]. We are interested in the SS propagation and neglect the ordinary sound mode. The normal layer displacement $u(x, y, z, t)$ by definition has only one component along the Z axis. In such a case, the generalized force density has only the Z component according to Eq. (3): $\vec{\Lambda} = (0, 0, \Lambda_z)$. Eq. (6) is specific for SmALC since it determines the condition of the smectic layer continuity [29–31].

The SmALC free energy density F in the presence of the external electric field $\vec{E}(x, y, z, t)$ has the form [29–31]

$$F = \frac{1}{2}B\left(\frac{\partial u}{\partial z}\right)^2 + \frac{1}{2}K\left(\frac{\partial^2 u}{\partial x^2} + \frac{\partial^2 u}{\partial y^2}\right)^2 - \frac{1}{2}\varepsilon_0\varepsilon_{ik}E_iE_k \tag{7}$$

Here $K \sim 10^{-11}N$ is the Frank elastic constant associated with the SmALC orientational energy inside layers, ε_0 is the free space permittivity, and ε_{ik} is the SmALC permittivity tensor including the terms defined by the smectic layer strains. The purely orientational second term in the free energy density F (7) can be neglected since for the typical values of the elastic constants B and K $K\left(k_\perp^S\right)^2 \ll B$ where $k_\perp^S$, the SS wave vector component is parallel to the layer plane. The permittivity tensor ε_{ik} is given by [30]

$$\begin{aligned} &\varepsilon_{xx} = \varepsilon_{yy} = \varepsilon_\perp + a_\perp\frac{\partial u}{\partial z}; \varepsilon_{zz} = \varepsilon_\parallel + a_\parallel\frac{\partial u}{\partial z}; \\ &\varepsilon_{xz} = \varepsilon_{zx} = -\varepsilon_a\frac{\partial u}{\partial x}; \varepsilon_{yz} = \varepsilon_{zy} = -\varepsilon_a\frac{\partial u}{\partial y}; \varepsilon_a = \varepsilon_\parallel - \varepsilon_\perp \end{aligned} \tag{8}$$

where $\varepsilon_{\parallel}, \varepsilon_{\perp}$ are the diagonal components of the permittivity tensor ε_{ik} along and perpendicular to the optical axis and $a_{\perp} \sim 1$, $a_{\parallel} \sim 1$ are the phenomenological dimensionless coefficients [29, 30]. SmALC is an optically uniaxial medium with the optical Z axis perpendicular to the smectic layer plane [29–31]. Combining Eqs. (1)–(8), we obtain the equation of motion for the smectic layer normal displacement $u(x,y,z,t)$ in the electric field $\vec{E}(x,y,z,t)$ [16, 17]:

$$
\begin{aligned}
&-\rho\nabla^2\frac{\partial^2 u}{\partial t^2} + \left[\alpha_1\nabla_\perp^2\frac{\partial^2}{\partial z^2} + \frac{1}{2}(\alpha_4+\alpha_{56})\nabla^2\nabla^2\right]\frac{\partial u}{\partial t} + B\nabla_\perp^2\frac{\partial^2 u}{\partial z^2} \\
&= \frac{\varepsilon_0}{2}\nabla_\perp^2\left[\frac{\partial}{\partial z}\left(a_\perp\left(E_x^2+E_y^2\right)+a_\parallel E_z^2\right) - 2\varepsilon_a\left(\frac{\partial}{\partial x}(E_xE_z)+\frac{\partial}{\partial y}(E_yE_z)\right)\right]
\end{aligned}
\tag{9}
$$

Here $\nabla^2{}_\perp u = \partial^2 u/\partial x^2 + \partial^2 u/\partial y^2$. In the absence of the external electric field, the homogeneous solution of the equation of motion (9) represents the SS wave with the dispersion relation [29]:

$$
\Omega_S = s_0\frac{k_\perp^S k_z^S}{k^S};\, s_0 = \sqrt{\frac{B}{\rho}}
\tag{10}
$$

Here, $\left(k_\perp^S\right)^2 = \left(k_x^S\right)^2 + \left(k_y^S\right)^2$ and Ω_S and s_0 are the SS frequency and velocity, respectively [29]. It is seen from Eq. (10) that the SS frequency $\Omega_S = 0$ for the propagation direction along the smectic layer plane and perpendicular to it. The decay constant Γ is given by

$$
\Gamma = \frac{1}{2\rho}\left[\alpha_1\frac{\left(k_\perp^S\right)^2\left(k_z^S\right)^2}{\left(k^S\right)^2} + \frac{1}{2}(\alpha_4+\alpha_{56})\left(k^S\right)^2\right]
\tag{11}
$$

If the viscosity terms responsible for the SS wave decay can be neglected, then the homogeneous part of Eq. (9) reduces to the SS wave equation with the dispersion relation (10) [29–31]:

$$
\rho\nabla^2\frac{\partial^2 u}{\partial t^2} = B\nabla_\perp^2\frac{\partial^2 u}{\partial z^2}
$$

We use equation of motion (9) for the evaluation of the light-enhanced dynamic grating $u(x,y,z,t)$.

3. SPP modes in a MIM waveguide with SmALC core

LC slab optical waveguide represents a LC layer of a thickness about 1 μm confined between two glass slides of lower refractive index than LC [14]. LC as a waveguide core provides the photonic signal modulation and switching by using the electro-optic or nonlinear optical effects of LC mesophases [35]. For instance, the large optical nonlinearities were implemented in order to create optical paths by photonic control of solitons in NLC [35]. Various electrode geometries may create due to the electro-optic effect periodically modulated LC core waveguides which can serve as efficient guided distributed Bragg reflectors with the tuning ranges of about 100–1550 nm optical wavelength range [35]. Plasmonic waveguides based on the manipulation and routing of SPPs can demonstrate a subwavelength beyond the diffraction limit together with large bandwidth and high operation rate typical for photonics [36]. The plasmonic devices can be integrated into nanophotonic chips due to their small scale and the compatibility with the VLSI electronic technology [36]. Plasmonic devices are the promising candidates for future integrated photonic circuits for broadband light routing, switching, and interconnecting [36]. It has been shown that different plasmonic structures can provide SPP light waveguiding determining the SPP mode properties [36]. MIM waveguide representing a dielectric sandwiched between two metal slabs attracted a research interest as a basic component of nanoscale plasmonic integrated circuits [37]. LC-tunable waveguides have been proposed as a core element of low-power variable attenuators, phase-shifters, switches, filters, tunable lenses, beam steers, and modulators [37, 38]. Typically NLCs have been used due to their strong optical anisotropy, responsivity to external electric and magnetic fields, and low power [37, 38]. Different types of NLC plasmonic waveguides have been proposed and investigated theoretically [36–38]. Recently, SmALCs attracted attention due to their layered structure and reconfigurable layer curvature [39]. The possibility of the dynamic variation of smectic layer configuration by external fields is intensively studied [39]. We investigated theoretically SLS in the optical slab waveguide with the SmALC core where the third-order optical nonlinearity mechanism was related to the smectic layer dynamic grating created by the interfering waveguide modes [27]. We also considered theoretically the MIM waveguide with the SmALC core [24, 26].

The structure of such a symmetric waveguide of the thickness 2d is shown in **Figure 3** [24, 26]. The plane of the waveguide is perpendicular to the SmALC optical axis Z. The SmALC in the waveguide core is homeotropically oriented, i.e., the smectic layers are parallel to the waveguide claddings $z = \pm d$, while the SmALC elongated molecules are mainly parallel to the Z axis [29]. Typically the waveguide dimension in the Y axis direction is much larger than d, and the dependence on the coordinate y in Eqs. (8) and (9) can be omitted. Than we obtain $u = u(x, z, t)$, $\nabla^2 u = \partial^2 u/\partial x^2 + \partial^2 u/\partial z^2$, $\nabla_\perp^2 u = \partial^2 u/\partial x^2$, $\left(k_\perp^S\right)^2 = \left(k_x^S\right)^2$, and the SmALC permittivity tensor (8) takes the form

$$\varepsilon_{xx} = \varepsilon_\perp + a_\perp \frac{\partial u}{\partial z};\ \varepsilon_{zz} = \varepsilon_\parallel + a_\parallel \frac{\partial u}{\partial z};\ \varepsilon_{xz} = \varepsilon_{zx} = -\varepsilon_a \frac{\partial u}{\partial x}; \varepsilon_a = \varepsilon_\parallel - \varepsilon_\perp \tag{12}$$

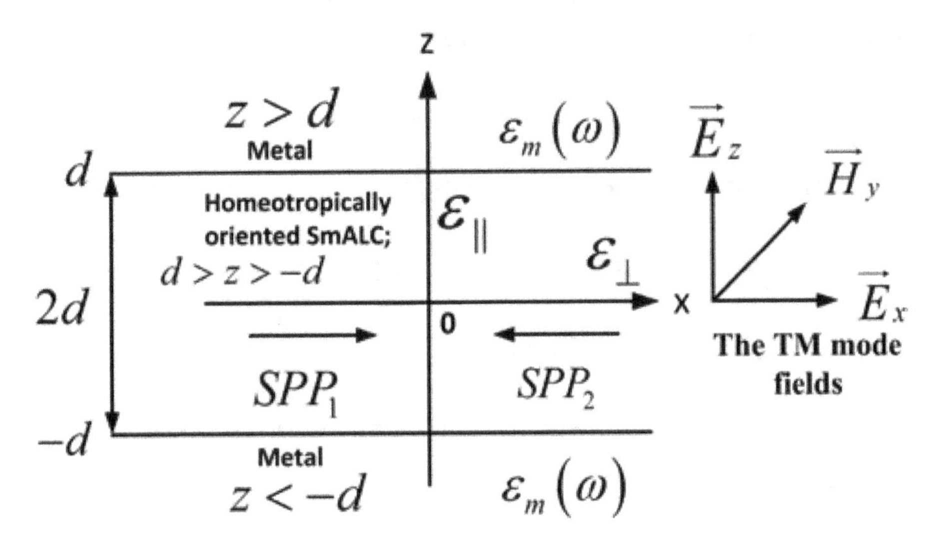

Figure 3.
The MIM waveguide with the homeotropically oriented SmALC core and counter-propagating SPPs.

The permittivity $\varepsilon_m(\omega)$ of the metal claddings is described by the Drude model [7, 8]:

$$\varepsilon_m(\omega) = 1 - \frac{\omega_p^2}{[\omega^2 + (i\omega/\tau)]} \tag{13}$$

where $\omega_p = \sqrt{n_0 e^2/(\varepsilon_0 m)}$ is the plasma frequency of the free electron gas; n_0 is the free electron density in the metal;e, m are the electron charge and mass, respectively; and ω, τ are the SPP angular frequency and lifetime, respectively [7, 8]. The electric field $\vec{E}(x,z,t)$ of the optical wave propagating in a nonlinear medium is described by the following wave equation including the nonlinear part of the electric induction $\vec{D}^{NL}$ [6]:

$$curlcurl\vec{E} + \mu_0 \frac{\partial^2 \vec{D}^L}{\partial t^2} = -\mu_0 \frac{\partial^2 \vec{D}^{NL}}{\partial t^2} \tag{14}$$

Here μ_0 is the free space permeability and $\vec{D}^L$ is the nonlinear part of the electric induction. The SPP can propagate in the plasmonic waveguide only as a transverse magnetic (TM) mode with the electric and magnetic fields given by $\vec{E}_{TM} = (E_x, 0, E_z)$; $\vec{H}_{TM} = (0, H_y, 0)$ [7]. In such a case, we obtain for $\vec{D}^L$ and $\vec{D}^{NL}$ in SmALC using Eq. (12)

$$D_x^L = \varepsilon_0 \varepsilon_\perp E_x;\ D_z^L = \varepsilon_0 \varepsilon_\parallel E_z \tag{15}$$

$$D_x^{NL} = \varepsilon_0 \left(a_\perp \frac{\partial u}{\partial z} E_x - \varepsilon_a \frac{\partial u}{\partial x} E_z \right);\ D_z^{NL} = \varepsilon_0 \left(a_\parallel \frac{\partial u}{\partial z} E_z - \varepsilon_a \frac{\partial u}{\partial x} E_x \right) \tag{16}$$

The linear part $\vec{D}_m^L$ of the electric induction in the metal claddings has the form: $\vec{D}_m^L = \varepsilon_0 \varepsilon_m(\omega) \vec{E}$ [7]. The SPP TM mode electric and magnetic fields for $|z| > d$ in the metal claddings $\vec{H}_{1,2}(x,z,t), \vec{E}_{1,2}(x,z,t)$ and for $|z| \leq d$ in SmALC $\vec{H}_{SA}(x,z,t), \vec{E}_{SA}(x,z,t)$ have the form [24, 26]

$$\vec{H}_{1,2}(x,z,t) = \frac{1}{2} \vec{a}_y H_{1,20} \exp\left(\mp k_z^m z + i k_x x - i\omega t\right) + c.c., |z| > d \tag{17}$$

$$\vec{E}_{1,2}(x,z,t) = \frac{1}{2} \left[\vec{a}_x E_{1,2x0} + \vec{a}_z E_{1,2z0} \right] \exp\left(\mp k_z^m z + i k_x x - i\omega t\right) + c.c., |z| > d \tag{18}$$

$$\vec{H}_{SA}(x,z,t) = \frac{1}{2} \vec{a}_y \left[A \exp\left(k_z^S z\right) + B \exp\left(-k_z^S z\right) \right] \exp\left(i k_x x - i\omega t\right) + c.c., |z| \leq d \tag{19}$$

$$\vec{E}_{SA}(x,z,t) = \frac{1}{2} \{ \vec{a}_x \frac{k_z^S}{i\omega \varepsilon_0 \varepsilon_\perp} \left[A \exp\left(k_z^S z\right) - B \exp\left(-k_z^S z\right) \right] \\ - \vec{a}_z \frac{k_x}{\omega \varepsilon_0 \varepsilon_\parallel} \left[A \exp\left(k_z^S z\right) + B \exp\left(-k_z^S z\right) \right] \} \exp i(k_x x - \omega t) + c.c.;\ |z| \leq d \tag{20}$$

Here c.c. stands for complex conjugate. The SPP fields (17)–(20) are confined in the Z direction. In the linear approximation substituting expressions (15), (18), and

(20) into the homogeneous part of the wave equation (14) for the claddings and SmALC core, respectively, we obtain the following expressions for the complex wave numbers k_z^m and k_z^S [24, 26]:

$$k_z^m = \sqrt{k_x^2 - \varepsilon_m(\omega)\omega^2/c^2} \tag{21}$$

$$k_z^S = \sqrt{k_x^2(\varepsilon_\perp/\varepsilon_\parallel) - \omega^2\varepsilon_\perp/c^2} \tag{22}$$

where c is the free space light velocity. The boundary conditions for the fields (17)–(20) at the interfaces $z = \pm d$ have the form [7, 8]

$$H_{1y}(z = d) = H_{SAy}(z = d);\ H_{2y}(z = -d) = H_{SAy}(z = -d) \tag{23}$$

$$E_{1x}(z = d) = E_{SAx}(z = d);\ E_{2x}(z = -d) = E_{SAx}(z = -d) \tag{24}$$

Substituting expressions (17)–(20) into Eqs. (23) and (24), we obtain the dispersion relation for the SPP TM modes in the MIM waveguide given by [24, 26]

$$\exp\left(-4k_z^S d\right) = \left(\frac{k_z^m}{\varepsilon_m(\omega)} + \frac{k_z^S}{\varepsilon_\perp}\right)^2 \left(\frac{k_z^m}{\varepsilon_m(\omega)} - \frac{k_z^S}{\varepsilon_\perp}\right)^{-2} \tag{25}$$

Dispersion relation obtained for the general case of different claddings [7] coincides with expression (25) for the symmetric structure with the same claddings. The results of the numerical solution of Eq. (25) for the typical values of the MIM waveguide parameters and the SPP frequencies ω corresponding to the optical wavelength range $\lambda_{opt} \sim 1 - 1.6$ μm and $2d \sim 1$ μm are presented in **Figures 4** and **5**.

These results show that $\mathrm{Re}\,k_z^S \sim 10^6 m^{-1} \gg \mathrm{Im}k_z^S \sim 10^4 m^{-1}$ and $\mathrm{Re}\,k_x \sim 10^7 m^{-1}$ $\gg \mathrm{Im}k_x \sim 10^3 m^{-1}$ [24, 26]. In such a case, the SPP oscillation length in the Z direction is defined by the relationship $2\pi\left(\mathrm{Im}k_z^S\right)^{-1} \sim 10^{-4} m \gg d \sim 10^{-6} m$, and $\mathrm{Im}k_z^S$ can be neglected inside the MIM waveguide, and $k_z^S \approx \mathrm{Re}\,k_z^S$ [24, 26]. The SPP propagation length in the X direction $L_{SPP} = (\mathrm{Im}k_x)^{-1} \sim 10^{-4} - 10^{-3} m \gg \lambda_{SPP} = 2\pi(\mathrm{Re}\,k_x)^{-1} < 10^{-6} m$ where λ_{SPP} is the SPP wavelength. Hence, at the optical wavelength-scale distances, $\mathrm{Im}k_x$ can be neglected, and $k_x \approx \mathrm{Re}\,k_x$ [24, 26].

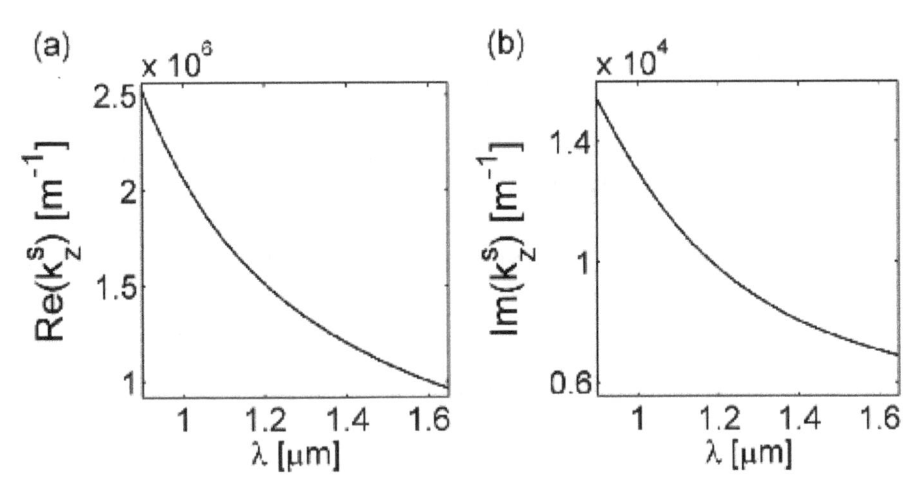

Figure 4.
The spectral dependence of $\mathrm{Re}\,k_z^S$ *(a) and* $\mathrm{Im}k_z^S$ *(b).*

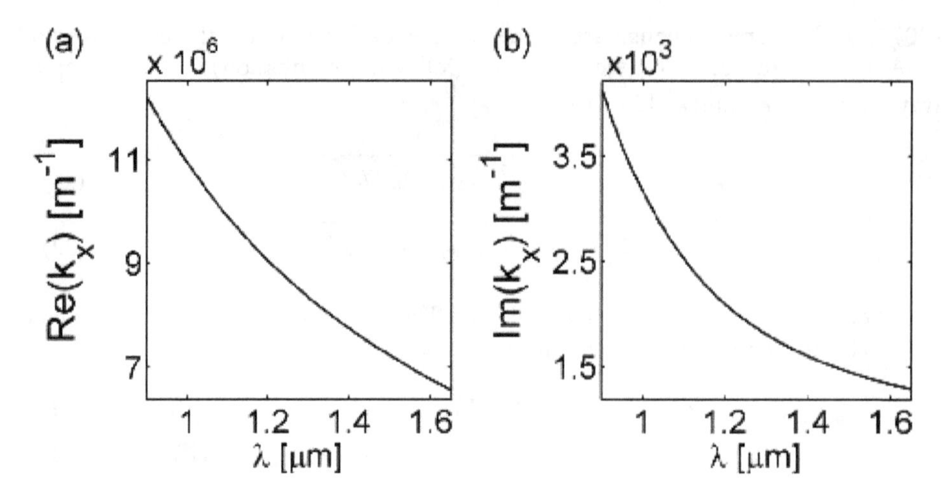

Figure 5.
The spectral dependence of $\mathrm{Re}\, k_x$ *(a) and* $\mathrm{Im} k_x$ *(b).*

Consequently, for a given optical frequency ω, a single localized TM mode can exist in the SmALC core of the MIM waveguide with the electric field $\vec{E}_{SA}(x,z,t)$ given by [24, 26]

$$\vec{E}_{SA} = E_0\left[\vec{a}_x \cosh\left(\left(\mathrm{Re}\, k_z^S\right)z\right) - \vec{a}_z i \frac{k_x\varepsilon_\perp}{k_z^S\varepsilon_\parallel}\sinh\left(\left(\mathrm{Re}\, k_z^S\right)z\right)\right] \times \exp\left[i((\mathrm{Re}\, k_x)x - \omega t)\right] + c.c. \tag{26}$$

The numerical estimations show that for the SPP modes with the close optical frequencies $\omega_{1,2} \sim 10^{15}\mathrm{s}^{-1}$ and the frequency difference $\Delta\omega = \omega_1 - \omega_2 \sim 10^8 s^{-1} \ll \omega_1$, the wave numbers of the both SPPs $k_{z1,2}^S$ and $k_{x1,2}$ are practically equal. As a result, only counter-propagating SPP modes can strongly interact in the MIM core creating the dynamic grating of smectic layers as it is seen from Eq. (9). The electric field of the counter-propagating SPP modes of the type (26) in the MIM waveguide SmALC core has the form

$$\vec{E}_{SA1,2} = E_{SA1,20}\left[\vec{a}_x \cosh\left(\left(\mathrm{Re}\, k_z^S\right)z\right) \mp \vec{a}_z i \frac{k_x\varepsilon_\perp}{k_z^S\varepsilon_\parallel}\sinh\left(\left(\mathrm{Re}\, k_z^S\right)z\right)\right] \times \exp\left[\pm i(\mathrm{Re}\, k_x)x - i\omega_{1,2}t\right] + c.c. \tag{27}$$

Substituting expression (27) into equation of motion (9), we obtain the expression of the smectic layer displacement localized dynamic grating $u(x,z,t)$:

$$u(x,z,t) = U_0 \sinh\left(2\left(\mathrm{Re}\, k_z^S\right)z\right)\exp\left[i(2(\mathrm{Re}\, k_x)x - \Delta\omega t)\right] + c.c. \tag{28}$$

Here

$$U_0 = -\frac{4\varepsilon_0 E_{SA10}E_{SA20}^*}{G\left(k_x, k_z^S, \Delta\omega\right)}(\mathrm{Re}\, k_x)^2\left(\mathrm{Re}\, k_z^S\right)h; \quad h = \left\{a_\perp - a_\parallel \frac{|k_x|^2\varepsilon_\perp^2}{\left|k_z^S\right|^2\varepsilon_\parallel^2} - 2\varepsilon_a \frac{(\mathrm{Re}\, k_x)^2\varepsilon_\perp}{\left(\mathrm{Re}\, k_z^S\right)^2\varepsilon_\parallel}\right\} \tag{29}$$

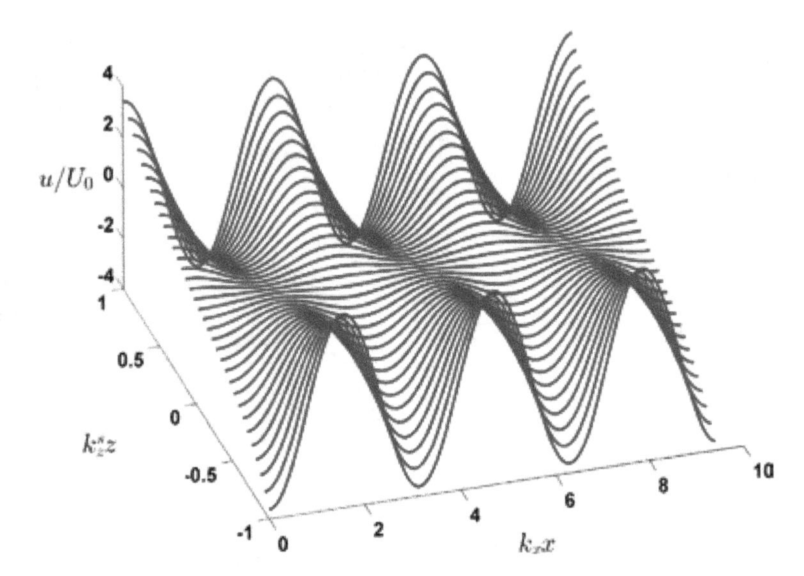

Figure 6.
The normalized smectic layer displacement $u(x, z, t = t_0)$ for the optical wavelength $\lambda_{opt} = 1.6\ \mu m$.

$$G\left(k_x, k_z^S, \Delta\omega\right) = 4\rho(\Delta\omega)^2\left[-(\operatorname{Re} k_x)^2 + \left(\operatorname{Re} k_z^S\right)^2\right]$$
$$-i\Delta\omega\{-\alpha_1(2\operatorname{Re} k_x)^2\left(2\operatorname{Re} k_z^S\right)^2 \tag{30}$$
$$+\frac{1}{2}(\alpha_4+\alpha_{56})\left[-(2\operatorname{Re} k_x)^2+\left(2\operatorname{Re} k_z^S\right)^2\right]^2\} - B(2\operatorname{Re} k_x)^2\left(2\operatorname{Re} k_z^S\right)^2$$

Expression (28) is the enhanced solution of Eq. (9). The homogeneous solution of Eq. (9) is overdamped for the typical values of SmALC parameters and $\Delta\omega \sim 10^8 s^{-1}$, and it can be neglected The normalized smectic layer displacement $u(x, z, t = t_0)/U_0$ for the optical wavelength $\lambda_{opt} = 1.6\mu m$ is shown in **Figure 6**. It is seen from **Figure 6** that the dynamic grating is localized inside the MIM waveguide in the Z direction and oscillates in the propagation direction X.

4. Nonlinear interaction of SPPs in the MIM waveguide

The light-enhanced dynamic grating (28) results in the nonlinear polarization defined by Eq. (16). In order to investigate the interaction of the counter-propagating SPPs (27), we should solve wave Eq. (14) including the nonlinear term $\vec{D}^{NL}$. We use the SVA approximation for the SPP electric field amplitudes $E_{SA1,20}(t) = |E_{SA1,20}(t)| \exp i\theta_{SA1,2}(t)$ where $|E_{SA1,20}(t)|$ and $\theta_{SA1,2}(t)$ are the SVA magnitudes and phases, respectively [6]. For the distances of the order of magnitude of the SPP wavelength $\lambda_{SPP} < 1\ \mu m$, the dependence of SAVs on the x coordinate can be neglected. We assume according to the SVA approximation that

$$\left|\frac{\partial^2 E_{SA1,20}}{\partial t^2}\right| \ll \omega_1 \left|\frac{\partial E_{SA1,20}}{\partial t}\right| \tag{31}$$

Substituting expressions (27) and (28) into Eqs. (16), we evaluate the nonlinear part $\vec{D}^{NL}$ of the electric induction in SmALC. Then, substituting relationships (15),

(16), and (27) into wave equation (14), taking into account the dispersion relation (22), neglecting the terms $\partial^2 E_{SA1,20}/\partial t^2$ according to condition (31), combining the phase-matched terms with the frequencies $\omega_{1,2}$, and dividing the real and imaginary parts, we derive the equations for the SVA magnitudes $|E_{SA1,20}(t)|$ and phases $\theta_{SA1,2}(t)$. They have the form

$$\frac{1}{\omega_{1,2}}\frac{\partial|E_{SA1,20}(t)|^2}{\partial t}F_1(z) = \mp\frac{8\varepsilon_0 \mathrm{Im}G\left(k_x,k_z^S,\Delta\omega\right)[\mathrm{Re}\,k_x]^2 h|E_{SA10}(t)|^2|E_{SA20}(t)|^2}{\varepsilon_\perp\left|G\left(k_x,k_z^S,\Delta\omega\right)\right|^2}F_2(z) \tag{32}$$

$$\frac{1}{\omega_{1,2}}\frac{\partial\theta_{1,2}}{\partial t}F_1(z) = -\frac{4\varepsilon_0\,\mathrm{Re}\,G\left(k_x,k_z^S,\Delta\omega\right)|E_{SA2,10}(t)|^2[\mathrm{Re}\,k_x]^2 h}{\varepsilon_\perp\left|G\left(k_x,k_z^S,\Delta\omega\right)\right|^2}F_2(z) \tag{33}$$

Here we assumed that the factor $\exp\,[\pm(\mathrm{Im}k_x)x]\approx 1$ for the distances $x \ll (\mathrm{Im}k_x)^{-1}$. The functions $F_{1,2}(z)$ describing the SPP mode localization inside the MIM waveguide are given by

$$F_1(z) = \cosh^2\left(k_z^S z\right) + \frac{|k_x|^2}{\left|k_z^S\right|^2}\sinh^2\left(k_z^S z\right) \tag{34}$$

$$\begin{aligned} F_2(z) = {} & \cosh^2\left(k_z^S z\right)\left[\cosh\left(2k_z^S z\right)\left(a_\perp\left(k_z^S\right)^2 + \varepsilon_a\frac{\varepsilon_\perp}{\varepsilon_\parallel}k_x^2\right) - \varepsilon_a\frac{\varepsilon_\perp}{\varepsilon_\parallel}k_x^2\right] \\ & -k_x^2\sinh^2\left(k_z^S z\right)\left[\cosh\left(2k_z^S z\right)\left(a_\parallel\frac{\varepsilon_\perp}{\varepsilon_\parallel} - \varepsilon_a\right) - \varepsilon_a\right] \end{aligned} \tag{35}$$

Here we neglected the small quantities $\mathrm{Im}k_x$ and $\mathrm{Im}k_z^S$ assuming that for $x \ll (\mathrm{Im}k_x)^{-1}$ and $|z| \le d$, we may use the relationships $k_x \approx \mathrm{Re}\,k_x$ and $k_z^S \approx \mathrm{Re}\,k_z^S$. We integrate both parts of Eqs. (32) and (33) over the MIM waveguide thickness $-d \le z \le d$ [40]. After the integration, Eqs. (32) and (33) take the form

$$\begin{aligned} \frac{1}{\omega_{1,2}}\frac{\partial|E_{SA1,20}(t)|^2}{\partial t} & = \mp\frac{8\varepsilon_0 \mathrm{Im}G\left(k_x,k_z^S,\Delta\omega\right)k_x^2 h\left(k_z^S\right)^2}{\varepsilon_\perp\left|G\left(k_x,k_z^S,\Delta\omega\right)\right|^2} \\ & \times|E_{SA10}(t)|^2|E_{SA20}(t)|^2 F_N\left(k_x,k_z^S\right) \end{aligned} \tag{36}$$

$$\frac{1}{\omega_{1,2}}\frac{\partial\theta_{1,2}}{\partial t} = -\frac{4\varepsilon_0\,\mathrm{Re}\,G\left(k_x,k_z^S,\Delta\omega\right)k_x^2 h\left(k_z^S\right)^2}{\varepsilon_\perp\left|G\left(k_x,k_z^S,\Delta\omega\right)\right|^2}|E_{SA2,10}(t)|^2 F_N\left(k_x,k_z^S\right) \tag{37}$$

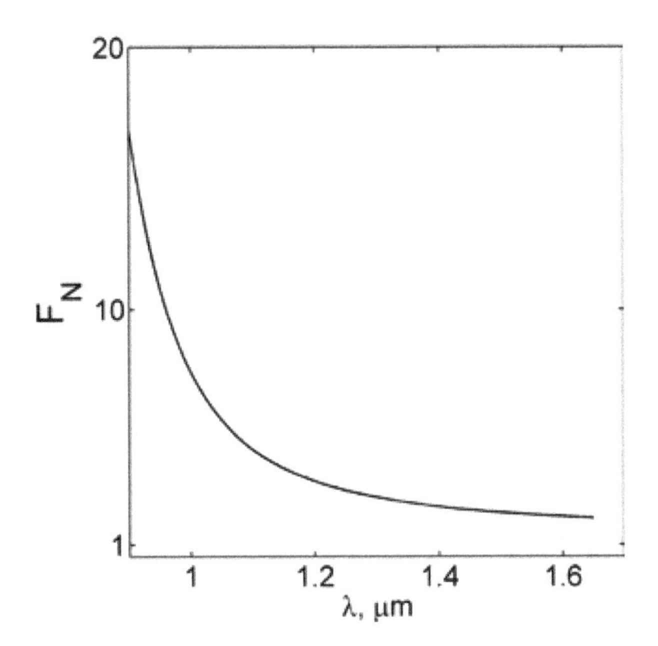

Figure 7.
The spectral dependence of the localization factor $F_N\left(k_x, k_z^S\right)$.

where the localization factor $F_N\left(k_x, k_z^S\right)$ is given by

$$\begin{aligned}
&F_N\left(k_x, k_z^S\right)\\
&=\{\frac{\left[\sinh\left(4k_z^S d\right)+4k_z^S d\right]}{4}\left[a_\perp - a_\parallel\frac{\varepsilon_\perp k_x^2}{\varepsilon_\parallel\left(k_z^S\right)^2}+\varepsilon_a\frac{k_x^2}{\left(k_z^S\right)^2}\left(1+\frac{\varepsilon_\perp}{\varepsilon_\parallel}\right)\right]\\
&+\sinh\left(2k_z^S d\right)\left(a_\perp + a_\parallel\frac{\varepsilon_\perp k_x^2}{\varepsilon_\parallel\left(k_z^S\right)^2}\right)-\left(2k_z^S d\right)\varepsilon_a\frac{k_x^2}{\left(k_z^S\right)^2}\left(1+\frac{\varepsilon_\perp}{\varepsilon_\parallel}\right)\}\\
&\times\left[\sinh\left(2k_z^S d\right)\left(1+\frac{k_x^2}{\left(k_z^S\right)^2}\right)+\left(2k_z^S d\right)\left(1-\frac{k_x^2}{\left(k_z^S\right)^2}\right)\right]^{-1}
\end{aligned} \tag{38}$$

The spectral dependence of the localization factor $F_N\left(k_x, k_z^S\right)$ is presented in **Figure 7**.

It is seen from **Figure 7** that $F_N\left(k_x, k_z^S\right)$ is varying by an order of magnitude in the range of the optical wavelengths essential for optical communications. The addition of Eq. (36) results in the following conservation condition [6]:

$$\frac{\partial}{\partial t}\left(\frac{|E_{SA10}(t)|^2}{\omega_1}+\frac{|E_{SA20}(t)|^2}{\omega_2}\right)=0 \tag{39}$$

We obtain from Eq. (39) the Manley-Rowe relation for the SVA magnitudes $|E_{SA1,20}(t)|^2$ [6]:

$$\frac{|E_{SA10}(t)|^2}{\omega_1} + \frac{|E_{SA20}(t)|^2}{\omega_2} = const = I_0 \tag{40}$$

We introduce the dimensionless quantities

$$I_{1,2}(t) = \frac{|E_{SA1,20}(t)|^2}{\omega_{1,2} I_0}; \;\; I_1(t) + I_2(t) = 1 \tag{41}$$

Substituting relationship (41) into Eq. (36), we obtain

$$\frac{\partial I_{1,2}}{\partial t} = \mp g I_1 I_2 \tag{42}$$

where the gain g has the form

$$g = \frac{8\varepsilon_0 \mathrm{Im} G\left(k_x, k_z^S, \Delta\omega\right) k_x^2 h \left(k_z^S\right)^2 \omega_1 \omega_2 I_0}{\varepsilon_\perp \left| G\left(k_x, k_z^S, \Delta\omega\right)\right|^2} F_N\left(k_x, k_z^S\right) \tag{43}$$

The spectral dependence of the gain g is shown in **Figure 8**. The solution of Eq. (41) has the form

$$I_1(t) = \frac{I_1(0)\exp(-gt)}{1 - I_1(0)[1 - \exp(-gt)]} \tag{44}$$

$$I_2(t) = \frac{1 - I_1(0)}{1 - I_1(0)[1 - \exp(-gt)]} \tag{45}$$

It is easy to see from Eqs. (44) and (45) that the solutions $I_{1,2}(t)$ satisfy the Manley-Rowe relation (40). Expressions (44) and (45) describe the energy

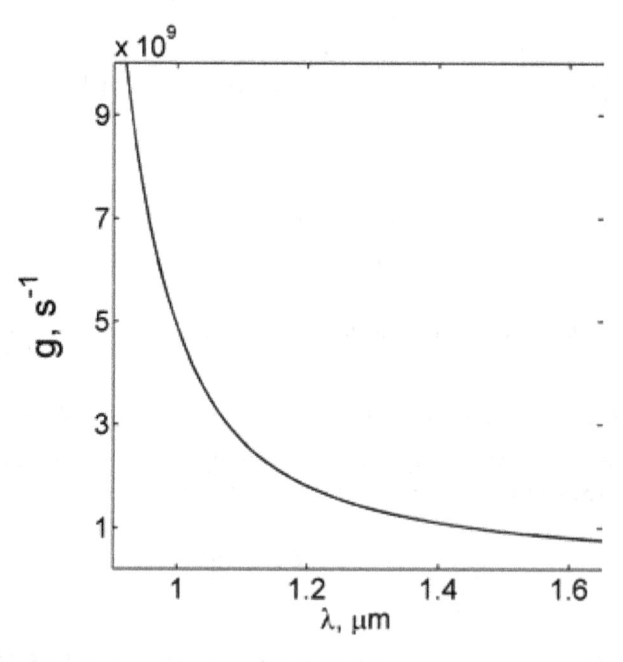

Figure 8.
The spectral dependence of the gain g for the SPP electric field amplitude $|E_{SA10}(t)| = 10^6 V/m$.

exchange between the SPPs interfering on the smectic layer dynamic grating. Indeed, $I_1 \to 0$ and $I_2 \to 1$ for $g > 0$ and $t \to \infty$. Actually, the SLS of the orientational type takes place [6]. The SPP_1 with the normalized intensity I_1 plays a role of the pumping wave, while the SPP_2 is a signal wave. The temporal dependence of $I_{1,2}(t)$ for the pumping wave amplitude $|E_{SA10}(t)| = 10^6 V/m$ is shown in **Figure 9**.

It is seen form **Figure 6** that for $I_1(0) > I_2(0)$, the characteristic time t_0 exists when $I_1(t_0) = I_2(t_0)$. Using expressions (44) and (45), we obtain

$$t_0 = \frac{1}{g} \ln \left[\frac{I_1(0)}{I_2(0)}\right] \tag{46}$$

Substitute expression (46) into Eqs. (44) and (45). Then they take the form

$$I_{1,2}(t) = \frac{1}{2}\left[1 \mp \tanh\left[\frac{g}{2}(t - t_0)\right]\right] \tag{47}$$

The time duration of the energy exchange between the SPPs is about $1ns$ as it is seen from **Figure 9**. Substituting relationships (43)–(45) into Eq. (37), we evaluate the phases $\theta_{SA1,2}(t)$. They are given by the following expressions:

$$\theta_{SA1}(t) - \theta_{SA1}(0) = -\frac{\operatorname{Re} G\left(k_x, k_z^S, \Delta\omega\right)}{2\operatorname{Im} G\left(k_x, k_z^S, \Delta\omega\right)} \times \ln\left[I_2(0)\exp(gt) + I_1(0)\right] \tag{48}$$

$$\theta_{SA2}(t) - \theta_{SA2}(0) = \frac{\operatorname{Re} G\left(k_x, k_z^S, \Delta\omega\right)}{2\operatorname{Im} G\left(k_x, k_z^S, \Delta\omega\right)} \times \ln\left[1 - I_1(0) + I_1(0)\exp(-gt)\right] \tag{49}$$

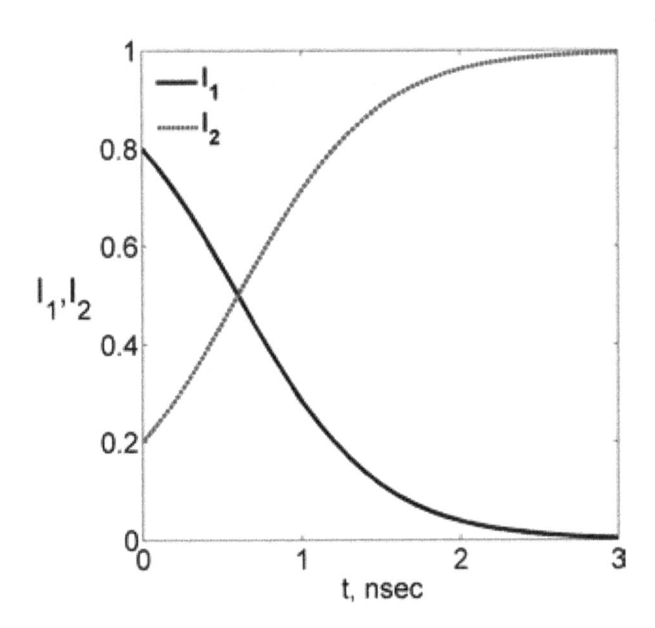

Figure 9.
The temporal dependence of the SPP normalized intensities $I_{1,2}(t)$ for pumping wave amplitude $|E_{SA10}(t)| = 10^6 V/m$.

The temporal dependence of the SPP SVA phases $\theta_{SA1,2}(t)$ is shown in **Figure 10**.

It is seen from expressions (48) and (49) that SLS of the SPPs in the MIM waveguide is accompanied by XPM. For the large time intervals $t \to \infty$, the phase of the pumping wave increases linearly:

$$\theta_{SA1}(t) - \theta_{SA1}(0) \to -\frac{\operatorname{Re} G\left(k_x, k_z^S, \Delta\omega\right)}{2\operatorname{Im} G\left(k_x, k_z^S, \Delta\omega\right)} gt \tag{50}$$

Such a behavior corresponds to the rapid oscillations of the depleted pumping wave amplitude. The signal wave phase for $t \to \infty$ tends to a constant value:

$$\theta_{SA2}(t) - \theta_{SA2}(0) \to \frac{\operatorname{Re} G\left(k_x, k_z^S, \Delta\omega\right)}{2\operatorname{Im} G\left(k_x, k_z^S, \Delta\omega\right)} \ln\left[1 - I_1(0)\right] \tag{51}$$

Substituting expressions (41) and (47) into Eq. (29), we obtain the explicit expression for the dynamic grating amplitude. It takes the form

$$U_0 = -\frac{2\varepsilon_0 I_0 k_x^2 k_z^S \sqrt{\omega_1 \omega_2} h}{G\left(k_x, k_z^S, \Delta\omega\right) \cosh\left[\frac{g}{2}(t - t_0)\right]} \exp i(\theta_{SA1} - \theta_{SA2}) \tag{52}$$

The temporal dependence of the amplitude (52) normalized absolute value $|U_0/U_{0\,\max}|$ is presented in **Figure 11**. Here

$$|U_{0\,\max}| = \frac{2\varepsilon_0 I_0 \left|k_x^2 k_z^S\right| \sqrt{\omega_1 \omega_2} |h|}{\left|G\left(k_x, k_z^S, \Delta\omega\right)\right|} \tag{53}$$

We evaluate now the hydrodynamic flow velocity in the MIM wave guide core. Substituting expression (28) into Eqs. (1) and (6), we obtain

$$v_z(x, z, t) = -i\Delta\omega U_0 \sinh\left(2k_z^S z\right) \exp\left[i(2k_x x - \Delta\omega t)\right] + c.c. \tag{54}$$

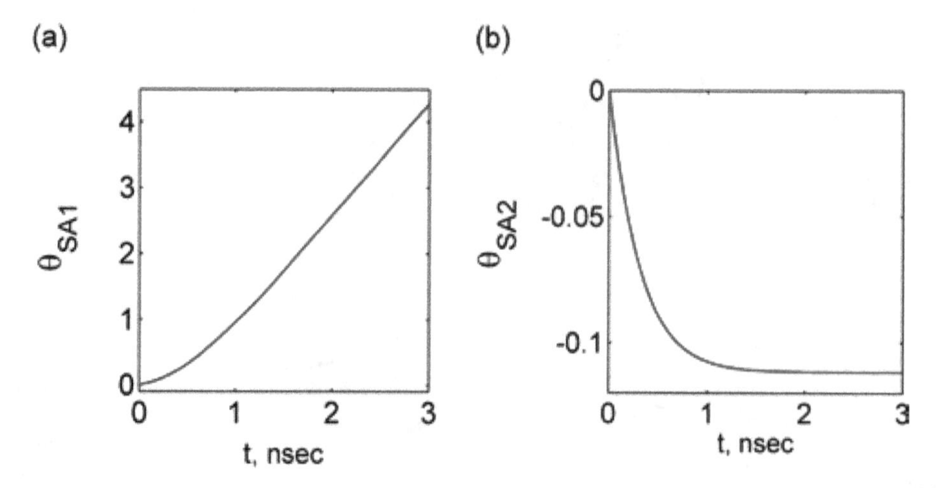

Figure 10.
The temporal dependence of the SPP SVA phases $\theta_{SA1}(t)$ (a) and $\theta_{SA2}(t)$ (b).

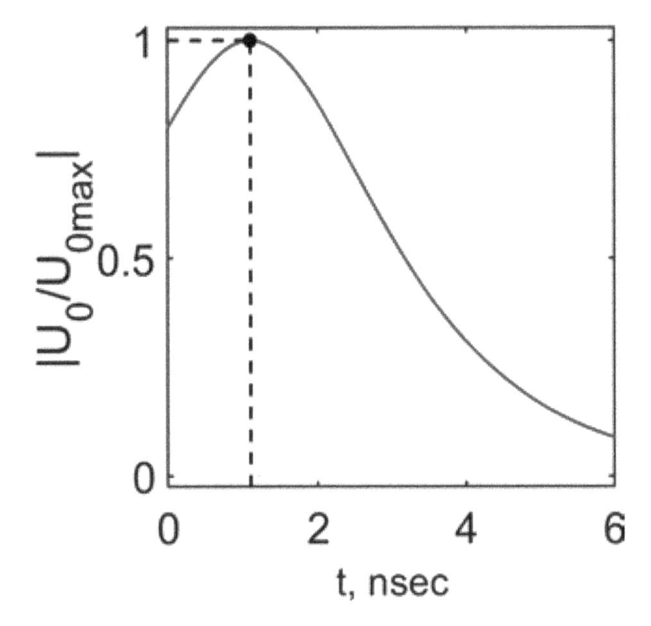

Figure 11.
The temporal dependence of the dynamic grating amplitude normalized absolute value $|U_0/U_{0\,\max}|$.

$$v_x(x, z, t) = \Delta\omega U_0 \frac{k_z^S}{k_x} \cosh\left(2k_z^S z\right) \exp\left[i(2k_x x - \Delta\omega t)\right] + c.c. \tag{55}$$

Expressions (28) and (52)–(55) and **Figure 11** show that the orientational and hydrodynamic excitations in SmALC core of the MIM waveguide enhanced by the SPPs are spatially localized and reach their maximum value during the time of the energy exchange between the interacting SPPs.

5. Conclusions

We investigated theoretically the nonlinear interaction of SPPs in the MIM waveguide with the SmALC core. The third-order nonlinearity mechanism is related to the smectic layer oscillations that take place without the change of the mass density. We solved simultaneously the equation of motion for the smectic layer normal displacement and the Maxwell equations for SPPs including the nonlinear polarization caused by the smectic layer strain. We evaluated the dynamic grating of the smectic layer displacement enhanced by the interfering SPPs. We evaluated the SVAs of the interacting SPPs. It has been shown that the SLS of the orientational type takes place. The pumping wave is depleted, while the signal wave is amplified up to the saturation level defined by the total intensity of the interacting waves. SLS is accompanied by XPM. The phase of the depleted pumping wave rapidly increases, while the phase of the amplified wave tends to a constant value. The SPP characteristic rise time is of the magnitude of 10^{-9} s for a feasible SPP electric field of 10^6 V/m. The smectic layer displacement and hydrodynamic velocity enhanced by SPPs are spatially localized and reach their maximum value during the time of the strong energy exchange between the interfering SPPs.

Author details

Boris I. Lembrikov*, David Ianetz and Yossef Ben Ezra
Department of Electrical Engineering and Electronics, Holon Institute of Technology (HIT), Holon, Israel

*Address all correspondence to: borisle@hit.ac.il

References

[1] Kauranen M, Zayats AV. Nonlinear plasmonics. Nature Photonics. 2012;**6**: 737-748. DOI: 10.1038/NPHOTON. 2012.244

[2] Willner AE, Yilmaz OF, Wang J, Wu X, Bogoni A, Zhang L, et al. Optically efficient nonlinear signal processing. IEEE Journal of Selected Topics in Quantum Electronics. 2011;**17**: 320-332. DOI: 10.1109/JSTQE.2010. 2055551

[3] Willner AE, Khaleghi S, Chitgarha MR, Yilmaz OF. All-optical signal processing. Journal of Lightwave Technology. 2014;**32**:660-680. DOI: 10.1109/JLT.2013.2287219

[4] Gai X, Choi D-Y, Madden S, Luther-Davies B. Materials and structures for nonlinear photonics. In: Wabnitz S, Eggleton BJ, editors. All-Optical Signal Processing. Heidelberg: Springer; 2015. pp. 1-33. DOI: 10.1007/978-3-319-14992-9

[5] Kik PG, Brongersma ML. Surface plasmon nanophotonics. In: Brongersma ML, Kik PG, editors. Surface Plasmon Nanophotonics. Dordrecht: Springer; 2007. pp. 1-9. ISBN: 978-1-4020-4349-9

[6] Shen YR. The Principles of Nonlinear Optics. Hoboken. New Jersey, USA: Wiley; 2003. p. 563. ISBN: 0-471- 43080-3

[7] Maier SA. Plasmonics: Fundamentals and Applications. New York: Springer; 2007. p. 223. ISBN: 978-0387-33150-8

[8] Sarid D, Challener W. Modern Introduction to Surface Plasmons. Cambridge: Cambridge University Press; 2010. 371 p. ISBN: 978-0-521-76717-0

[9] Sederberg S, Firby CJ, Greig SR, Elezzabi AY.Integrated nanoplasmonic waveguides for magnetic, nonlinear, and strong-field devices. Nanophotonics. 2017;**6**:235-257. DOI: 10.1515/ nanoph-2016-0135

[10] Lapine M. Colloquium: Nonlinear metamaterials. Reviews of Modern Physics. 2014;**86**:1093-1123. DOI: 10.11 03/RevModPhys.86.1093

[11] Zheludev NI, Kivshar YS. From metamaterials to metadevices. Nature Materials. 2012;**11**:917-924. DOI: 10.10 38/NMAT3431

[12] Khoo IC. Nonlinear optics, active plasmonics and metamaterials with liquid crystals. Progress in Quantum Electronics. 2014;**38**:77-117. DOI: 10.10 16/j.pquantelec.2014.03.001

[13] Khoo IC. Nonlinear optics of liquid crystalline materials. Physics Reports. 2009;**471**:221-267. DOI: 10.1016/j. physr ep.2009.01.001

[14] Khoo I-C. Liquid Crystals. 2nd ed. Hoboken, New Jersey, USA: Wiley; 2007. p. 368. ISBN 978-0-471-75153-3

[15] Giallorenzi TG, Weiss JA, Sheridan JP. Light scattering from smectic liquid-crystal waveguides. Journal of Applied Physics. 1976;**47**: 1820-1826. DOI: 10.1063/1.322898

[16] Kventsel GF, Lembrikov BI. Two-wave mixing on the cubic non-linearity in the smectic A liquid crystals. Liquid Crystals. 1994;**16**:159-172. ISSN: 0267-82 92

[17] Kventsel GF, Lembrikov BI. Stimulated light scattering in smectic A liquid crystals. Liquid Crystals. 1995;**19**: 21-37. ISSN: 0267-8292

[18] Kventsel GF, Lembrikov BI. Self-focusing and self-trapping in smectic A liquid crystals. Molecular Crystals and

Liquid Crystals. 1995;**262**:629-643. ISSN: 1542-1406

[19] Kventsel GF, Lembrikov BI. The four-wave mixing and the hydrodynamic excitations in smectic A liquid crystals. Molecular Crystals and Liquid Crystals. 1995;**262**:5 91-627. ISSN: 1542-1406

[20] Kventsel GF, Lembrikov BI. Second sound and nonlinear optical phenomena in smectic A liquid crystals. Molecular Crystals and Liquid Crystals. 1996;**282**: 145-189. ISSN: 1542-1406

[21] Lembrikov BI. Light interaction with smectic A liquid crystals: Nonlinear effects. HAIT Journal of Science and Engineering. 2004;**1**:306-347. ISSN: 1565-4990

[22] Lembrikov BI, Ben-Ezra Y. Surface plasmon polariton (SPP) interactions at the interface of a metal and smectic liquid crystal. In: Proceedings of the 17th International Conference on Transparent Optical Networks (ICTON 2015); July 5-9, 2015; Budapest, Hungary (We.C4.4). 2015. pp. 1-4

[23] Lembrikov BI, Ben-Ezra Y, Ianetz D. Stimulated scattering of surface plasmon polaritons (SPPs) in smectic A liquid crystal. In: Proceedings of the 18th International Conference on Transparent Optical Networks (ICT-ON- 2016); July 10-14, 2016; Trento, Italy (We.B4.2). 2016. pp. 1-4

[24] Lembrikov BI, Ianetz D, Ben-Ezra Y. Metal/insulator/metal (MIM) plasmonic waveguide containing a smectic A liquid crystal (SALC) layer. In: Proceedings of the 19th International Conference on Transparent Optical Networks (ICTON 2017); July 2-6, 2017; Girona, Catalonia, Spain (Tu.A4.3). 2017. pp. 1-4

[25] Lembrikov BI, Ianetz D, Ben-Ezra Y. Nonlinear optical phenomena in silicon-smectic A liquid crystal (SALC) waveguiding structures. In: Proceedings of the 20th Int'l. Conf. on Transparent Optical Networks (ICTON 2018), 1-5 July 2018 Bucharest, Romania (Mo. D4.1). 2018. pp. 1-4

[26] Lembrikov BI, Ianetz D, Ben-Ezra Y. Nonlinear optical phenomena in smectic A liquid crystals. In: Choudhury PK, editor. Liquid Crystals—Recent Advancements in Fundamental and Device Technologies. Rijeka, Croatia: InTechOpen; 2018. pp. 131-157. DOI: 10.5772/intechopen.70997

[27] Lembrikov BI, Ianetz D, Ben-Ezra Y. Nonlinear optical phenomena in a silicon-smectic A liquid crystal (SALC) waveguide. Materials. 2019;**12**:2086-1-2086-17. DOI: 10.3390/ma12132086

[28] Lembrikov BI, Ianetz D, Ben-Ezra Y. Light enhanced flexoelectric polarization in waveguiding structures with a smectic A liquid crystal (SALC) layer. In: Proceedings of the 20th International Conference on Transparent Optical Networks (ICTON 2019); 9-13 July 2019; Angers, France, Th.B4.3. 2019. pp. 1-4

[29] De Gennes PG, Prost J. The Physics of Liquid Crystals. 2nd ed. New York, USA: Oxford University Press; 1993. 597 p. ISBN: 978-0198517856

[30] Stephen MJ, Straley JP. Physics of liquid crystals. Reviews of Modern Physics. 1974;**46**:617-704. ISSN: 0034-6861

[31] Chandrasekhar S. Liquid Crystals. 2nd ed. Cambridge: Cambridge University Press; 1992. p. 480. ISBN: 978-0521427418

[32] Liao Y, Clark NA, Pershan PS. Brillouin scattering from smectic liquid crystals. Physical Review Letters. 1973; **30**:639-641. DOI: 10.1103/PhysRevLett.30.639

[33] Ricard L, Prost J. “Second sound” propagation and the smectic response function. Journal DE Physique Colloque C3. 1979;**40**(supplement 4): C3-83-C3-86. DOI: 10.1051/jphyscol:1979318

[34] Ricard L, Prost J. Critical behavior of second sound near the smectic A nematic phase transition. Journal De Physique. 1981;**42**:861-873. DOI: 10.1051//jphys:0198100420 6086100

[35] Zografopoulos DC, Asquini R, Kriezis EE, d’Alessandro A, Becceherelli R. Guided-wave liquid-crystal photonics. Lab on a Chip. 2012; **12**:3598-3610. DOI: 10.1039/c21c40514h

[36] Zografopoulos DC, Becceherelli R, Tasolamprou AC, Kriezis EE. Liquid-crystal tunable waveguides for integrated plasmonic components. Photonics and Nanostructures-Fundamentals and Applications. 2013; **11**:73-84. DOI: 10.1016/j.photonics.2012.08.004

[37] Zografopoulos DC, Beccherelli R. Liquid-crystal-tunable metal-insulator-plasmonic waveguides and Bragg resonators. Journal of Optics. 2013;**15**: 1-5. DOI: 10.1088/2040-8978/15/5/0550 09

[38] Prokopidis KP, Zografopoulos DC, Kriezis EE. Rigorous broadband investigation of liquid-crystal plasmonic structures using finite-difference time-domain dispersive-anisotropic models. Journal of the Optical Society of America B. 2013;**30**(10):2722-2730. DOI: 10.1364/JOSAB.30.002722

[39] Ma L-L, Hu W, Zheng Z-G, Wu S-B, Chen P, Li Q, et al. Light-activated liquid crystalline hierarchical architecture toward photonics. Advanced Optical Materials. 2019; **1900393**:1-19. DOI: 10.1002/adom.201900393

[40] Suhara T, Fujimura M. Waveguide Nonlinear-Optic Devices. Berlin, Germany: Springer; 2003. p. 321. ISBN: 3-540-01527-2

3

Nanomanipulation with Designer Thermoplasmonic Metasurface

Chuchuan Hong, Sen Yang and Justus Chukwunonso Ndukaife

Abstract

Plasmonic nanoantennas provide an efficient platform to confine electromagnetic energy to the deeply subwavelength scales. The resonant absorption of light by the plasmonic nanoantennas provides the means to engineer heat distribution at the micro and nanoscales. We present thermoplasmonic metasurfaces for on-chip trapping, dynamic manipulation and sensing of micro and nanoscale objects. This ability to rapidly concentrate objects on the surface of the metasurface holds promise to overcome the diffusion limit in surface-based optical biosensors. This platform could be applied for the trapping and label-free sensing of viruses, biological cells and extracellular vesicles such as exosomes.

Keywords: metasurface, nano-antenna, thermoplasmonic, optofluidic, nano-tweezer

1. Introduction

As an important category of metamaterials [1–4], which are artificial structures designed to achieve unique properties not occurring in nature, electromagnetic metasurfaces are able to provide unique electromagnetic responses for complete control over the phase, amplitude or polarization of light in a two-dimensional platform [5–9]. There are numerous kinds of applications that can be enabled by metasurfaces including holograms, metalens, near-eye displays, vortex beam generators and compact Spatial Light Modulators (SLM) [9–12] (**Figure 1**).

A meta-atom is the building block of a metasurface and it is usually a resonator made of either plasmonic or dielectric materials. In this section, we will focus on plasmonic metasurfaces to emphasize the significance of plasmonic metasurfaces for diverse applications, including on-chip optical trapping and optofluidic control. We will also present thermoplasmonic metasurfaces capable of inducing microfluidic motion in microchannels based on electrothermoplasmonic (ETP) effect, and discuss their role in on-chip nanoparticle trapping and manipulation [16–19].

Noble metals such as gold (Au) or silver (Ag) are typically used as plasmonic materials at visible or near-IR frequencies. Due to their ability to sustain plasmon oscillation, a well-designed metallic nanostructure can confine light in a very tiny volume, which generally creates a smaller mode volume than dielectric structures, especially at the resonance frequency. Hence, plasmonic resonators can tightly confine light in a tiny region so that they largely enhance the local electric field intensity. Furthermore, due to the Ohmic losses, the plasmonic materials generate heat under illumination. The enhanced local electric field results in efficient local

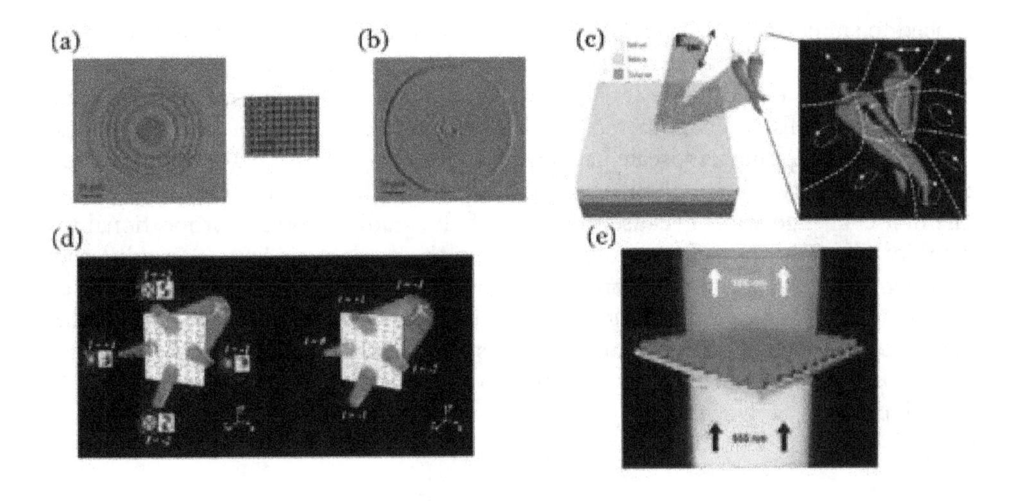

Figure 1.
Various applications based on metasurfaces. (a) SEM image of metasurface lens composed of silicon nano-post array. (b) SEM image of a vortex beam generator made of silicon nano-post array. (c) A plasmonic metasurface hologram insensitive to light polarization states. (d) A vortex beam opto-multiplexer and demultiplexer made of gold at terahertz. (e) A titanium dioxide metasurface enhancing third harmonic generation to produce ultraviolet light.

heating. Although, several research works in the literature have considered the loss in plasmonics as detrimental, it is now known that the loss in plasmonics can be beneficial to fast on-chip nano-particle manipulation [17, 20]. The loss-induced heating effect has given rise to the burgeoning field of thermoplasmonic metasurface.

2. Optical trapping and optical tweezers

Optical tweezer, which employs a tightly focused laser to trap microscopic objects have proven to be a versatile tool for many scientific researches such as in biophysics [21] and was recently recognized with a 2018 Physics Nobel prize. The initial experiment to trap particles with a laser beam used two counter-propagating loosely focused beams to localize the particles at a node of the generated standing wave, and this was reported by Arthur Ashkin and his colleagues in 1970 [22]. Subsequently, a single beam optical tweezer, which employs a tightly focused laser beam to achieve three-dimensional manipulation and trapping of microscale parti-cles was demonstrated in 1986 [23]. Dielectric particles are trapped when the refractive index of the particles is higher than the refractive index of the surround-ing medium.

Generally, the optical force is decomposed into two parts, the gradient force and the scattering force. The scattering force is also called radiation pressure and acts in the direction of light propagation. The gradient force is the significant part in single beam optical trapping procedure, because it is oriented perpendicular to the axis and push particles towards a region with higher optical intensity. Thus, the gradient force ensures that a particle is trapped in an optical tweezer. The total optical force induced on a particle can be determined by Maxwell's Stress Tensor method (MST). When the size of the particle is much smaller than the wavelength of trapping light, the particl e can be considered as a dipole, so that the dipole approximation can be applied to simplifying the calculation of the optical force [24]. The detailed derivation of MST and dipole approximation can be obtained in Ref. [25].

In biological applications, target particles range in size from micrometer size scales such as cells and nanometer length scales such as viruses, protein molecules and vesicles. Trapping nanoscale particles is challenging in free-space optical tweezers. There are two main challenges faced by researchers using optical tweezers. The first challenge arises because the value of the gradient force is proportional to the third power of the radius of the particles in the quasi-static limit. Secondly, the optical gradient force is proportional to the gradient of light field intensity, and it is thus limited because the diffraction limit limits the achievable gradient in the light field intensity. One approach explored by Ashkin to increase the trapping stability for nanoscale objects in free-space optical tweezers involves the use of very high optical powers. However, this results in serious damage to the objects being trapped, a process termed opticution.

3. Plasmonic optical tweezers

In order to overcome the limit on particle size that can be trapped in conventional optical tweezers, plasmonic optical tweezers have been proposed Novotny et al. [17, 26–33]. Plasmonic optical tweezers employ plasmonic nanoantenna, which can squeeze light to nanoscale volumes comparable to the size of the target particles, thus significantly enhancing the gradient force applied on nanoscale particles. Unlike the light-matter interaction between dielectric materials and electromagnetic wave, plasmonic materials uniquely react to the light field through free-electron-photon coupling to generate a specific type of surface wave called surface plasmon polariton (SPP). Furthermore, a subwavelength plasmonic particle will efficiently couple to propagating light to generate localized surface plasmon resonance (LSPR) and further enhance the electric field due to this resonance. These phenomena permit light to be strongly localized at deep subwavelength scales. Besides, surface plasmon structures offer additional advantages on lab-on-a-chip manipulation due to its compatibility with integrated photonics devices [34, 35].

There are two main challenges with the use of plasmonic nanoantennas for near-field nano-optical tweezers. These challenges include the Ohmic loss in the materials, which invariably results in loss-induced heating, and the need to rely on slow Brownian diffusion to transport particles towards the illuminated nanoantenna. The Ohmic loss in plasmonic materials at visible or near-IR is inevitable. It will not only hamper the efficiency of particle trapping due to the need to generate enough trapping potential energy to overcome the Brownian motion and possible thermophoretic force, but the heat generated by Ohmic loss would be problematic in many experiments. Though the light field is tightly confined near an illuminated plasmonic nanoantenna leading to very high field intensity enhancement [16], which are advantageous for particle trapping, the damage from temperature rise due to this field enhancement affects the experimental trapping stability in several ways. Experimentally, bioparticles are vulnerable in an environment with increas-ing temperature. Furthermore, excessive photothermal heating could deform the plasmonic nanoantenna [36]. In aqueous solutions, this heating effect would even generate bubbles by boiling the water and disrupt the entire experiment. Wang et al. demonstrated that by integrating a heat sink in plasmonic tweezers made up of layers of high thermal conductivity materials, the adverse effects arising from the heating effect can be mitigated [37].

Rather than treating the photothermal heating effect as detrimental in plasmonic nanotweezer experim ents, pl asmonic heating can be harnessed to enable new functionalities in near-field nano-optical trapping. Ndukaife et al. proposed and demonstrated that thermal effect due heating loss can actually promote the trapping process thereby eliminating the need to rely on slow Brownian diffusion to deliver particles towards the plasmonic nanotweezer [18, 20]. This new platform, referred to as electrothermoplasmonic tweezers harnesses the localized photothermal heating from an illuminated plasmonic nanoantenna to establish a thermal gradient in the fluid. The thermal gradient results in a gradient in the permittivity and electrical conductivity of the fluid. An applied AC electrical field acts on these gradients to result in rapid microfluidic flow motion, which transports particles towards the plasmonic hotspot. This typical fluid motion is called electrothermoplasmonic (ETP) flow. Under this condition, heating effect actually speeds up the loading procedure and converts the slow traditional diffusion-based particle loading of conventional plasmonic nano-tweezers into a fast and directional particle loading in electrothermoplasmonic tweezers. The technique works with small temperature rises of a few degrees and only the thermal gradient needs to be optimized, thereby making it suitable for handling biological objects.

By all means, plasmonic optical tweezers are promising and especially useful for manipulation of extremely small particles. By careful thermal engineering to mitigate excessive heating effect, plasmonic optical tweezers can be utilized to enable many applications.

4. Electrothermoplasmonic, electro-osmotic and thermophoretic effects

As mentioned in the previous section, the heating effect from plasmonic nanoantennas can be utilized to enable new capabilities in near-field nano-optical trapping. The physics nature behind this kind of trapping comprises the interplay of multiple physical phenomena. In this section, we describe several forces that can be experienced by objects in electrothermoplasmonic tweezers in the presence of optical illumination and applied AC electric field. To understand how the thermal effect influences the trapping process, we firstly look at the thermophoresis phenomenon, which is the motion of particles or molecules in the presence of thermal gradients. Unlike the normal diffusion process due to Brownian motion, thermophoresis is induced by a temperature gradient ∇T to cause the drifting of particles. In most cases, particles prefer to move towards the region of lower temperature, away from the plasmonic nanoantenna. This trend is called "thermophobic" behavior or positive thermophoresis. However, this motion can be reversed in certain instances whereby particles will move towards the region of higher temperature, which is called "thermophilic" behavior, also known as negative thermophoresis.

Quantitatively, in a diluted suspension (particle weight fraction w << 1), the mass flow J can be written as [38, 39]:

$$J \approx -D\nabla c - cD_T\nabla T, \tag{1}$$

where D and D_T are the Brownian diffusion coefficient and thermodiffusion coefficient, respectively.c denotes the concentration. In diluted concentrations, thermodiffusion velocity is generally assumed to be linearly dependent on the temperature gradient ∇T with the thermodiffusion coefficient D_T:

$$\vec{v} = -D_T\nabla T \tag{2}$$

Under steady-state, the thermodiffusion is balanced by ordinary diffusion and the concentration coefficients are related in an exponential law:

$$c/c_0 = \exp\left[-(D_T/D)(T - T_0)\right] \tag{3}$$

Finally, the thermophoretic force induced by thermal gradients on a particle in the fluid is given by [40]:

$$F_{therm} = -K_b T(z) S_T \nabla T \tag{4}$$

Here, K_b stands for Boltzmann's constant and $T(z)$ corresponds to the temperature at a given position z. Soret coefficient S_T is defined as the ratio: $S_T = D_T/D$, which depicts how strong the thermodiffusion is at steady state. The Soret coefficient is influenced by many factors including temperature, size of the particles, surface charge of the particles and ions in the solvent. From Eq. (4), we concluded that the thermophoretic force is proportional to Soret coefficient but with opposite sign. It is useful to know that Soret coefficient is also temperature dependent and flips its sign from positive to negative by decreasing temperature, as expressed in Eq. (5) [38, 39]. S_T^{∞} represents a high-T thermophobic limit and T^* is the temperature where S_T switches sign, and T^0 embodies the strength of temperature effects. At that moment, force applied onto particles switches its orientation from "thermophobic" to "thermophilic" behavior.

$$S_T(T) = S_T^{\infty}\left[1 - \exp\left(\frac{T^* - T}{T^0}\right)\right] \tag{5}$$

The direction of the thermophoretic force can be tuned from a repulsive to an attractive force by tuning the interfacial permittivity of the electrical double layer (EDL) surrounding the particles in solution. The EDL exists on the surface of an object when it is exposed to a fluid. The object could be a small charged particle floating inside the liquid or a surface coated with fluid. The charged object attracts counterions from the solution to screen the surface charge. This layer is called the Stern layer. This charged object consequently attracts ions from the liquid with opposite charge via Coulombic force to electrically screen the first layer and the particle. The second layer is called diffuse layer. The potential difference across the diffuse layer is defined as Zeta potential, ζ. The charges in the diffuse layer are not as tightly anchored to the particle as the first layer and the thickness can be affected by tangential stress [41].

Based on Anderson's model [42], the thermophoretic mobility of the particle is associated with its Zeta potential.

$$D_T = -\frac{\varepsilon}{2\eta T}\frac{2\Lambda_1}{2\Lambda_1 + \Lambda_p}\left(1 + \frac{\partial ln\varepsilon}{\partial lnT}\right)\zeta^2 \tag{6}$$

where ε is the solvent permittivity, and Λ_1 and Λ_p are the thermal conductivities of the solvent and the particle, respectively. In bulk water, for example, the differential permittivity change with temperature $\frac{\partial ln\varepsilon}{\partial lnT} = -1.4$ at room temperature. In building up the model of ET flow numerical analysis, this term can be taken into account to ensure the accuracy of the results [16]. Inside an EDL, however, the value of $\frac{\partial ln\varepsilon}{\partial lnT}$ can reach +2.4, which is crucial to reverse the Soret coefficient and induce a negative thermophoresis behavior. Hence, one way to induce negative thermophoresis behavior is to ensure that an EDL has been established, by charging the surface of the particles. Adding surfactants, such as Cetyltrimethylammonium Chloride (CTAC), for example, into particle solution can help to build up an EDL on the particles [27].

The photothermal heating of the fluid by the plasmonic nanoantennas can also be utilized to induce strong electro-convection fluidic motion, which is called electrothermal (ET) effect. When the temperature inside fluid is no longer uniform due to a thermal gradient, a gradient in the permittivity and the conductivity of the fluid will induced as well. In such a system, a local free charge distribution must be present if Gauss's Law and charge conservation are to be satisfied simulta-neously. When an AC electric field is applied, both free and bounded local charge density responds to the applied electric field so that a non-zero body force is generated on the fluid:

$$f_{ET} = \rho_e E - \frac{1}{2}|E|^2 \nabla \varepsilon_m \tag{7}$$

After perturbative expansion in the limit of small temperature gradient, the force density is expressed as [43]:

$$f_{ET} = \frac{1}{2} Re \left\{ \frac{\varepsilon(\alpha - \beta)(\nabla T.E)E^*}{\sigma + i\omega\varepsilon} + \frac{1}{2}\varepsilon\alpha|E|^2 \nabla T \right\} \tag{8}$$

where $\alpha = 1/_{\varepsilon}\frac{d\varepsilon}{dT}$ and $\beta = 1/_{\sigma}\frac{d\sigma}{dT}$. ε stands for permittivity and σ stands for conductivity. ω is the AC frequency. So far, the body force is decomposed into two parts: the first term on the right-hand side of Eq. (7) is the Coulombic force, and it increases as AC frequency goes down. It is worth mentioning that this electrother-mal flow is able to generate a fast flow to rapidly transport particles from a long distance of several hundreds of microns to the hot spot. One no longer need to wait for particles to diffuse through Brownian motion into the target region to be successfully trapped.

Based on this concept, a novel plasmonic nano-tweezer called electrothermo-plasmonic tweezers using a single plasmonic nanoantenna was invented [18, 44]. In this platform, a plasmonic nanoantenna is illuminated causing the surrounding fluid near the nanoantenna to be slightly heated up due to the Ohmic loss nature of plasmonic materials. Localized heating of fluid medium induces local gradients in the fluid's electrical conductivity and permittivity. An applied AC electric field acts on these gradients to induce an electrothermal microfluidic flow, which acts to transport particles towards the illuminated plasmonic nanoantenna.

The electrothermal flow from plasmon-induced heating can be modeled through the solution of the electromagnetic wave-equation, heat equation and the Navier-Stokes equation. Mathematically, we can solve the wave equation to find the electric field in the vicinity of the plasmonic nanoantenna:

$$\nabla \times (\nabla \times E) - k_0^2 \epsilon(\mathbf{r})\mathbf{E} = 0 \tag{9}$$

The heat source density generated by plasmonic heating is calculated using the electric field distribution and the induced current density representing the energy loss:

$$\mathrm{q(r)} = \frac{1}{2} Re\,(\mathrm{J} \cdot \mathbf{E}^*) \tag{10}$$

The temperature field distribution is determined by solving the heat equation below:

$$\nabla \cdot [-\kappa \nabla \mathrm{T}(\mathbf{r}) + \rho c_p \mathrm{T}(\mathbf{r})\mathrm{u}(\mathbf{r})] = q(\mathbf{r}) \tag{11}$$

Finally, the Navier-Stokes equation is solved to find the velocity of the fluid:

$$\rho_0[\mathbf{u}(\mathbf{r}) \cdot \nabla]\mathbf{u}(\mathbf{r}) + \nabla p(\mathbf{r}) - \eta\nabla^2\mathbf{u}(\mathbf{r}) = \mathbf{F}, \tag{12}$$

where **F** here is the electrothermal force described in the prior section [43]. We have also established the theoretical framework between the ETP flow velocity with laser power and external AC electric field. It is noted that the ETP flow velocity scales linearly with laser power and quadratically with AC electric field amplitude [16].

Electro-osmotic flow is another category of body flow happening in a microfluidic channel [45]. Electro-osmosis occurs when tangential electric field acts on the loosely bound charges in the EDL along the channel walls. Because of the existence of electrical double layer, near the charged channel walls the tangential electric field will act on the EDL charges to induce an electro-osmotic flow with slip velocity given by:

$$u_s = -\frac{\varepsilon_w \zeta}{\eta} E_{\parallel} \tag{13}$$

where $E_{\parallel}$ is the tangential component of the bulk electric field, η is the fluid viscosity and ζ is the zeta potential. When there is a plasmonic structure existing inside a microfluid channel on the channel wall, this plasmonic structure perturbs the AC electric field, resulting in a non-zero tangential component of the AC electric field. For instance, Jamshidi or Hwang and their colleagues integrated photoconductive material as the electrode of applied bias inside a microfluidic channel to design a so-called "NanoPen" device for dynamical manipulation on particles using low-power laser illumination [46, 47].

This perturbed external electric field not only contributes to electro-osmosis, but could also induce a dielectrophoretic force on the suspended particles [48]. Particles experience dielectrophoresis only when the external field is non-uniform and the force that particles feel, the DEP force, does not depend on the polarity of the AC field. DEP can be observed both in AC or DC electric field. There is a crucial evaluator for DEP force direction called Clausius-Mossoti factor.

$$K = \frac{\varepsilon_2 - \varepsilon_1}{\varepsilon_2 + 2\varepsilon_1} \tag{14}$$

In terms of Clausius-Mossoti factor, the DEP force is expressed as:

$$F_{DEP} = 2\pi\varepsilon_1 R^3 K \nabla E_0^2 \tag{15}$$

From which, we understand DEP force is proportional to particle volume and magnitude of K. The direction of DEP force is along the gradient of the electric field intensity and its sign depends on the sign of Clausius-Mossoti factor.

In AC field and lossy medium, the Clausius-Mossoti factor is frequency dependent and it is given by:

$$K = \frac{\varepsilon_2^* - \varepsilon_1^*}{\varepsilon_2^* + 2\varepsilon_1^*}, \varepsilon_1^* = \varepsilon_1 + \frac{\sigma_1}{j\omega} \text{ and } \varepsilon_2^* = \varepsilon_2 + \frac{\sigma_2}{j\omega} \tag{16}$$

The Clausius-Mossoti factor determines whether the DEP force is attractive or repulsive ant it depends on the frequency of AC field and electric properties of particles and medium.

So far, we have briefly introduced the main mechanisms which could occur in plasmonic nanotweezers. The physical phenomena could be harnessed to introduce new capabilities in plasmonic nanotweezers. A recent article [49] has proposed the use of DEP force to promote particle transport towards plasmonic resonators.

5. Thermoplasmonic nanohole metasurface

In this section, we discuss thermoplasmonic nanotweezer based on nanohole metasurface [50], which enables high-throughput large-ensemble nanoparticle assembly in a lab-on-a-chip platform. As mentioned in previous sections, optical metasurfaces achieve control over the properties of light. Besides the optical response, thermoplasmonic metasurfaces allow to engineer the thermal response at micro and nanoscale. Upon illumination of metasurface, the combination of optical and thermal effects enables robust large-ensemble many-particle trapping.

The nanohole metasurface comprises of an array of subwavelength nanoholes in a 125 nm thick gold film and illuminated with a laser of 1064 nm wavelength. This nanohole metasurface serves as a plasmonic resonator supporting both the localized surface plasmon resonance (LSPR) and Bloch mode surface plasmon polariton (SPP). Due to the plasmon response, electromagnetic field is confined near the rims of the nanohole and the region in between nanohole, as depicted in **Figure 2**. This enhances the photothermal conversion efficiency and hence higher temperature rise under illumination of the structured plasmonic film. As predicted, when an external AC field is applied onto the nanohole metasurface, a long-range electrothermo-plasmonic (ETP) flow is generated inside the fluid, bringing particles from long distance towards the hot spot. Simultaneously, because of the existence of nanohole array, the AC electric field in the channel is no longer uniform but tangential components are created, which induces AC electroosmotic flow. The numerical simulation results of electroosmotic flow are depicted in the **Figure 3**.

For the experimental demonstration, the gold metasurface is fabricated using template stripping method on a silicon wafer. Using standard nanofabrication procedures, an array of nanoholes are fabricated on a silicon wafer, which is the same dimension and scale as the nanohole array on gold film. A 125 nm gold film is then deposited onto this silicon template using an electron beam evaporation process. Finally, a UV-sensitive curing-agent was uniformly spread onto the gold film and covered by an ITO-coated glass substrate with the same size as the silicon template.

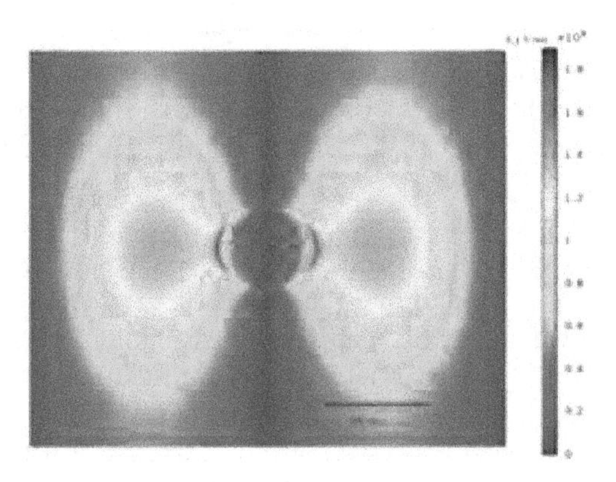

Figure 2.
Electric field distribution of one of the single nanoholes by numerical simulation.

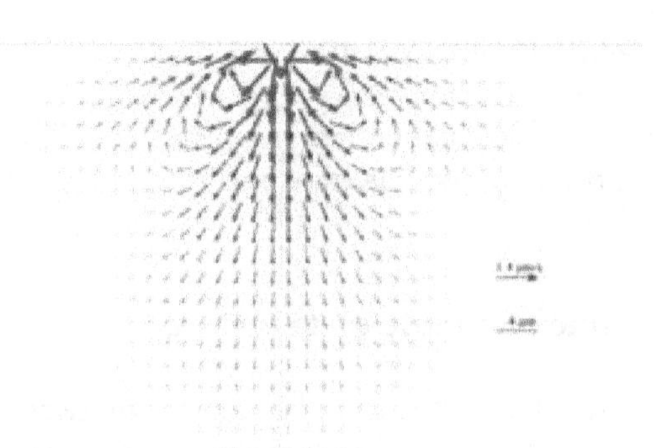

Figure 3.
Velocity profile of the induced AC electro-osmosis flow from numerical simulation.

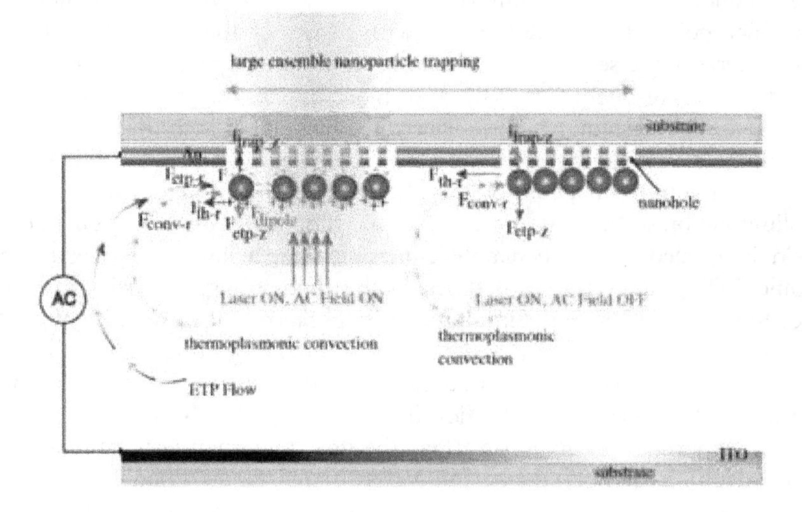

Figure 4.
Illustration of microfluidic chip coupling with nanohole plasmonic metasurface.

After UV illumination is applied to harden the curing-agent, the gold film layer is stripped off and transferred to a glass substrate by gently inserting a blade in between the gold film and silicon template due to the weak attachment between gold and silicon. The silicon template can be cleaned by soaking in hydrogen peroxide and sulfuric acid solution in the ratio 3:1 to remove the organics on the surface and any remaining gold flakes on the template. Furthermore, the residual gold remaining on the template can be removed thoroughly by a gold etchant. The template can be reused multiple times.

To make the gold nanohole metasurface into a microfluidic chip, another ITO-coated glass coverslip is placed above the nanohole array sample, with a 120 μm dielectric spacer in between. For the trapping experiments, a dilute solution of 1 μm PS beads was injected into the channel. AC electric field is applied in between the ITO-coated glass cover slip and gold film. A 1064 nm laser illumination is focused on the nanohole array. The mechanism of particle trapping in the thermoplasmonic nanohole metasurface channel is depicted in **Figure 4**. The strong photo-induced heating in combination with an applied AC electric field creates the ETP flow that enables long-range capture and transport of particles towards the laser position. The particles brought close to the nanohole array are trapped by the optical gradient force. In the lateral direction, the particle-particle separation distance is tuned by the dipole-dipole repulsion force between the particles [51, 52].

Under the microscope, when both laser illumination and AC electric field is turned on, particles are seen moving very directionally towards the nanohole metasurface. By selectively turning on and off AC electric field, particle-particle spacing can be controlled dynamically, due to the polarization of the particles induced by AC electric field. Briefly, particles are polarized by the AC field and a dipole-dipole repulsive force creates an in-plane interparticle separation between them. The ETP flow is measured experimentally using microparticle image velocimetry and its radial velocity is shown in **Figure 5(a)**.

Experimental images of trapped particles on the surface of the nanohole array are depicted in **Figure 5(b)**. When both the laser illumination and the AC electric field are applied, the particles are trapped with a certain interparticle spacing between them due to the in-plane dipole-dipole repulsion force. When the AC field is turned off, with the laser still on, the AC electric field-induced dipole-dipole repulsion force disappears, and the assembly becomes more compact.

They also show that the ETP flow induced by the array of nanohole is higher than that can be achieved with a single nanohole or an unpatterned gold film. The plasmonic resonance induced by patterned nanohole truly enhances the electro-thermal effect. These results obtained from micro-particle image velocimetry analysis are depicted in **Figure 6**.

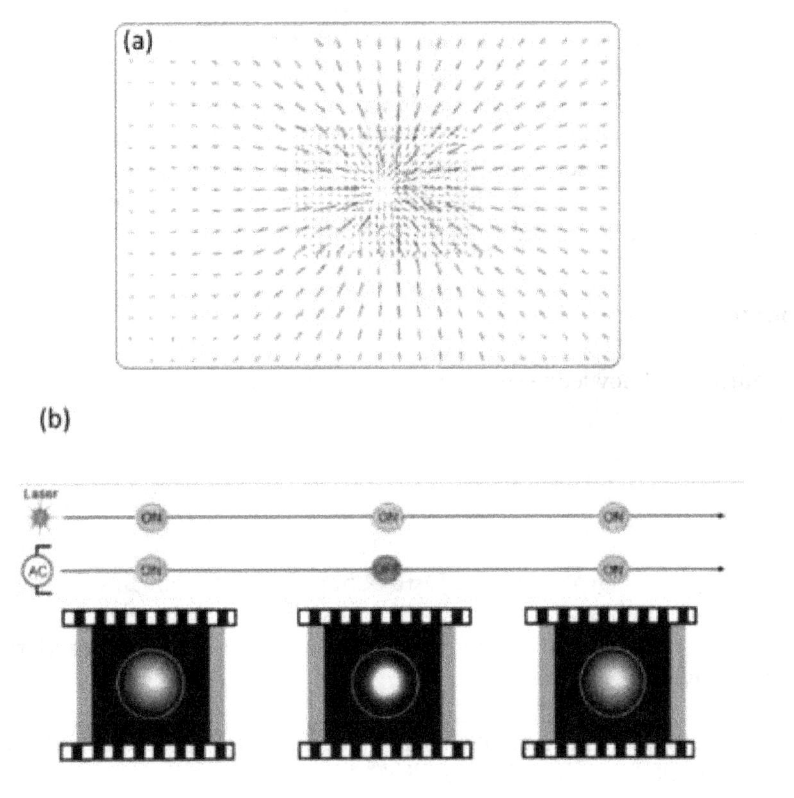

Figure 5.
(a) Experimentally measured radial velocity of the ETP flow (b) sequence of trapping and particle assembling of 200 nm diameter polystyrene beads on the surface of the plasmonic nanohole array when AC field is alternatively turned on and off.

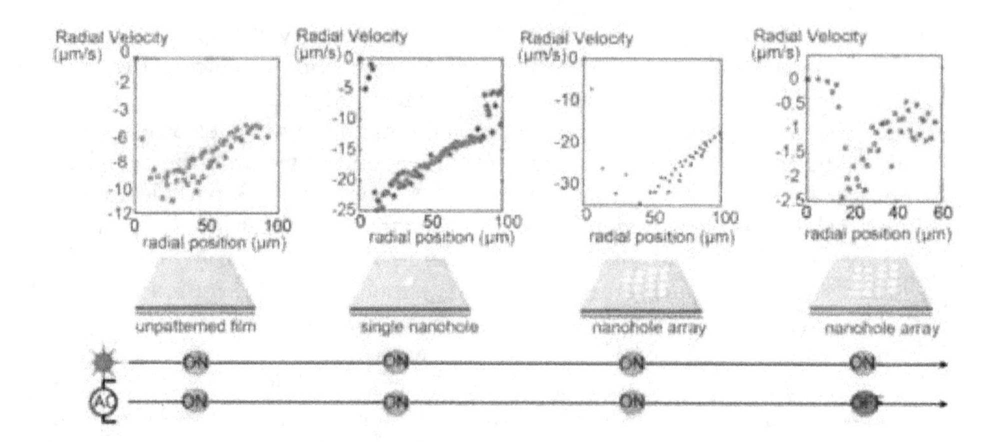

Figure 6.
The nanohole array enables an enhanced ETP flow that is higher than the velocity induced when the planer film or a single nanohole is excited.

6. Conclusion

We have introduced the importance of plasmonic metasurface for applications in optical trapping, nano-tweezers and tiny particle manipulation, by demonstrating a nanoparticle trapping approach that utilizes a thermoplasmonic nanohole metasurface. The application of laser illumination and a.c. electric field results in new physical effects such as electrothermoplasmonic flow and a.c. electroosmosis that can work in concert with optical gradient forces to enable to new advanced features in micro and nanoparticle manipulation. The recent reports in the literature as articulated in this section shows that the intrinsic loss in plasmonic systems is not always detrimental but could work in synergy with the high electric field enhancement to realize advanced lab-on-a-chip devices in nanomanufacturing, nanophotonics, life science and quantum optics.

Acknowledgements

The authors acknowledge support from NSF ECCS-1933109 and Vanderbilt University.

Author details

Chuchuan Hong[1,2], Sen Yang[2,3] and Justus Chukwunonso Ndukaife[1,2*]

1 Department of Electrical Engineering and Computer Science, Vanderbilt University, Nashville, TN, USA

2 Vanderbilt Institute of Nanoscale Science and Engineering, Vanderbilt University, Nashville, TN, USA

3 Interdisciplinary Materials Science, Vanderbilt University, TN, USA

*Address all correspondence to: justus.ndukaife@vanderbilt.edu

References

[1] Liu Y, Zhang X. Metamaterials: A new frontier of science and technology. Chemical Society Reviews. 2011; **40** (5): 2494

[2] Smith DR. Metamaterials and negative refractive index. Science. 2004;**305**(5685):788-792

[3] Zheludev NI, Kivshar YS. From metamaterials to metadevices. Nature Materials. 2012;**11**(11):917-924

[4] Jahani S, Jacob Z. All-dielectric metamaterials. Nature Nanotechnology. 2016;**11**(1):23-36

[5] Yu N, Capasso F. Flat optics with designer metasurfaces. Nature Materials. 2014;**13**(2):139-150. Available from: http://www.nature.com/articles/nmat3839

[6] Arbabi E, Arbabi A, Kamali SM, Horie Y, Faraji-Dana M, Faraon A. MEMS-tunable dielectric metasurface lens. Nature Communications. 2018;**9** (1):812

[7] Lin D, Fan P, Hasman E, Brongersma ML. Dielectric gradient metasurface optical elements. Science. 2014;**345**(6194):298-302. Available from: http://www.ncbi.nlm.nih.gov/pub med/25035488

[8] Ni X, Kildishev AV, Shalaev VM. Metasurface holograms for visible light. Nature Communications. 2013; **4** (1): 2807

[9] Zhan A, Colburn S, Trivedi R, Fryett TK, Dodson CM, Majumdar A. Low-contrast dielectric metasurface optics. ACS Photonics. 2016;**3**(2): 209-214

[10] Zheng G, Mühlenbernd H, Kenney M, Li G, Zentgraf T, Zhang S. Metasurface holograms reaching 80% efficiency. Nature Nanotechnology. 2015;**10**(4):308-312

[11] Hong C, Colburn S, Majumdar A. Flat metaform near-eye visor. Applied Optics. 2017;**56**(31):8822

[12] Yue F, Wen D, Xin J, Gerardot BD, Li J, Chen X. Vector vortex beam generation with a single plasmonic metasurface. ACS Photonics. 2016;**3**(9): 1558-1563

[13] Deng Z-L, Deng J, Zhuang X, Wang S, Li K, Wang Y, et al. Diatomic metasurface for vectorial holography. Nano Letters. 2018;**18**(5):2885-2892

[14] Zhao H, Quan B, Wang X, Gu C, Li J, Zhang Y. Demonstration of orbital angular momentum multiplexing and demultiplexing based on a metasurface in the terahertz band. ACS Photonics. 2018;**5**(5):1726-1732

[15] Semmlinger M, Zhang M, Tseng ML, Huang T-T, Yang J, Tsai DP, et al. Generating third harmonic vacuum ultraviolet light with a TiO2 metasurface. Nano Letters. 2019;**19**(12): 8972-8978

[16] Hong C, Yang S, Ndukaife JC. Optofluidic control using plasmonic TiN bowtie nanoantenna. Optical Materials Express. 2019;**9**(3):953. Available from: https:// www. osapublishing.org/abstrac t.cfm?URI=ome-9-3-953

[17] Ndukaife JC, Kildishev AV, Agwu Nnanna AG, Wereley S, Shalaev VM, Boltasseva A. Plasmon-assisted optoelectrofluidics. In: Cleo. Washington, D.C.: OSA; 2015. p. AW3K.5

[18] Ndukaife JC, Kildishev AV, Nnanna AGA, Shalaev VM, Wereley ST, Boltasseva A. Long-range and rapid transport of individual nano-objects by a hybrid electrothermoplasmonic nanotweezer. Nature Nanotechnology. 2016;**11**(1):53-59

[19] Ndukaife JC, Mishra A, Guler U, Nnanna AGA, Wereley ST, Boltasseva A. Photothermal heating enabled by plasmonic nanostructures for electrokinetic manipulation and sorting of particles. ACS Nano. 2014; **8**(9):9035-9043

[20] Ndukaife JC, Shalaev VM, Boltasseva A. Plasmonics–turning loss into gain. Science. 2016;**351**(6271): 334-335

[21] Grier DG. A revolution in optical manipulation. Nature. Aug 2003;**424** (6950):810-816

[22] Ashkin A. Acceleration and trapping of particles by radiation pressure. Physical Review Letters. 1970;**24**(4): 156-159

[23] Ashkin A, Dziedzic JM, Bjorkholm JE, Chu S. Observation of a single-beam gradient force optical trap for dielectric particles. Optics Letters. 1986;**11**(5):288

[24] Wang K, Crozier KB. Plasmonic trapping with a gold nanopillar. ChemPhysChem. 2012;**13**(11): 2639-2648

[25] Collinge MJ, Draine BT. Discrete-dipole approximation with polarizabilities that account for both finite wavelength and target geometry. Journal of the Optical Society of America. A. 2004;**21**(10):2023

[26] Gao D, Ding W, Nieto-Vesperinas M, Ding X, Rahman M, Zhang T, et al. Optical manipulation from the microscale to the nanoscale: Fundamentals, advances and prospects. Light: Science & Applications. 2017; **6**(9):e17039-e17039

[27] Lin L, Wang M, Peng X, Lissek EN, Mao Z, Scarabelli L, et al. Opto-thermoelectric nanotweezers. Nature Photonics. 2018;**12**(4):195-201

[28] Roxworthy BJ, Johnston MT, Lee-Montiel FT, Ewoldt RH, Imoukhuede PI, Toussaint KC. Plasmonic optical trapping in biologically relevant media. PLoS One. 2014;**9**(4):e93929

[29] Crozier KB. Quo vadis, plasmonic optical tweezers? Light: Science & Applications. 3 Dec 2019;**8**(1):35

[30] Kravets VG, Kabashin AV, Barnes WL, Grigorenko AN. Plasmonic surface lattice resonances: A review of properties and applications. Chemical Reviews. 2018;**118**(12):5912-5951

[31] Juan ML, Righini M, Quidant R. Plasmon nano-optical tweezers. Nature Photonics. Jun 2011;**5**(6):349-356

[32] Shoji T, Tsuboi Y. Plasmonic optical tweezers toward molecular manipulation: Tailoring plasmonic nanostructure, light source, and resonant trapping. Journal of Physical Chemistry Letters. 2014;**5**(17):2957-2967

[33] Donner JS, Baffou G, McCloskey D, Quidant R. Plasmon-assisted optofluidics. ACS Nano. 2011;**5**(7): 5457-5462

[34] Harter T, Muehlbrandt S, Ummethala S, Schmid A, Nellen S, Hahn L, et al. Silicon–plasmonic integrated circuits for terahertz signal generation and coherent detection. Nature Photonics. 2018;**12**(10):625-633

[35] Gramotnev DK, Bozhevolnyi SI. Plasmonics beyond the diffraction limit. Nature Photonics. 2010;**4**(2):83-91

[36] Roxworthy BJ, Bhuiya AM, Inavalli VVGK, Chen H, Toussaint KC. Multifunctional plasmonic film for recording near-field optical intensity. Nano Letters. 2014;**14**(8):4687-4693

[37] Wang K, Schonbrun E, Steinvurzel P, Crozier KB. Trapping and

rotating nanoparticles using a plasmonic nano-tweezer with an integrated heat sink. Nature Communications. 2011; **2**(1):469

[38] Iacopini S, Piazza R. Thermophoresis in protein solutions. Europhysics Letters. 2003;**63**(2):247-253

[39] Duhr S, Braun D. Why molecules move along a temperature gradient. Proceedings of the National Academy of Sciences. 2006;**103**(52):19678-19682

[40] Lamhot Y, Barak A, Peleg O, Segev M. Self-trapping of optical beams through thermophoresis. Physical Review Letters. 2010;**105**(16):163906

[41] Lu GW, Gao P. Emulsions and microemulsions for topical and transdermal drug delivery. In: Handbook of Non-Invasive Drug Delivery Systems. Elsevier; 2010. pp. 59-94. Available from: https://linkinghub.elsevier.com/retrieve/pii/B9780815520252100034

[42] Derjaguin BV, Churaev NV, Muller VM. Forces near interfaces. In: Surface Forces. Boston, MA: Springer US; 1987. pp. 1-23

[43] Ramos A, Morgan H, Green NG, Castellanos A. Ac electrokinetics: A review of forces in microelectrode structures. Journal of Physics D: Applied Physics. 1998;**31**(18):2338-2353

[44] Melcher JR. Electric fields and moving media. IEEE Transactions on Education. 1974;**17**(2):100-110

[45] SQUIRES TM, BAZANT MZ. Induced - charge electro - osmosis. Journal of Fluid Mechanics. 2004;**509**: 217-252

[46] Jamshidi A, Neale SL, Yu K, Pauzauskie PJ, Schuck PJ, Valley JK, et al. NanoPen: Dynamic, low-power, and light-actuated patterning of nanoparticles. Nano Letters. 2009;**9**(8): 2921-2925

[47] Hwang H, Park J-K. Rapid and selective concentration of microparticles in an optoelectrofluidic platform. Lab on a Chip. 2009;**9**(2):199-206

[48] Elitas M, Martinez-Duarte R, Dhar N, McKinney JD, Renaud P. Dielectrophoresis-based purification of antibiotic-treated bacterial subpopulations. Lab on a Chip. 2014; **14**(11):1850-1857

[49] Zaman MA, Padhy P, Hansen PC, Hesselink L. Dielectrophoresis-assisted plasmonic trapping of dielectric nanoparticles. Physical Review A. 2017; **95**(2):023840

[50] Ndukaife JC, Xuan Y, Nnanna AGA, Kildishev AV, Shalaev VM, Wereley ST, et al. High-resolution large-ensemble nanoparticle trapping with multifunctional thermoplasmonic nanohole metasurface. ACS Nano. 2018; **12**(6):5376-5384

[51] Mittal M, Lele PP, Kaler EW, Furst EM. Polarization and interactions of colloidal particles in ac electric fields. The Journal of Chemical Physics. 2008; **129**(6):064513

[52] Work AH, Williams SJ. Characterization of 2D colloids assembled by optically-induced electrohydrodynamics. Soft Matter. 2015;**11**(21): 4266-4272

4

Scattering from Multilayered Graphene-Based Cylindrical and Spherical Particles

Shiva Hayati Raad, Zahra Atlasbaf and Mauro Cuevas

Abstract

This chapter discusses various approaches for calculating the modified Mie-Lorenz coefficients of the graphene-based multilayered cylindrical and spherical geometries. Initially, the Kubo model of graphene surface conductivity is discussed. Then, according to it, the formulations of scattering from graphene-based conformal particles are extracted. So, we have considered a graphene-wrapped cylinder and obtained its scattering coefficients by considering graphene surface currents on the shell. Later, a layered nanotube with multiple stacked graphene-dielectric interfaces is introduced, and for analyzing the plane wave scattering, graphene surface conductivity is incorporated in the transfer matrix method (TMM). Unlike the previous section, the dielectric model of graphene material is utilized, and the boundary conditions are applied on an arbitrary graphene interface, and a matrix-based formulation is concluded. Then, various examples ranging from super-scattering to super-cloaking are considered. For the scattering analysis of the multilayered spherical geometries, recurrence relations are introduced for the corresponding modified Mie-Lorenz coefficients by applying the boundary conditions at the interface of two adjacent layers. Later, for a sub-wavelength nanoparticle with spherical morphology, the full electrodynamics response is simplified in the electrostatic regime, and an equivalent circuit is proposed. Various practical examples are included to clarify the importance of scattering analysis for graphene-based layered spheres in order to prove their importance for developing novel optoelectronic devices.

Keywords: multilayered, graphene, spherical, cylindrical, nanoparticles, modified Mie-Lorenz theory, 2D material

1. Introduction

Cylindrically layered structures have various exotic applications. For instance, a metal-core dielectric-shell nano-wire has been proposed for the cloaking applications in the visible spectrum. The functionality of this structure is based on the induction of antiparallel currents in the core and shell regions, and the design procedure is the so-called scattering cancelation technique [1]. Experimental realization of a hybrid gold/silicon nanowire photodetector proves the practicality of these structures [2]. As an alternative approach for achieving an invisible cloak, cylindrically wrapped impedance surfaces are designed by a periodic arrangement of metallic patches, and the approach is denominated as mantle cloaking [3].

Conversely, cylindrically layered structures can be designed in a way that they exhibit a scattering cross-section far exceeding the single-channel limit. This phenomenon is known as super-scattering and has various applications in sensing, energy harvesting, bio-imaging, communication, and optical devices [4, 5]. Moreover, a cylindrical stack of alternating metals and dielectrics behaves as an anisotropic cavity and exhibits a dramatic drop of the scattering cross-section in the transition from hyperbolic to elliptic dispersion regimes [6, 7]. The Mie-Lorenz theory is a powerful, an exact, and a simple approach for designing and analyzing the aforementioned structures.

Multilayered spherical structures have also attracted lots of interests in the field of optical devices. A dielectric sphere made of a high index material supports electric and magnetic dipole resonances which results in peaks in the extinction cross-section [8]. Moreover, by covering the dielectric sphere with a plasmonic metal shell, an invisible cloak is realizable, which is useful for sensors and optical memories [9]. By stacking multiple metal-dielectric shells, an anisotropic medium for scattering shaping can be achieved [10].

From the above discussions, it can be deduced that tailoring the Mie-Lorenz resonances in the curved particles results in developing novel optical devices. In this chapter, we are going to extend the realization of various optical applications based on the excitations of localized surface plasmons (LSP) in graphene-wrapped cylindrical and spherical particles. To this end, initially we introduce a brief discussion of modeling graphene material based on corresponding surface conductivity or dielectric model. Later, we extract the modified Mie-Lorenz coefficients for some curved structures with graphene interfaces. The importance of developed formulas has been proven by providing various design examples. It is worth noting that graphene-wrapped particles with a different number of layers have been proposed previously as refractive index sensors, waveguides, super-scatterers, invisible cloaks, and absorbers [11–15]. Our formulation provides a unified approach for the plane wave and eigenmode analysis of graphene-based optical devices.

Graphene is a 2D carbon material in a honeycomb lattice that exhibits extraordinary electrical and mechanical properties. In order to solve Maxwell's equations in the presence of graphene, two approaches are applied by various authors, and we will review them in the following paragraphs. It should be noted that although we are discussing the graphene planar model, we will use the same formulas for the curved geometries when the number of carbon atoms exceeds 10^4, letting us neglect the effect of defects. Therefore, the radii of all cylinders and spheres are considered to be greater than 5 nm [16]. Moreover, bending the graphene does not have any considerable impact on the properties of its surface plasmons, except for a small downshift of the frequency. **Figure 1** shows the propagation of the graphene surface plasmons on the S-shaped and G-shaped curves [17].

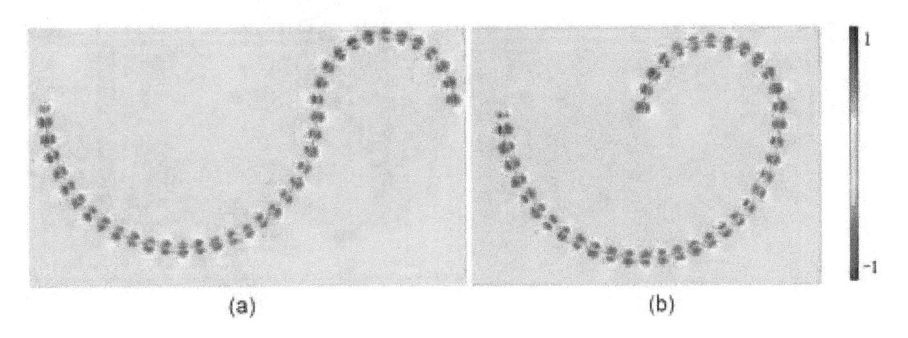

Figure 1.
Propagation of graphene surface plasmons on curved structures: (a) S-shaped and (b) G-shaped [17].

Since graphene material is atomically thin, in order to consider its impact on the electromagnetic response of a given structure, boundary conditions at the interface can be simply altered. To this end, graphene surface currents that are proportional to its surface conductivity should be accounted for ensuring the discontinuity of tangential magnetic fields. In the infrared range and below, we can describe the graphene layer with a complex-valued surface conductivity σ which may be modeled using the Kubo formulas [18, 19]. The intraband and interband contributions of graphene surface conductivity under local random phase approximations read as [18]:

$$\sigma_{\text{intra}}(\omega) = \frac{2ie^2 k_B T}{\pi\hbar^2(\omega + i\Gamma)} \ln\left[2\cosh\left(\frac{\mu_c}{2k_B T}\right)\right] \quad (1)$$

$$\sigma_{\text{inter}} \approx \frac{e^2}{4\hbar}\left[\frac{1}{2} + \frac{1}{\pi}\arctan\left(\frac{\hbar\omega - 2\mu_c}{2k_B T}\right) - \frac{i}{2\pi}\ln\frac{(\hbar\omega + 2\mu_c)^2}{(\hbar\omega - 2\mu_c)^2 + (2k_B T)^2}\right] \quad (2)$$

The parameters $\hbar$, e, μ_c, Γ, and T are reduced Plank's constant, electron charge, chemical potential, charge carriers scattering rate, and temperature, respectively. The above equations are valid for the positive valued chemical potentials. Moreover, graphene-based structures can be analyzed by assigning a small thickness of $\Delta \approx 0.35\,nm$ to the graphene interface and later approaching it to the zero. In this method, by defining the volumetric conductivity as $\sigma_{g,V} = \sigma_g/\Delta$ and using it in

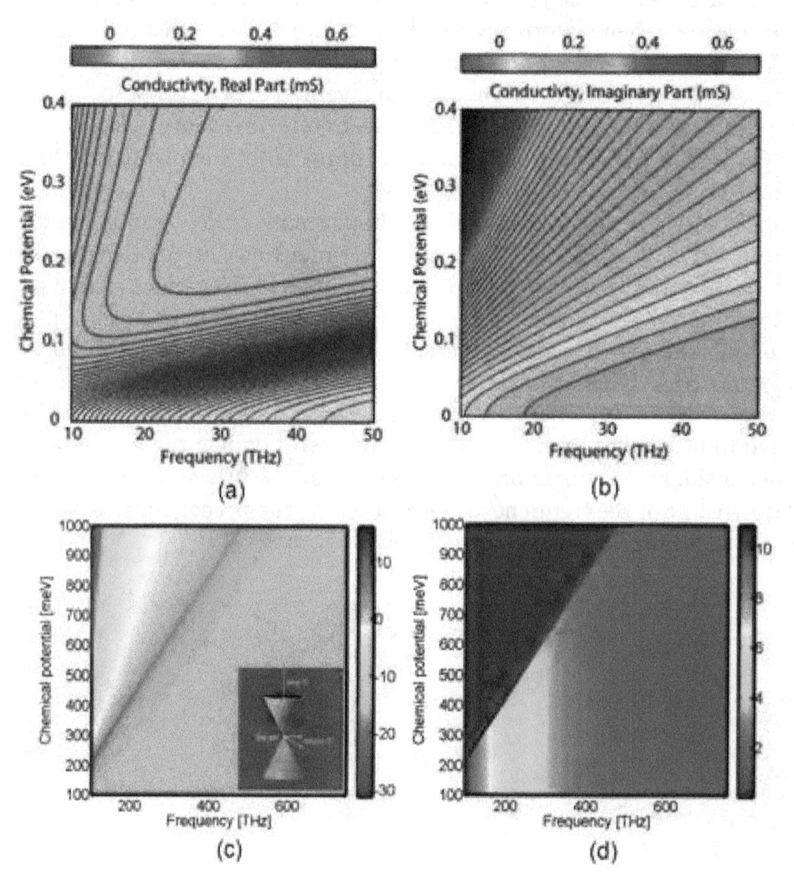

Figure 2.
(a) and (b) the real and imaginary parts of graphene surface conductivity [20] and (c) and (d) the real and imaginary parts of graphene equivalent permittivity [21].

Maxwell's curl equations, the equivalent complex permittivity of the layer can be obtained as [20]:

$$\varepsilon_{g,eq} = \left(-\frac{\sigma_{g,i}}{\omega\Delta} + \varepsilon_0\right) + i\left(\frac{\sigma_{g,r}}{\omega\Delta}\right) \tag{3}$$

where subscripts r and i represent the real and imaginary parts of the surface conductivity, respectively. Both models will be used in the following sections.

Figure 2(a) and **(b)** shows the real and imaginary parts of graphene surface conductivity at the temperature of T = 300°K. The real part of the conductivity accounts for the losses, while the positive valued imaginary parts represent the plasmonic properties [20]. Moreover, the real and imaginary parts of the graphene equivalent bulk permittivity are shown in **Figure 2(c)** and **(d)**. The negative valued real relative permittivity represents the plasmonic excitation, and the imaginary part of the permittivity represents the losses [21]. It should be noted that all of the formulas of this chapter are adapted with $\exp(-i\omega t)$ time-harmonic dependency.

2. Graphene-coated cylindrical tubes

In this section, the modified Mie-Lorenz coefficients of a single-layered graphene-coated cylindrical tube will be extracted. The formulation is expanded into the multilayered graphene-based tubes through exploiting the TMM method, and later, various applications of the analyzed structures, including emission and radiation properties, complex frequencies, super-scattering, and super-cloaking, will be explained.

2.1 Scattering from graphene-coated wires

Let us consider a graphene-wrapped infinitely long cylindrical tube. The structure is shown in **Figure 3(a)**, and it is considered that a TE^z-polarized plane wave illuminates the cylinder. In general, TE and TM waves are coupled in the cylindrical geometries. For the normally incident plane waves, they become decoupled, and they can be treated separately. For simplicity, we consider the normal incidence of plane waves where the wave vector k is perpendicular to the cylinder axis.

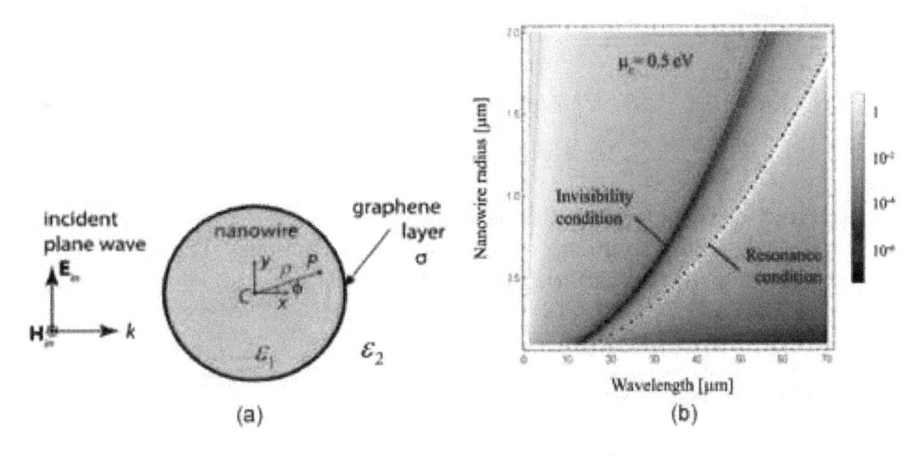

Figure 3.
(a) A single-layered graphene-coated cylinder under TE^z plane wave illumination and (b) corresponding scattering efficiency for ε_1 = 3.9 and μ_c = 0.5 eV. The normalization factor in this figure is the diameter of the cylinder [23].

In order to obtain the modified Mie-Lorenz coefficients, the incident, scattered, and internal electromagnetic fields are expanded in terms of cylindrical coordinates special functions which are, respectively, the Bessel functions and exponentials in the radial and azimuthal directions. In order to exploit a terse mathematical notation, the vector wave functions are introduced as [22]:

$$\boldsymbol{M}_n = k\left(in\frac{Z_n(k\rho)}{k\rho}\hat{\rho} - Z_n'(k\rho)\hat{\phi}\right)e^{in\phi} \tag{4}$$

$$\boldsymbol{N}_n = kZ_n(k\rho)e^{in\phi}\hat{z} \tag{5}$$

The complete explanation of the above vector wave functions and their self and mutual orthogonally relations can be found in the classic electromagnetic books [22]. In the above equation, Z_n is the solution of the Bessel differential equation, and n is its order. It is clear that in the environment, the radial field contains the Hankel function of the first kind in order to account for the radiation condition at infinity, while in the medium region, the Bessel function is utilized to satisfy the finiteness condition at the origin of the structure.

In the graphene-based cylindrical structures, the plasmonic state is achieved via illuminating a TE^z wave to the structure. Therefore, for the normal illumination, the incident, scattered, and dielectric electromagnetic fields are shown with the superscripts $l = in, sca, d$, respectively, and they read as [23]:

$$\boldsymbol{H}_l = \frac{H_0}{k_l}\sum_{n=-\infty}^{\infty} A_n i^n \boldsymbol{N}_n(k_l\rho) \tag{6}$$

$$\boldsymbol{E}_l = \frac{E_0}{k_l}\sum_{n=-\infty}^{\infty} B_n i^n \boldsymbol{M}_n(k_l\rho) \tag{7}$$

where H_0 and E_0 are the magnitudes of the incident electric and magnetic fields, respectively, and they are related via the intrinsic impedanc e of the free space. The wavenumber in the region l is denoted by k_l. The coefficients $[A_n, B_n]$ are, respectively, $[1, 1]$, $[a_n, b_n]$, and $[c_n, d_n]$ for the incident, scattered, and core regions. Moreover, a_n and b_n are the well-known Mie-Lorenz coefficients, which are called the modified Mie-Lorenz coefficients for the scattering analysis of graphene-based structures.

The boundary conditions at the graphene interface at $\rho = R_1$ are the continuity of the tangential electric fields along with the discontinuity of tangential magnetic fields. Therefore:

$$\hat{\phi}.\boldsymbol{E}_d = \hat{\phi}.(\boldsymbol{E}_{sca} + \boldsymbol{E}_{in}) \tag{8}$$

$$\hat{z}.\boldsymbol{H}_d = \hat{z}.(\boldsymbol{H}_{sca} + \boldsymbol{H}_{in}) + \hat{\phi}.\boldsymbol{E}_d\sigma \tag{9}$$

By applying the boundary conditions in the expanded fields, the linear system of equations for extracting the unknowns can be readily obtained. The solution of the extracted equations for the scattering coefficients leads to:

$$a_n = \frac{k_1 J_n'(k_1R_1)\left[\varepsilon_2 J_n(k_2R_1) - i\frac{k_2\sigma}{\omega\varepsilon_0}J_n'(k_2R_1)\right] - k_2\varepsilon_1 J_n(k_1R_1)J_n'(k_2R_1)}{k_2 H_n^{(1)'}(k_2R_1)\left[\varepsilon_1 J_n(k_1R_1) + i\frac{k_1\sigma}{\omega\varepsilon_0}J_n'(k_1R_1)\right] - k_1\varepsilon_2 H_n^{(1)}(k_2R_1)J_n'(k_1R_1)} \tag{10}$$

$$b_n = \frac{k_2\varepsilon_1\left[H_n^{(1)'}(k_2R_1)J_n(k_2R_1) - H_n^{(1)}(k_2R_1)J_n'(k_2R_1)\right]}{k_2 H_n^{(1)'}(k_2R_1)\left[\varepsilon_1 J_n(k_1R_1) + i\frac{k_1\sigma}{\omega\varepsilon_0}J_n'(k_1R_1)\right] - k_1\varepsilon_2 H_n^{(1)}(k_2R_1)J_n'(k_1R_1)} \tag{11}$$

The same procedure can be repeated for the TM^z illumination. The normalized scattering cross-section (NSCS) reads as:

$$\mathrm{NSCS} = \sum_{n=-\infty}^{\infty} |a_n|^2 \tag{12}$$

where the normalization factor is the single-channel scattering limit of the cylindrical structures. In order to have some insight into the scattering performance of graphene-wrapped wires, the scattering efficiency for $\varepsilon_1 = 3.9$ and $\mu_c = 0.5$ eV is plotted in **Figure 3(b)** by varying the radius of the wire. As the figure illustrates, a peak valley line shape occurs in each wavelength. They correspond to invisibility and scattering states and will be further manipulated in the next sections to develop some novel devices. The excitation frequency of the plasmons is the complex poles of the extracted coefficients [24] which will be discussed in the next subsection. Interestingly, the scattering states of graphene-coated dielectric cores are polarization-dependent. By using a left-handed metamaterial as a core, this limitation can be obviated [25].

2.1.1 Eigenmode problem and complex frequencies

As in any resonant problem, additional information can be obtained by studying the solutions to the boundary value problem in the absence of external sources (eigenmode approach). Although, from a formal point of view, this approach has many similar aspects with those developed in previous sections, the eigenmode problem presents an additional difficulty related to the analytic continuation in the complex plane of certain physical quantities. Due to the fact that the electromagnetic energy is thus leaving the LSP (either by ohmic losses or by radiation towards environment medium), the LSP should be described by a complex frequency where the imaginary part takes into account the finite lifetime of such LSP. The eigenmode approach is not new in physics, but its appearance is associated to any resonance process (at an elementary level could be an RLC circuit), where the complex frequency is a pole of the analytical continuation to the complex plane of the response function of the system (e.g., the current on the circuit). Similarly, in the eigenmode approach presented here, the complex frequencies correspond to poles of the analytical continuation of the multipole terms (Mie-Lorenz coefficients) in the electromagnetic field expansion.

In order to derive complex frequencies of LSP modes in terms of the geometrical and constitutive parameters of the structure, we use an accurate electrodynamic formalism which closely follows the usual separation of variable approach devel-oped in Section 2.1. We can obtain a set of two homogeneous equations for the m–th LSP mode [26]:

$$\begin{aligned} &\frac{k_2}{\varepsilon_2} H_n^{(1)'}(x_2) a_n - \frac{k_1}{\varepsilon_1} J_n'(x_1) b_n = 0 \\ &H_n^{(1)}(x_2) a_n - J_n(x_1) b_n = \frac{\sigma k_1}{\omega \varepsilon_0 \varepsilon_1} b_n J_n'(x_1) \end{aligned} \tag{13}$$

where the prime denotes the first derivative with respect to the argument of the function and $x_j = k_j R_1$ $(j = 1, 2)$. For this system to have a nontrivial solution, its determinant must be equal to zero, a condition which can be written as:

$$D_n = h_n(x_2) - j_n(x_1) + \sigma \omega i R_1 j_n(x_1) h_n(x_2) = 0 \tag{14}$$

where $j_n(x) = \frac{J_n'(x)}{xJ_n(x)}$, $h_n(x) = \frac{H_n^{(1)'}(x)}{xH_n^{(1)}(x)}$. This condition is the dispersion relation of LSPs, and it determines the complex frequencies in terms of all the parameters of the wire cylinder.

2.1.2 Non-retarded dispersion relation

When the size of the cylinder is small compared to the eigenmode wavelength, i.e., $\lambda_n = \frac{2\pi c}{\omega_n} >> R_1$, where c is the speed of light in free space, Eq. (13) can be approximated by using the quasistatic approximation as follows. Using the small argument asymptotic expansions for Bessel and Hankel functions, the functions $j_n(x) \approx \frac{n}{x^2}$ and $h_n(x) \approx -\frac{n}{x^2}$ [27]. Thus, the dispersion relation (14) adopts the form:

$$i\omega\varepsilon_0 \frac{\varepsilon_1 + \varepsilon_2}{\sigma} = \frac{n}{R_1} \tag{15}$$

Taking into account that in the non-retarded regime the propagation constant of the plasmon propagating along perfectly flat graphene sheet can be approximated by:

$$k_{sp} = i\omega\varepsilon_0 \frac{\varepsilon_1 + \varepsilon_2}{\sigma}, \tag{16}$$

it follows that the dispersion relation (14) for LSPs in dielectric cylinders wrapped with a graphene sheet can be written as:

$$k_{sp} 2\pi R_1 = 2\pi n \tag{17}$$

where n is the LSP multipole order. The dispersion Eq. (17), known as Bohr condition, states that the n–th LSP mode of a graphene-coated cylinder accommodates along the cylinder perimeter exactly n oscillation periods of the propagating surface plasmon corresponding to the flat graphene sheet.

For large doping ($\mu_c \gg k_B T$) and relatively low frequencies ($\hbar\omega \ll \mu_c$), the intraband contribution to the surface conductivity (1), the Drude term, plays the leading role. In this case, the non-retarded dispersion equation Eq. (14) is written as:

$$\varepsilon_1 + \varepsilon_2 = \frac{e^2 \mu_c n}{\hbar^2 \pi \varepsilon_0 R_1 \omega(\omega + i\gamma_c)} \tag{18}$$

which can be analytically solved for the plasmon eigenfrequencies,

$$\omega_n = \sqrt{\frac{\omega_0^2 n^2}{\varepsilon_1 + \varepsilon_2} - \left(\frac{\gamma_c}{2}\right)^2} - i\frac{\gamma_c}{2} \approx \frac{n\omega_0}{\sqrt{\varepsilon_1 + \varepsilon_2}} - i\frac{\gamma_c}{2}, \tag{19}$$

where $\omega_0^2 = \frac{e^2 \mu_c}{\pi \varepsilon_0 \hbar^2 R_1}$ is the effective plasma frequency of the graphene coating. It is worth noting that the real part of ω_n is proportional to $\sqrt{\mu_c}$, and as a consequence, the net effect of the chemical potential increment is to increase the spectral position of the resonance peaks when the structure is excited with a plane wave or a dipole emitter [28, 29].

In the following example, we consider a graphene-coated wire with a core radius $R_1 = 30$ nm, made of a non-magnetic dielectric material of permittivity $\varepsilon_1 = 2.13$

n	FR (μm^{-1})	AA (μm^{-1})
1	$0.8868268 - i0.4105015 \times 10^{-3}$	$0.8873162 - i0.2532113 \times 10^{-3}$
2	$1.254676 - i0.2532181 \times 10^{-3}$	$1.254855 - i0.2532113 \times 10^{-3}$
3	$1.536735 - i0.2531644 \times 10^{-3}$	$1.536877 - i0.2532113 \times 10^{-3}$
4	$1.774508 - i0.2531757 \times 10^{-3}$	$1.774632 - i0.2532113 \times 10^{-3}$

Table 1.
Resonance frequencies ω_n for the first four eigenmodes ($1 \leq n \leq 4$), $R_1 = 30$ nm, $\mu_c = 0.5$ eV, $\gamma_c = 0.1$ meV, $\varepsilon_1 = 2.13$, $\mu_1 = 1$, $\varepsilon_2 = 1$, and $\mu_2 = 1$ [26].

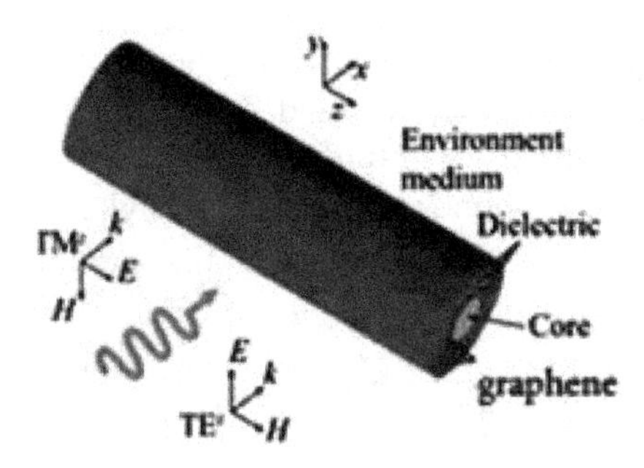

Figure 4.
Multilayered cylindrical structure consisting of alternating graphene-dielectric stacks under plane wave illumination. The 2D graphene shells are represented volumetrically for the sake of illustration [31].

immersed in the vacuum. The graphene parameters are $\mu_c = 0.5$ eV, $\gamma_c = 0.1$ meV, and $T = 300° K$. **Table 1** shows the first four eigenfrequencies calculated by solving the full retarded (FR) dispersion Eq. (14) (second column) and by using the analytical approximation (AA) given by Eq. (19) (third column). Since the radius of the wire is small compared with the eigenmode wavelengths, good agreement is obtained between the complex FR and AA ω_n values, even when the AA assigns $Im\ \omega_n = -\gamma_c/2 \approx 0.25 \times 10^{-3} \mu m^{-1}$ to all multipolar plasmon modes.

2.2 Multilayered graphene-based cylindrical structures

In this section, multilayered cylindrical tubes with multiple graphene interfaces are of interest. In order to ease the derivation of the unknown expansion coefficients, matrix-based TMM formulation is generalized to the tubes with several graphene interfaces. Initially, consider a layered cylinder constructed by the staked ordinary materials under TE^z plane wave illumination, as shown in **Figure 4**. The total magnetic field at the environment can be expressed as the superposition of incident and scattered waves as in Section 2.1. The unknown expansion coefficients of the scattered wave can be determined by means of the T_n matrix defined as [30]:

$$T_n = [D_{n,C}(R_1)]^{-1} \cdot \left\{ \prod_{q=1}^{N} D_{n,q}(R_q) \cdot [D_{n,q}(R_{q+1})]^{-1} \right\} \cdot D_{n,N+1}(R_{q+1}) \tag{20}$$

where C represents the core layer In the above equation the dynamical matrix $D_{n,q}$ of each region is constructed based on its constitutive and geometrical parameters distinguished through the subscript q (q = 1, 2, ..., N). We have:

$$D_{n,q}(x) = \begin{bmatrix} J_n(x_1) & Y_n(x_1) \\ z_q^{-1}J'_n(x_1) & z_q^{-1}Y'_n(x_1) \end{bmatrix} \tag{21}$$

The argument of the above special functions is $x_1 = k_q x$, and the TE^z wave impedance equals $z_q^{-1} = \sqrt{\varepsilon_q}$. After generating T_n matrix for the structure, a_n coefficients can be calculated as:

$$a_n = \frac{T_{n,21}}{T_{n,21} + iT_{n,22}} \tag{22}$$

In order to incorporate the graphene surface conductivity in the above formulas, let us consider each graphene interface as a thin dielectric with the equivalent complex permittivity defined in Eq. (3) and utilize the TMM formulation in the limiting case of a small radius at the graphene interface with the wave number of k_g, i.e., $R_{q+1} - R_q = t_g$. At each boundary, using the Taylor expansion as $J_n(k_g R_q) = J_n(k_g R_{q+1}) - t_g k_g J'_n(k_g R_{q+1})$ in the T_n matrix, the graphene interface can be represented by the following matrices. We have:

$$T_g^{TE} = \begin{bmatrix} 1 & -i\sigma\eta_0 \\ 0 & 1 \end{bmatrix} \tag{23}$$

$$T_g^{TM} = \begin{bmatrix} 1 & 0 \\ i\sigma\eta_0 & 1 \end{bmatrix} \tag{24}$$

where the free-space impedance η_0 equals 377 ohms. Once T_n matrix is generated, the modified Mie-Lorenz coefficients and thus scattering cross-section are readily attainable. In the following subsections, the above equations will be used to design some novel optoelectronic devices.

2.2.1 Application in mantle cloaking

Widely tunable scattering cancelation is feasible by using patterned graphene-based patch meta-surface around the dielectric cylinder as shown in **Figure 5**. The surface impedance of the graphene patches can be simply and accurately calculated by closed-form formulas, to be inserted in the modified Mie-Lorenz theory [32].

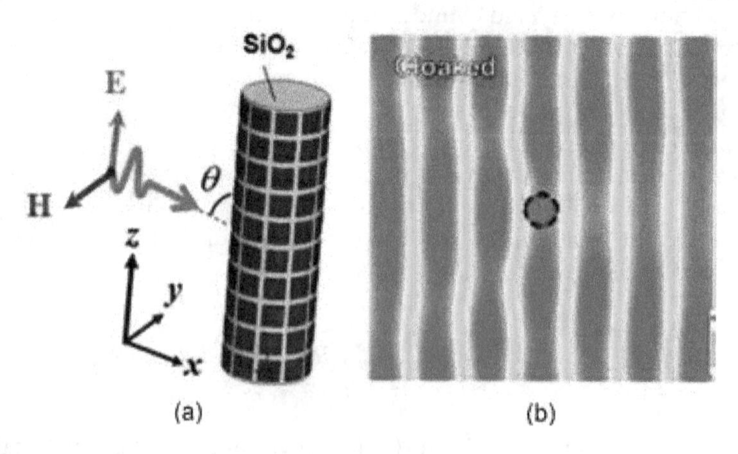

Figure 5.
(a) Electromagnetic cloaking of a dielectric cylinder using graphene meta-surface and (b) corresponding electric field distribution [32].

2.2.2 Application in super-scattering

Let us consider a triple shell graphene-based nanotube under plane wave illumination, as shown in **Figure 6(a)**. This structure is used to design a dual-band super-scatterer in the infrared frequencies. To this end, modified Mie-Lorenz coefficients of various scattering channels should have coincided with the proper choice of geometrical and optical parameters. In order to construct the T_n matrix for this geometry, one needs to multiply nine 2×2 dynamical matrices, which is mathematically complex for analytical scattering manipulation. Therefore, the associated planar structure, shown in **Figure 6(b)**, is used to develop the dispersion engineering method as a quantitative design procedure of the super-scatter. The separations of the free-standing graphene layers are $d_1 = d_2 = 45$ nm in the planar structure, and the transmission line model is used to analyze it. Moreover, the chemical potential of lossless graphene material is $\mu_c = 0.2$ eV in all layers. The dispersion diagram of the planar structure is illustrated in **Figure 7(a)**, which predicts the presence of three plasmonic resonances in each scattering channel of the tube at around the frequencies that fulfill $\beta R_{eff} = n$, where R_{eff} is the mean of the radii of all layers and β is the propagation constant of the plasmons in the planar structure. This condition is known as Bohr's quantization formula [30], and its validity for our specific structure is proven by means of the previously developed formulas in **Figure 7(b)**. Eqs. (12), (22), and (23) are used to obtain this figure.

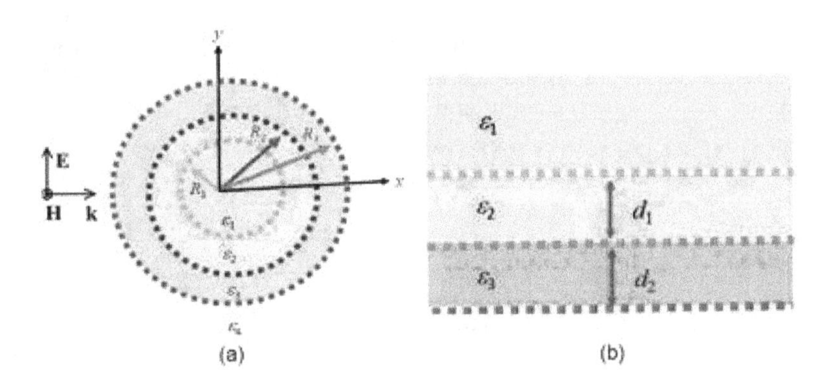

Figure 6.
(a) Multilayered cylindrical nanotube with three graphene shells and (b) associated planar structure [30]. R_1 is denoted with R_c in the text.

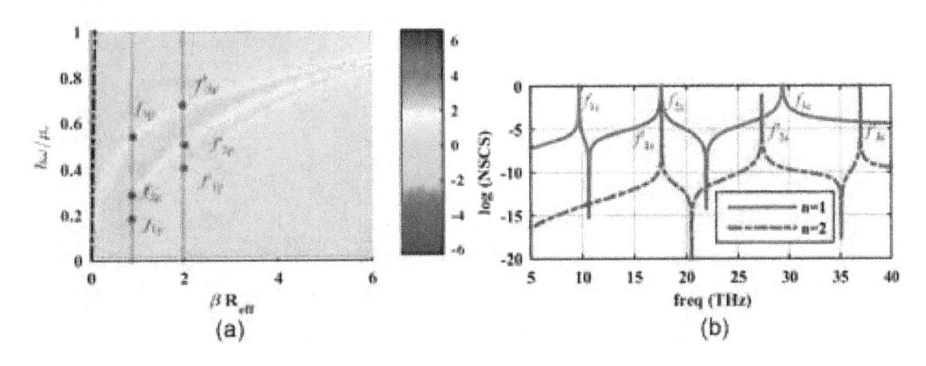

Figure 7.
(a) Dipole and quadruple Mie-Lorenz scattering coefficients for the tube of **Figure 6** *and (b) dispersion diagram of the associated planar structure [30]. f_{1p}, f_{2p}, and f_{3p} are the plasmonic resonances of the dipole mode predicted by the planar configuration. The prime denotes the same information for the quadruple mode. f_{1c}, f_{2c}, and f_{3c} are the same information calculated by the exact modified Mie-Lorenz theory of the multilayered cylindrical structure.*

In order to design a dual-band super-scatterer, the plasmonic resonances of two scattering channels have coincided by fine-tuning the results of the Bohr's model. The optimized geometrical and constitutive parameters are R_c = 45.45 nm, d_1 = 45.05 nm, d_2 = 43.23 nm, ε_1 = 3.2, ε_2 = 2.1, ε_3 = 2.2, and ε_4 = 1. **Figure 8** shows the NSCS and magnetic field distribution for the dual operating bands of the structure. It is clear that NSCS exceeds the single-channel limit by the factor of 4, and in the corresponding magnetic field, there is a large shadow around the nanometer-sized cylinder at each operating frequency. Other designs are also feasible by altering optical and geometri-cal parameters. Furthermore, the far-field radiation pattern is a hybrid dipole-quadrupole due to simultaneous excitation of the first two channels. It should be noted that an inherent characteristic of the super-scatterer design using plasmonic graphene material is extreme sensitivity to the parameters. Moreover, in the presence of losses, the scattering amplitudes do not reach the single-channel limit anymore, and this restricts the practical applicability of the concepts to low-frequency windows.

2.2.3 Application in simultaneous super-scattering and super-cloaking

As another example, the dispersion diagram of **Figure 7(a)** along with Foster's theorem has been used to conclude that each scattering channel of the triple shell tube contains two zeros which are lying between the plasmonic resonances, predicted by the Bohr's model. Later, we have coincided the zeros and poles of the first two scattering channels in order to observe super-scattering and super-cloaking simultaneously [33]. The optimized material and geometrical parameters are ε_c = 3.2, ε_1 = ε_2 = 2.1, R_c = 45.45 nm, d_1 = 46.25 nm, and d_2 = 46.049 nm. The NSCS curves corresponding to the super-cloaking and super-scattering regimes are illustrated in

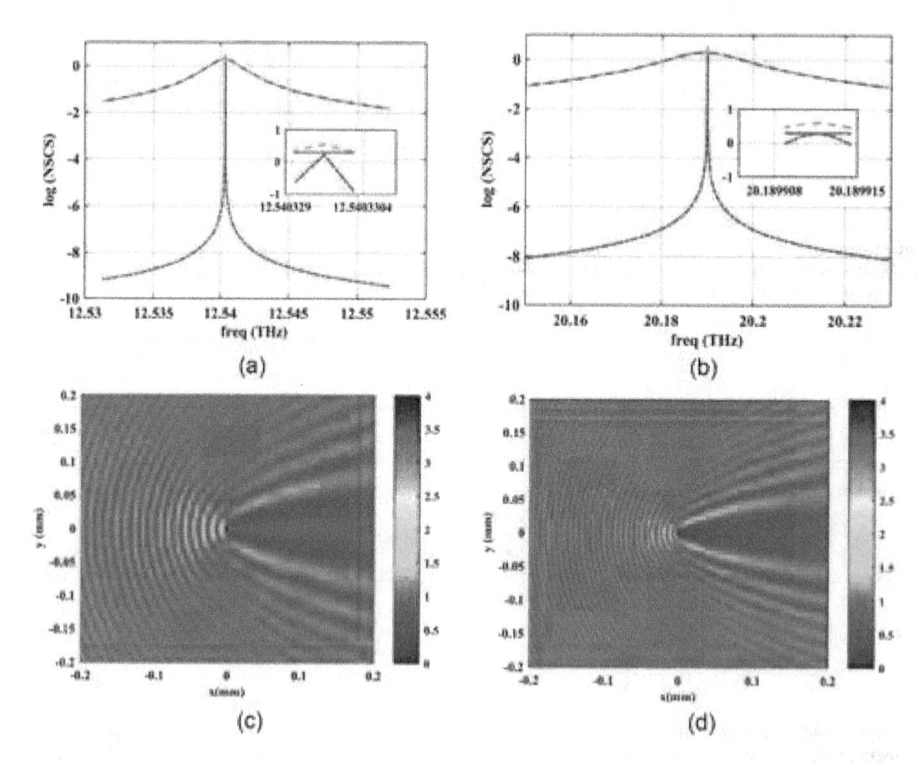

Figure 8.
(a) and (b) The NSCS of dual-band super-scatterer respectively, in the first and second operating frequencies and (c) and (d) corresponding magnetic field distributions [30].

Figure 9(a) and **(b)**, as well as the expected phenomenon, is clearly observed. The corresponding magnetic field distributions, shown in **Figure 9(c)** and **(d)**, also manifest the reduced and enhanced scatterings in the corresponding operating bands, respectively. Similar to the dual-band super-scatterer of the previous section, the performance of this structure is very sensitive to the optical, material, and geometrical parameters. By further increasing the number of graphene shells, other plasmonic resonances and zeros can be achieved for the manipulation of the optical response.

3. Graphene-coated spherical structures

In this section, multilayered graphene-coated particles with spherical morphology are investigated, and corresponding modified Mie-Lorenz coefficients are extracted by expanding the incident, scattered, and transmitted electromagnetic fields in terms of spherical harmonics. It is clear that by increasing the number of graphene layers, further degrees of freedom for manipulating the optical response can be achieved. For the simplicity of the performance optimization, an equivalent RLC circuit is proposed in the quasistatic regime for the sub-wavelength plasmons, and various practical examples are presented.

3.1 Multilayered graphene-based spherical structures

In this section, the most general graphene-based structure with N dielectric layers, as shown in **Figure 10**, is considered, and plane wave scattering is analyzed through extracting recurrence relations for modified Mie-Lorenz coefficients. It should be noted that since, in the TMM method, multiple matrix inversions are

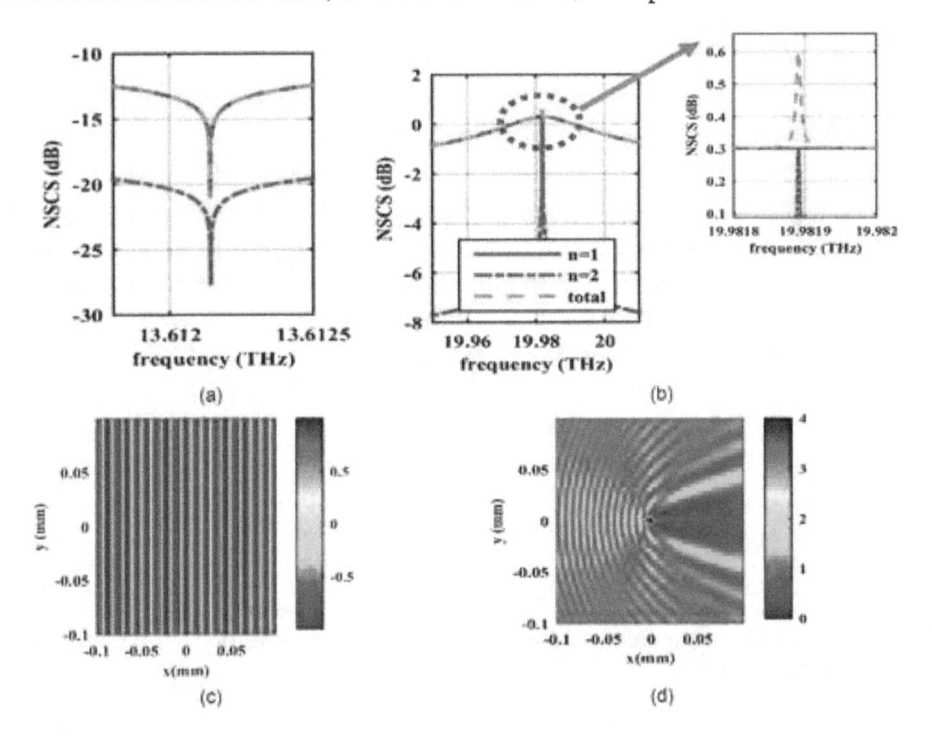

Figure 9.
Simultaneous super-scattering and super-cloaking using the structure of **Figure 6**. *NSCS for (a) super-cloaking and (b) super-scattering regimes and corresponding magnetic field distributions, respectively, in (c) and (d) [33].*

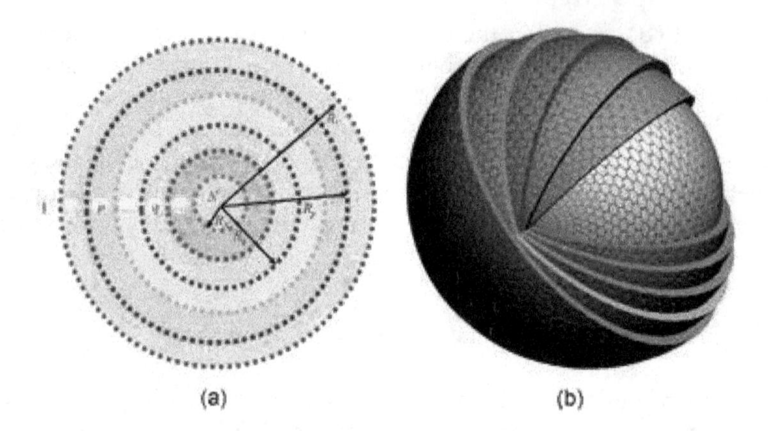

Figure 10.
Spherical graphene-dielectric stack (a) 2D and (b) 3D views [34]. Please note that the numbering of the layers is started from the outermost layer in order to preserve the consistency with the reference paper [35].

necessary, unlike the cylindrically layered structures of the previous section, the spherical geometries are analyzed through recurrence relations. Also, scattering from a single graphene-coated sphere has been formulated elsewhere [16], and it can be simply attained as the special case of our formulation.

The scattering analysis is very similar to that of the single-shell sphere [16], unless the Kronecker delta function is used in the expansions in order to find the electromagnetic fields of any desired layer with terse expansions. Therefore [34]:

$$\boldsymbol{E}_i = E_0 \sum_n \sum_m i^n \frac{2n+1}{n(n+1)} \left\{ \boldsymbol{M}_{mn}^{(1)} - i\boldsymbol{N}_{mn}^{(1)} \right\} \tag{25}$$

$$\begin{aligned} \boldsymbol{E}_{scat}^{p} &= E_0 \sum_n \sum_m i^n \frac{2n+1}{n(n+1)} \times \\ &\left\{ \left(1-\delta_p^N\right) B_H^p \boldsymbol{M}_{mn}^{(1)} - i\left(1-\delta_p^N\right) B_V^p \boldsymbol{N}_{mn}^{(1)} + \left(1-\delta_p^1\right) D_H^p \boldsymbol{M}_{mn} - i\left(1-\delta_p^1\right) D_V^p \boldsymbol{N}_{mn} \right\} \end{aligned} \tag{26}$$

By considering z_n as either j_n or $h_n^{(1)}$, which stand for the spherical Bessel and Hankel functions of the first kind with order n, respectively, and P_n^m as the associated Legendre function of order (n, m), the vector wave functions are defined as follows:

$$\boldsymbol{M}_{mn}\left(k_p\right) = z_n\left(k_p r\right) e^{im\phi} \left[\frac{im}{\sin\theta} P_n^m(\cos\theta)\hat{\boldsymbol{\theta}} - \frac{dP_n^m(\cos\theta)}{d\theta}\hat{\boldsymbol{\phi}} \right] \tag{27}$$

$$\begin{aligned} \boldsymbol{N}_{mn}\left(k_p\right) &= \frac{n(n+1)}{k_p r} z_n\left(k_p r\right) P_n^m(\cos\theta) e^{im\phi}\hat{\boldsymbol{r}} + \\ &\frac{1}{k_p r}\frac{d\left[r z_n\left(k_p r\right)\right]}{dr} \left[\frac{dP_n^m(\cos\theta)}{d\theta}\hat{\boldsymbol{\theta}} + \frac{im}{\sin\theta} P_n^m(\cos\theta)\hat{\boldsymbol{\phi}} \right] e^{im\phi} \end{aligned} \tag{28}$$

where super-indices (1) in the vector wave functions show that the Hankel functions are used in the field expansions. The boundary conditions at the interface of adjacent layers read as:

$$\hat{\boldsymbol{r}} \times \boldsymbol{E}^{(p)} = \hat{\boldsymbol{r}} \times \boldsymbol{E}^{(p+1)} \tag{29}$$

$$\frac{1}{i\omega\mu_{p+1}} \hat{\boldsymbol{r}} \times \nabla \times \boldsymbol{E}^{(p+1)} - \frac{1}{i\omega\mu_p} \hat{\boldsymbol{r}} \times \nabla \times \boldsymbol{E}^{(p)} = \sigma_{(p+1)p} \hat{\boldsymbol{r}} \times \left(\hat{\boldsymbol{r}} \times \boldsymbol{E}^{(p)} \right) \tag{30}$$

Therefore, the linear system of equations resulting from the above conditions is:

$$\begin{bmatrix} \xi_n^{pp} B_H^p \\ \partial \xi_n^{pp} B_V^p \end{bmatrix} + \begin{bmatrix} \psi_n^{pp}\left(D_H^p + \delta_p^1\right) \\ \partial \psi_n^{pp}\left(D_V^p + \delta_p^1\right) \end{bmatrix} = \begin{bmatrix} \xi_n^{(p+1)p} B_H^{(p+1)} \\ \partial \xi_n^{(p+1)p} B_V^{(p+1)} \end{bmatrix} + \begin{bmatrix} \psi_n^{(p+1)p} D_H^{(p+1)} \\ \partial \psi_n^{(p+1)p} D_V^{(p+1)} \end{bmatrix} \tag{31}$$

$$\frac{k_{p+1}}{\mu_{p+1}}\left(\begin{bmatrix} \partial \xi_n^{(p+1)p} B_H^{(p+1)} \\ \xi_n^{(p+1)p} B_V^{(p+1)} \end{bmatrix} + \begin{bmatrix} \partial \psi_n^{(p+1)p} D_H^{(p+1)} \\ \psi_n^{(p+1)p} D_V^{(p+1)} \end{bmatrix}\right) - \frac{k_p}{\mu_p}\left(\begin{bmatrix} \partial \xi_n^{pp} B_H^p \\ \xi_n^{pp} B_V^p \end{bmatrix} + \begin{bmatrix} \partial \psi_n^{pp}\left(D_H^p + \delta_p^1\right) \\ \psi_n^{pp}\left(D_V^p + \delta_p^1\right) \end{bmatrix}\right) =$$
$$-i\omega\sigma_{(p+1)p}\left(\begin{bmatrix} \xi_n^{pp} B_H^p \\ \partial \xi_n^{pp} B_V^p \end{bmatrix} + \begin{bmatrix} \psi_n^{pp}\left(D_H^p + \delta_p^1\right) \\ \partial \psi_n^{pp}\left(D_V^p + \delta_p^1\right) \end{bmatrix}\right) \tag{32}$$

where $\psi_n^{pq} = j_n\left(k_p R_q\right)$, $\xi_n^{pq} = h_n^{(1)}\left(k_p R_q\right)$, $\partial \psi_n^{pq} = \frac{1}{\rho} \mathrm{d}\left[\rho j_n(\rho)\right]\big|_{\rho=k_p R_q}$, and $\partial \xi_n^{pq} = \frac{1}{\rho} \mathrm{d}\left[\rho h_n^{(1)}(\rho)\right]\Big|_{\rho=k_p R_q}$ (d is defined as a symbol for the derivative with respect to the radial component). By rearranging the above equations, the coefficients of the layer $(p + 1)$ can be written in terms of the coefficients of the layer p as:

$$\begin{bmatrix} B_{H,V}^{(p+1)} \\ D_{H,V}^{(p+1)} \end{bmatrix} = \begin{bmatrix} \frac{1}{T_{Fp}^{H,V}} & \frac{R_{Fp}^{H,V}}{T_{Fp}^{H,V}} \\ \frac{R_{Pp}^{H,V}}{T_{Pp}^{H,V}} & \frac{1}{T_{Pp}^{H,V}} \end{bmatrix} \begin{bmatrix} B_{H,V}^p \\ D_{H,V}^p + \delta_p^1 \end{bmatrix} \tag{33}$$

where the sub/superscripts H and V represent the TE and TM waves, respectively. The directions of propagation of these waves are realized thought the subscripts F (outgoing waves) and P (incoming waves). The effective reflection coefficients are extracted as:

$$R_{Fp}^H = \frac{k_{p+1}\mu_p \partial \psi_n^{(p+1)p} \psi_n^{pp} - k_p \mu_{p+1} \partial \psi_n^{pp} \psi_n^{(p+1)p} + g \psi_n^{pp} \psi_n^{(p+1)p}}{k_{p+1}\mu_p \partial \psi_n^{(p+1)p} \xi_n^{pp} - k_p \mu_{p+1} \partial \xi_n^{pp} \psi_n^{(p+1)p} + g \xi_n^{pp} \psi_n^{(p+1)p}} \tag{34}$$

$$R_{Pp}^H = \frac{k_{p+1}\mu_p \partial \xi_n^{(p+1)p} \xi_n^{pp} - k_p \mu_{p+1} \partial \xi_n^{pp} \xi_n^{(p+1)p} + g \xi_n^{pp} \xi_n^{(p+1)p}}{k_{p+1}\mu_p \partial \xi_n^{(p+1)p} \psi_n^{pp} - k_p \mu_{p+1} \partial \psi_n^{pp} \xi_n^{(p+1)p} + g \psi_n^{pp} \xi_n^{(p+1)p}} \tag{35}$$

$$R_{Fp}^V = \frac{k_{p+1}\mu_p \psi_n^{(p+1)p} \partial \psi_n^{pp} - k_p \mu_{p+1} \psi_n^{pp} \partial \psi_n^{(p+1)p} + g \partial \psi_n^{pp} \partial \psi_n^{(p+1)p}}{k_{p+1}\mu_p \psi_n^{(p+1)p} \partial \xi_n^{pp} - k_p \mu_{p+1} \xi_n^{pp} \partial \psi_n^{(p+1)p} + g \partial \xi_n^{pp} \partial \psi_n^{(p+1)p}} \tag{36}$$

$$R_{Pp}^V = \frac{k_{p+1}\mu_p \xi_n^{(p+1)p} \partial \xi_n^{pp} - k_p \mu_{p+1} \xi_n^{pp} \partial \xi_n^{(p+1)p} + g \partial \xi_n^{pp} \partial \xi_n^{(p+1)p}}{k_{p+1}\mu_p \xi_n^{(p+1)p} \partial \psi_n^{pp} - k_p \mu_{p+1} \psi_n^{pp} \partial \xi_n^{(p+1)p} + g \partial \psi_n^{pp} \partial \xi_n^{(p+1)p}} \tag{37}$$

Moreover, it can be readily shown that the transmission coefficients read as:

$$T_{Fp}^H = \frac{k_{p+1}\mu_p \left(\partial \psi_n^{(p+1)p} \xi_n^{(p+1)p} - \psi_n^{(p+1)p} \partial \xi_n^{(p+1)p}\right)}{k_{p+1}\mu_q \partial \psi_n^{(p+1)p} \xi_n^{pp} - k_p \mu_{p+1} \partial \xi_n^{pp} \psi_n^{(p+1)p} + g \xi_n^{pp} \psi_n^{(p+1)p}} \tag{38}$$

$$T_{Pp}^{H} = \frac{k_{p+1}\mu_p\left(\partial\xi_n^{(p+1)p}\psi_n^{(p+1)p} - \partial\psi_n^{(p+1)p}\xi_n^{(p+1)p}\right)}{k_{p+1}\mu_p\partial\xi_n^{(p+1)p}\psi_n^{pp} - k_p\mu_{p+1}\partial\psi_n^{pp}\xi_n^{(p+1)p} + g\,\psi_n^{pp}\xi_n^{(p+1)p}} \tag{39}$$

$$T_{Fp}^{V} = \frac{k_{p+1}\mu_p\left(\psi_n^{(p+1)p}\partial\xi_n^{(p+1)p} - \xi_n^{(p+1)p}\partial\psi_n^{(p+1)p}\right)}{k_{p+1}\mu_p\psi_n^{(p+1)p}\partial\xi_n^{pp} - k_p\mu_{p+1}\xi_n^{pp}\partial\psi_n^{(p+1)p} + g\,\partial\xi_n^{pp}\partial\psi_n^{(p+1)p}} \tag{40}$$

$$T_{Pp}^{V} = \frac{k_{p+1}\mu_p\left(\xi_n^{(p+1)p}\partial\psi_n^{(p+1)p} - \psi_n^{(p+1)p}\partial\xi_n^{(p+1)p}\right)}{k_{p+1}\mu_p\,\xi_n^{(p+1)p}\partial\psi_n^{pp} - k_p\mu_{p+1}\psi_n^{pp}\partial\xi_n^{(p+1)p} + g\,\partial\psi_n^{pp}\partial\xi_n^{(p+1)p}} \tag{41}$$

where $g = i\omega\sigma_{(p+1)p}\mu_p\mu_{p+1}$. By using $B_{H,V}^{N} = D_{H,V}^{1} = 0$, the recurrence relations can be started, and the field expansion coefficients in any desired layer can be obtained. The extinction efficiency is related to the external modified Mie-Lorenz coefficients via:

$$Q_{ext} = \frac{2\pi}{k^2}\Re\sum_{n=1}^{\infty}(2n+1)\left(B_V^1 + B_H^1\right) \tag{42}$$

where symbol $\Re$ represents the real part of the summation. In order to verify the extracted coefficients, the extinction efficiencies of three graphene-coated structures is provided in **Figure 11**. In the graphical representation of the structures, the dashed lines illustrate graphene interfaces, while the solid line shows a PEC core. The optical and geometrical parameters are R_1 = 200 nm, R_2 = 100 nm, R_3 = 50 nm, μ_c = 0.3 eV,T = 300° K, and τ = 0.02 ps. The analytical results are compared with the numerical results of CST 2017 commercial software, and good agreement is achieved. Moreover, the analytical formulation provides a fast and accurate tool for the scattering shaping of various spherical geometries.

In order to realize the priority of the closed-form analytical formulation with respect to the numerical analysis, the simulation times of both methods are included in **Table 2**. Considerable time reduction using the exact solution is evident. Moreover, since 3D meshing and perfectly matched layers are not required in this method, it is efficient in terms of memory as well.

3.1.1 Quasistatic approximation and RLC model

Based on the results of Section 3.1, the modified Mie-Lorenz coefficients of the graphene-based spherical particles form infinite summations in terms of spherical Bessel and Hankel functions. In general, graphene plasmons are excited in the subwavelength regime, and only the leading order term of the summation is sufficient for achieving the results with acceptable precision. In this regime, the polynomial expansion of the special functions can also be truncated in the first few terms [22]. Later, the extracted modified Mie-Lorenz coefficients can be rewritten in the form of the polynomials. To further simplify the real-time monitoring and performance optimization of the graphene-coated nanoparticles, an equivalent RLC circuit can be proposed by representing the rational functions in the continued fraction form as [36]:

$$Y_{TE/TM} = Y_0\frac{1}{Z_1 + \frac{1}{Z_2 + \frac{1}{Z_3 + \dots}}} \tag{43}$$

The equivalent circuit corresponding to the above representation is shown in **Figure 12**.

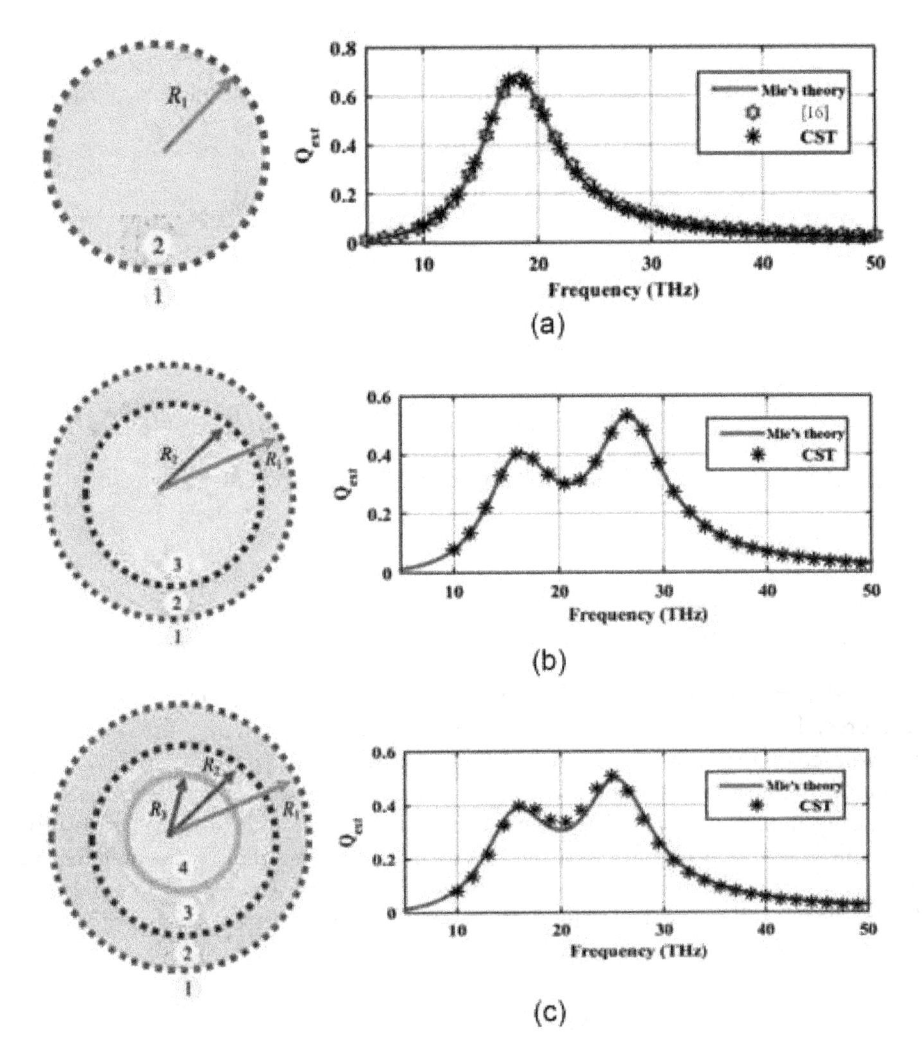

Figure 11.
The extinction efficiencies of graphene-based particles with different number of layers: (a) two, (b) three, and (c) four [34].

Structure	Simulation time	
	Analytical	CST
Figure 11(a)	0.053214 s	32 h, 50 m, 18 s
Figure 11(b)	0.045831 s	33 h, 45 m, 25 s
Figure 11(c)	0.151555 s	33 h, 34 m, 55 s

Table 2.
Comparing the simulation time of CST and our codes [34].

The continued fraction representation for the TM coefficients is:

$$b_1 = \frac{x^3}{-\sigma'_2 {\sigma'}_3^{-1} + \frac{x^2}{-{\sigma'}_1^{-1} {\sigma'}_2^{-2} {\sigma'}_3^{2} + \frac{x}{{\sigma'}_1^{2} {\sigma'}_2^{4} {\sigma'}_3^{-4} + O'(x)}}} \tag{44}$$

where $\sigma'_0 = {\sigma'}_3^{-1}\sigma'_2$, $\sigma'_1 = {\sigma'}_2^{-1}\sigma'_5 - {\sigma'}_2^{-2}\sigma'_3\sigma'_4$, $\sigma'_2 = i\frac{m}{3}(2 + 2g + m^2)$, $\sigma'_3 = \frac{2m}{9}(-1 + 2g + m^2)$, $\sigma'_4 = \frac{im}{3}\left(-1 + g + \frac{2gm^2}{5} + \frac{9m^2}{10}\right)$, $\sigma'_5 = \frac{4gm}{45}(1 + m^2)$.

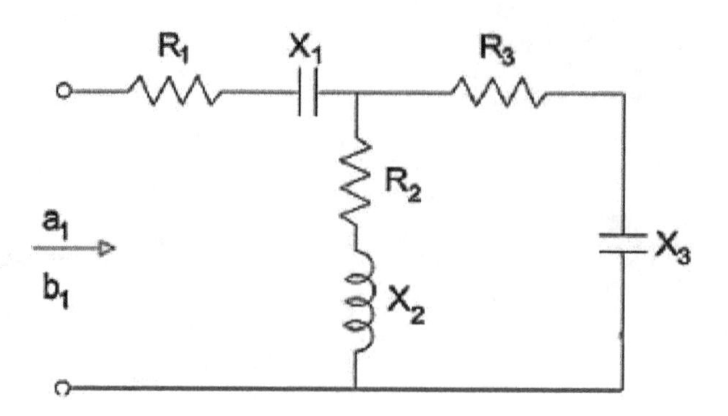

Figure 12.
The proposed equivalent circuit for the scattering analysis of electrically small graphene-coated spheres [36].

The elements of the equivalent circuit for the TM coefficients read as:

$$Y'_0 = x^3 \quad , \quad Z'_1 = -\sigma'_2 \sigma'^{-1}_3 \quad , \quad Z'_2 = \frac{-\sigma'^{-1}_1 \sigma'^{-2}_2 \sigma'^{2}_3}{x^2} \quad , \quad Z'_3 = x \sigma'^{2}_1 \sigma'^{4}_2 \sigma'^{-4}_3 \tag{45}$$

3.1.2 Application in emission

In order to illustrate the application of Mie analysis for the graphene-wrapped structures, let us consider vertical and horizontal dipoles in the proximity of a graphene-coated sphere, as shown in **Figure 13**. Although in the Mie analysis, the excitation is considered to be a plane wave, by using the scattering coefficients, the total decay rates can be calculated for the dipole emitters, and it can be proven that the localized surface plasmons of the graphene-wrapped spheres can enhance the total decay rate, which is connected to the Purcell factor [16, 37]. The amount of electric field enhancement for the radial-oriented and tangential oscillating dipoles with the distance of x_d, respectively, read as:

$$\frac{E}{E_0} = 1 + \sum_{n=1}^{\infty} i^{n+1} b_n (2n+1) P_n^1(0) \frac{h_n^{(1)}(x_d)}{x_d} \tag{46}$$

$$\frac{E}{E_0} = 1 + \sum_{n=1}^{\infty} i^n \frac{(2n+1)}{n(n+1)} \left\{ a_n P'_n(0) h_n^{(1)}(x_d) + i b_n P_n^1(0) \frac{\left[x_d h_n^{(1)}(x_d)\right]'}{x_d} \right\} \tag{47}$$

Figure 13(b) shows the local field enhancement for the average orientation of the dipole emitter in the vicinity of the sphere with R_1 = 20 nm, coated by a graphene material with the chemical potential of μ_c = 0.1 eV. As the figure shows, an enhanced electric field in the order of $\sim 10^4$ is obtained for the dipole distance of 1 nm with averaged orientation, and it decreases as the dipole moves away from the sphere.

3.1.3 Application in super-scattering

The possibility of a super-scatterer design using graphene-coated spherical particles is illustrated in **Figure 14**. The design parameters are ε_1 = 1.44, R_1 = 0.24 μm, and μ_c = 0.3 eV. The structure can be simply analyzed by the modified Mie-Lorenz coefficients. The general design concepts are similar to their cylindrical counterparts, namely, dispersion engineering using the associated planar structure, as shown in the inset of the figure. Due to the excitation of TM surface plasmons, the normalized extinction cross-section is five times greater than the bare dielectric

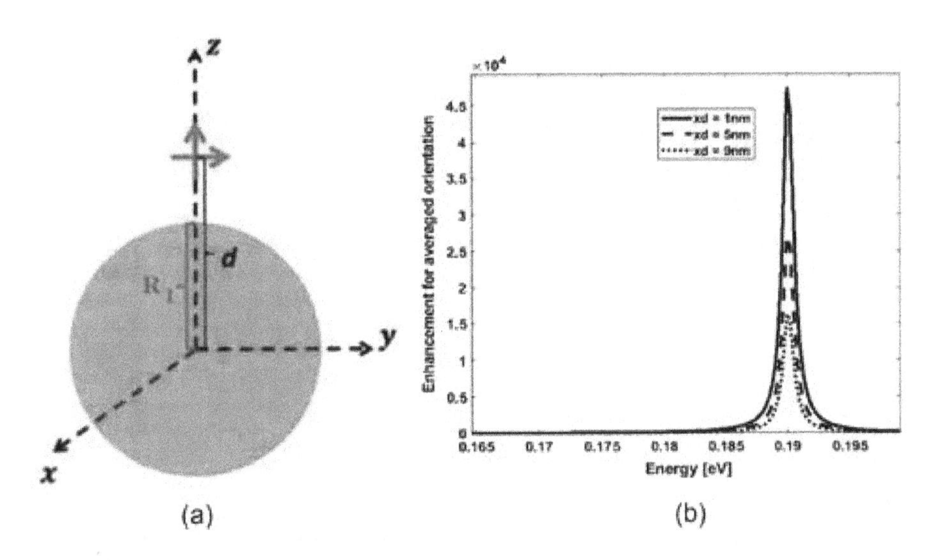

Figure 13.
(a) Vertical and horizontal dipole emitters in the proximity of the graphene-coated sphere and (b) the local field enhancement for various dipole distances with averaged orientation [37].

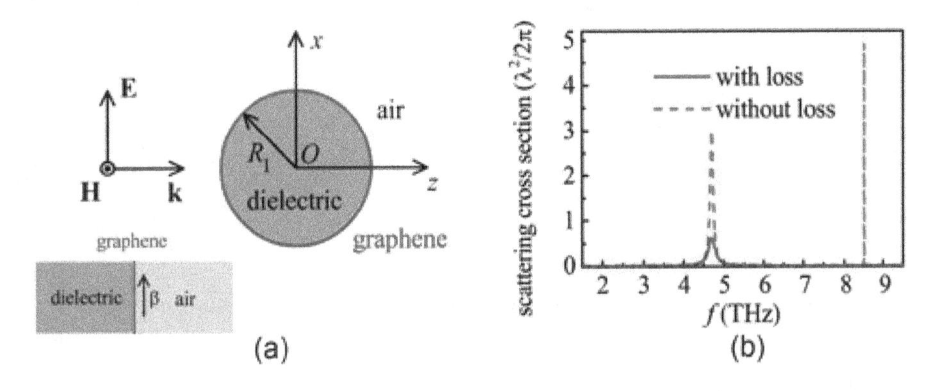

Figure 14.
(a) Atomically thin super-scatterer and associated planar structure shown in the inset and (b) corresponding normalized scattering cross-sections by considering lossless and lossy graphene shells [38].

sphere. Moreover, similar to the cylindrical super-scatterers, by considering a small amount of loss for the graphene coating by assigning $\Gamma = 0.11\, meV$, the performance is considerably degraded [38].

3.1.4 Application in wide-band cloaking

By pattering graphene-based disks with various radii around a dielectric sphere, it is feasible to design a wide-band electromagnetic cloak at infrared frequencies. The geometry of this structure is illustrated in **Figure 15**. In order to analyze the proposed cloak by the modified Mie-Lorenz theory, the polarizability of the disks can be inserted in the equivalent conductivity method. The extracted equivalent surface conductivity can be used to tune the surface reactance of the sphere for the purpose of cloaking [39].

3.1.5 Application in multi-frequency cloaking

The other application that can be adapted to our proposed formulation of multilayered spherical structures is multi-frequency cloaking. As **Figure 16** shows, by

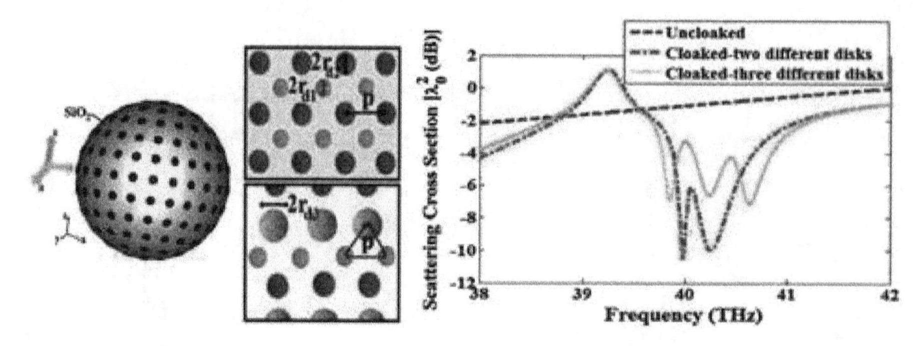

Figure 15.
Wide-band cloaking using graphene disks with varying radii [39].

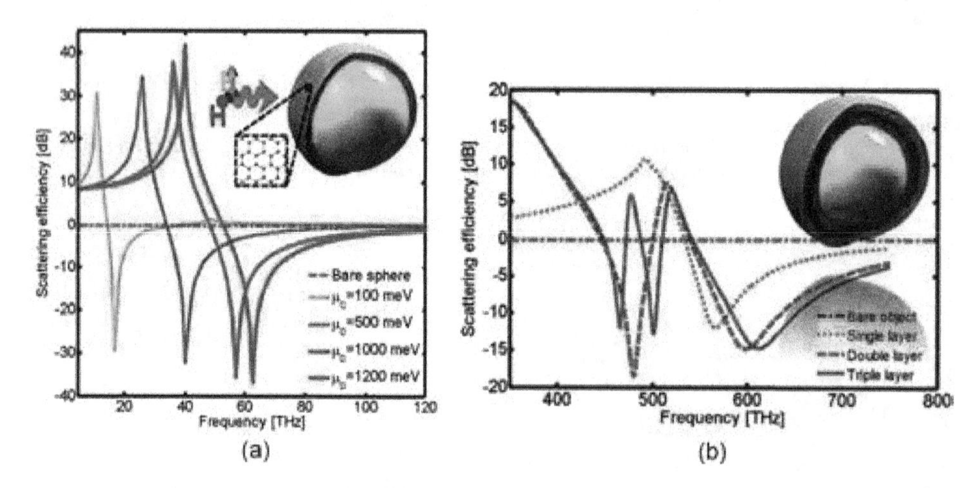

Figure 16.
(a) Single and (b) multi-frequency cloaking using single/multiple graphene shells around a spherical particle [21].

proper design, a single graphene coating can eliminate the dipole resonace in a single reconfigurable frequency. The radius of the sphere is R_1 = 100 nm and its core permittivity is $\varepsilon_1 = 3$. It can be concluded that double graphene shells can suppress the scattering in the dual frequencies since each graphene shell with different geometrical and optical properties can support localized surface plasmon reso-nances in a specific frequency. By further increase of the graphene shells, other frequency bands can be generated. **Figure 16(b)** shows the cloaking performance of a spherical particle with multiple graphene shells. The radii of the spheres are 107.5, 131.5, and 140 nm, and the corresponding chemical potentials are 900, 500, and 700 meV, respectively. The permittivity of the dielectric filler is 2.1 [21].

3.1.6 Application in electromagnetic absorption

As another example, a dielectric-metal core-shell spherical resonator (DMCSR) with the resonance frequency lying in the near-infrared spectrum is considered. In order to increase the optical absorption, the outer layer of the structure is covered with graphene. The localized surface plasmons of graphene are mainly excited in the far-infrared frequencies and in the near-infrared and visible range; it behaves like a dielectric. By hybridizing the graphene with a resonator, its optical absorption can be greatly enhanced. **Figure 17** shows the performance of the structure for various core radii [15].

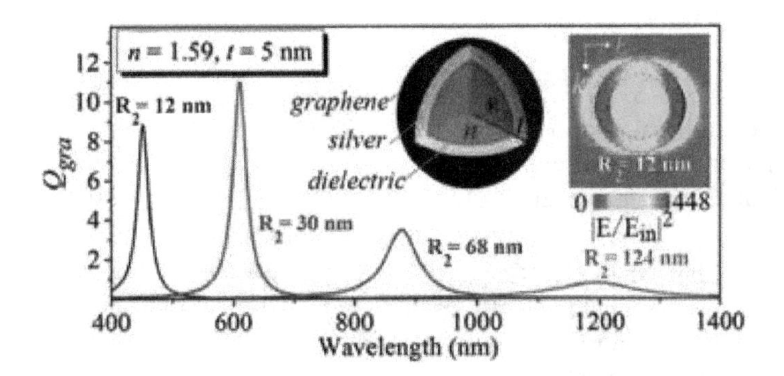

Figure 17.
Strong tunable absorption using a graphene-coated spherical resonator with fixed dielectric core refractive index of n and silver shell thickness of t [15].

The provided examples are just a few instances for scattering analysis of graphene-based structures. Based on the derived formulas, other novel optoelectronic devices based on graphene plasmons can be proposed. Moreover, since assemblies of polarizable particles fabricated by graphene exhibit interesting properties such as enhanced absorption, negative permittivity, giant near-field enhancement, and large enhancements in the emission and the radiation of the dipole emitters [40–43], the research can be extended to the multiple scattering theory.

Author details

Shiva Hayati Raad[1], Zahra Atlasbaf[1*] and Mauro Cuevas[2]

1 Department of Electrical and Computer Engineering, Tarbiat Modares University, Tehran, Iran

2 Faculty of Engineering and Information Technology, University of Belgrano, Buenos Aires, Argentina

*Address all correspondence to: atlasbaf@modares.ac.ir

References

[1] Kim K-H, No Y-S. Subwavelength core/shell cylindrical nanostructures for novel plasmonic and metamaterial devices. Nano Convergence. 2017;**4**(1): 1-13

[2] Fan P, Chettiar UK, Cao L, Afshinmanesh F, Engheta N, Brongersma ML. An invisible metal–semiconductor photodetector. Nature Photonics. 2012;**6**(6):380

[3] Padooru YR, Yakovlev AB, Chen P-Y, Alù A. Analytical modeling of conformal mantle cloaks for cylindrical objects using sub-wavelength printed and slotted arrays. Journal of Applied Physics. 2012;**112**(3):034907

[4] Ruan Z, Fan S. Superscattering of light from subwavelength nanostructures. Physical Review Letters. 2010;**105** (1):013901

[5] Qian C et al. Experimental observation of superscattering. Physical Review Letters. 2019;**122**(6):063901

[6] Naserpour M, Zapata-Rodríguez CJ. Tunable scattering cancellation of light using anisotropic cylindrical cavities. Plasmonics. 2017;**12**(3):675-683

[7] Díaz-Aviñó C, Naserpour M, Zapata-Rodríguez CJ. Conditions for achieving invisibility of hyperbolic multilayered nanotubes. Optics Communications. 2016;**381**:234-239

[8] Garcia-Etxarri A et al. Strong magnetic response of submicron silicon particles in the infrared. Optics Express. 2011;**19**(6):4815-4826

[9] Monticone F, Argyropoulos C, Alù A. Layered plasmonic cloaks to tailor the optical scattering at the nanoscale. Scientific Reports. 2012;**2**:912

[10] Liu W, Lei B, Shi J, Hu H. Unidirectional superscattering by multilayered cavities of effective radial anisotropy. Scientific Reports. 2016;**6**: 34775

[11] Velichko EA. Evaluation of a graphene-covered dielectric microtube as a refractive-index sensor in the terahertz range. Journal of Optics. 2016; **18**(3):035008

[12] Correas-Serrano D, Gomez-Diaz JS, Alù A, Melcón AÁ. Electrically and magnetically biased graphene-based cylindrical waveguides: Analysis and applications as reconfigurable antennas. IEEE Transactions on Terahertz Science and Technology. 2015;**5**(6):951-960

[13] Li R et al. Design of ultracompact graphene-based superscatterers. IEEE Journal of Selected Topics in Quantum Electronics. 2016;**23**(1):130-137

[14] Bernety HM, Yakovlev AB. Cloaking of single and multiple elliptical cylinders and strips with confocal elliptical nanostructured graphene metasurface. Journal of Physics: Condensed Matter. 2015;**27**(18):185304

[15] Wan M et al. Strong tunable absorption enhancement in graphene using dielectric-metal core-shell resonators. Scientific Reports. 2017;**7**(1):32

[16] Christensen T, Jauho A-P, Wubs M, Mortensen NA. Localized plasmons in graphene-coated nanospheres. Physical Review B. 2015;**91**(12):125414

[17] Xiao T-H, Gan L, Li Z-Y. Graphene surface plasmon polaritons transport on curved substrates. Photonics Research. 2015;**3**(6):300-307

[18] Falkovsky LA. Optical properties of graphene and IV–VI semiconductors. Physics-Uspekhi. 2008;**51, 9**:887

[19] Hanson GW. Dyadic Green's functions and guided surface waves for

a surface conductivity model of graphene. Journal of Applied Physics. 2008;**103**(6):064302

[20] Vakil A. Transformation Optics Using Graphene: One-Atom-Thick Optical Devices Based on Graphene. Dissertations-University of Pennsylvania; 2012

[21] Farhat M, Rockstuhl C, Bağcı H. A 3D tunable and multi-frequency graphene plasmonic cloak. Optics Express. 2013;**21**(10):12592-12603

[22] Bohren CF, Huffman DR. Absorption and Scattering of Light by Small Particles. New York: John Wiley & Sons; 2008

[23] Naserpour M, Zapata-Rodríguez CJ, Vuković SM, Pashaeiadl H, Belić MR. Tunable invisibility cloaking by using isolated graphene-coated nanowires and dimers. Scientific Reports. 2017; **7** (1): 12186

[24] Riso M, Cuevas M, Depine RA. Tunable plasmonic enhancement of light scattering and absorption in graphene-coated subwavelength wires. Journal of Optics. 2015;**17**(7):075001

[25] Pashaeiadl H, Naserpour M, Zapata-Rodríguez CJ. Scattering of electromagnetic waves by a graphene-coated thin cylinder of left-handed metamaterial. Optik. 2018;**159**:123-132

[26] Cuevas M, Riso MA, Depine RA. Complex frequencies and field distributions of localized surface plasmon modes in graphene-coated subwavelength wires. Journal of Quantitative Spectroscopy and Radiative Transfer. 2016;**173**:26-33

[27] Abromowitz M, Stegun IA, editors. Handbook of Mathematical Functions with Formulas, Graphs, and Mathematical Tables. Vol. 55. US Government Printing Office; 1948

[28] Cuevas M. Graphene coated subwavelength wires: A theoretical investigation of emission and radiation properties. Journal of Quantitative Spectroscopy and Radiative Transfer. 2017;**200**:190-197

[29] Cuevas M. Enhancement, suppression of the emission and the energy transfer by using a graphene subwavelength wire. Journal of Quantitative Spectroscopy and Radiative Transfer. 2018;**214**:8-17

[30] Raad SH, Zapata-Rodríguez CJ, Atlasbaf Z. Multi-frequency super-scattering from sub-wavelength graphene-coated nanotubes. JOSA B. 2019;**36**(8):2292-2298

[31] Díaz-Aviñó C, Naserpour M, Zapata-Rodríguez CJ. Optimization of multilayered nanotubes for maximal scattering cancellation. Optics Express. 2016;**24**(16):18184-18196

[32] Chen P-Y, Soric J, Padooru YR, Bernety HM, Yakovlev AB, Alù A. Nanostructured graphene metasurface for tunable terahertz cloaking. New Journal of Physics. 2013;**15**(12):123029

[33] Raad SH, Zapata-Rodríguez CJ, Atlasbaf Z. Graphene-coated resonators with frequency-selective super-scattering and super-cloaking. Journal of Physics D: Applied Physics. 2019; **52**(49):495101

[34] Raad SH, Atlasbaf Z, Rashed-Mohassel J, Shahabadi M. Scattering from Graphene-based multilayered spherical structures. IEEE Transactions on Nanotechnology. 2019;**18**:1129-1136

[35] Li L-W, Kooi P-S, Leong M-S, Yee T-S. Electromagnetic dyadic Green's function in spherically multilayered media. IEEE Transactions on Microwave Theory and Techniques. 1994;**42**(12):2302-2310

[36] Raad SH, Atlasbaf Z. Equivalent RLC ladder circuit for scattering by graphene-coated nanospheres. IEEE Transactions on Nanotechnology. 2019; **18**:212-219

[37] Sijercic E, Leung P. Enhanced terahertz emission from quantum dot by graphene-coated nanoparticle. Applied Physics B. 2018;**124**(7):141

[38] Li R, Lin X, Lin S, Liu X, Chen H. Atomically thin spherical shell-shaped superscatterers based on a Bohr model. Nanotechnology. 2015;**26**(50):505201

[39] Shokati E, Granpayeh N, Danaeifar M. Wideband and multi-frequency infrared cloaking of spherical objects by using the graphene-based metasurface. Applied Optics. 2017; **56**(11):3053-3058

[40] Raad SH, Atlasbaf Z. Tunable optical absorption using Graphene covered Core-Shell Nano-spheres. In: Iranian Conference on Electrical Engineering (ICEE). IEEE; 2018. pp. 98-102

[41] Raad SH, Atlasbaf Z. Tunable optical meta-surface using graphene-coated spherical nanoparticles. AIP Advances. 2019;**9**(7):075224

[42] Raad SH, Atlasbaf Z, Zapata-Rodríguez CJ. Multi-frequency near-field enhancement with graphene-coated nano-disk homo-dimers. Optics Express. 2019;**27**(25):37012-37024

[43] Cuevas M. Theoretical investigation of the spontaneous emission on graphene plasmonic antenna in THz regime. Superlattices and Microstructures. 2018;**122**:216-227

5

Wireless Optical Nanolinks with Yagi-Uda and Dipoles Plasmonic Nanoantennas

Karlo Queiroz da Costa, Gleida Tayanna Conde de Sousa, Paulo Rodrigues Amaral, Janilson Leão Souza, Tiago Dos Santos Garcia and Pitther Negrão dos Santos

Abstract

In this work, we present a theoretical analysis of wireless optical nanolinks formed by plasmonic nanoantennas, where the antennas considered are Yagi-Uda and cylindrical nanodipoles made of Au. The numerical analysis is performed by the finite element method and linear method of moments, where the transmission power and the near electric field are investigated and optimized for three nanolinks: Yagi-Uda/Yagi-Uda, Yagi-Uda/dipole and dipole/dipole. The results show that all these case can operate with good transmission power at different frequencies by adjusting the impedance matching in the transmitting antennas and the load impedance of the receiving antennas.

Keywords: nanoplasmonic, nanoantennas, wireless optical nanolink, power transmission

1. Introduction

The electromagnetic scattering of metals in optical frequency region possesses special characteristics. At these frequencies, there are electron oscillations in the metal called plasmons with distinct resonant frequencies, which produce strongly enhanced near fields at the metal surface. This effect can be analyzed using Lorentz-Drude model of the complex dielectric constant. The science of the electromagnetic optical response of metal nanostructures is known as plasmonics or nanoplasmonics [1, 2].

One subarea of nanoplasmonics is the field of optical nanoantennas, which are metal nanostructures used to transmit or receive optical fields [3–5]. This definition is similar to that of conventional radio frequency (RF) and microwave antennas. The main difference between these two regimes (RF-microwave and optical) is due to physical properties of the metals at optical frequencies where they cannot be considered as perfect conductors because of the plasmonic effects [2]. Comprehen-sive reviews on optical antennas have been presented in [6–11]. In these works, the authors described recent developments in calculation of such antennas, their appli-cations and challenges in their design. In **Figure 1**, we present some examples of fabricated nanoantennas.

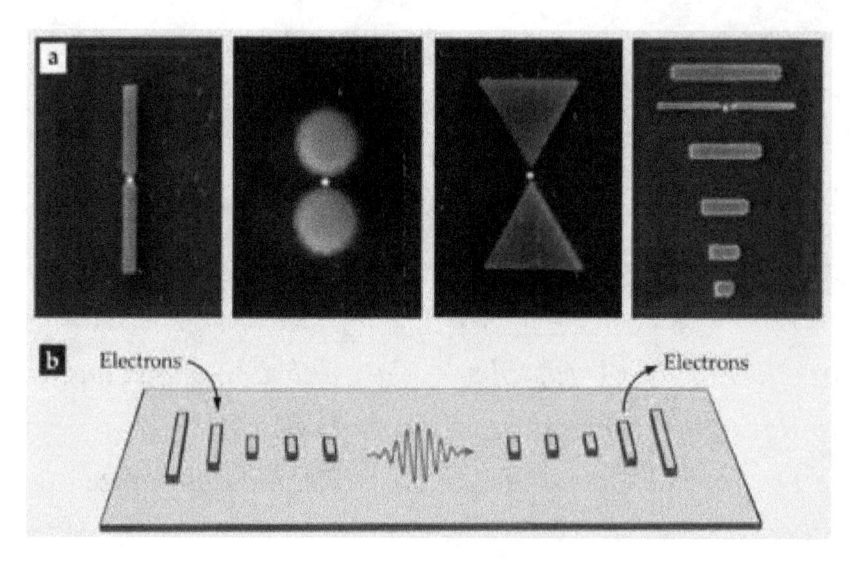

Figure 1.
(a) Scanning electron micrographs of various optical nanoantennas. (b) Application example of Yagi-Uda nanoantennas in wireless optical nanolink, where the nanoantennas perform transduction between electrical current and optical radiation [5].

Optical nanoantennas have received great interest in recent years in the scientific community due to their ability to amplify and confine optical fields beyond the light diffraction limit [6]. With this characteristic, it is possible to apply in several areas, such as nanophotonics, biology, chemistry, computer science, optics and engineering, among others [6, 7, 12]. In addition, these studies were expanded due to the development of computational numerical methods and innovations in nanofabrication techniques, such as electron beam lithography, colloidal lithography and ion beam lithography [9].

Optical wireless nanolinks with nanoantennas can be used to efficiently communicate between devices, significantly reducing the losses that occur in wired communication. Nanolinks with different geometries of nanoantennas were investigated in the literature [13–17]. In [13] the authors propose a broadband nanolink formed by dipole-loop antennas. The results showed that using this nanolink with dipole-loop antennas instead of conventional dipoles, it is possible to increase the operating bandwidth of the system to the range of 179.1–202.5 THz, which is within the optical range of telecommunications. In [14], a wireless nanolink formed by dipole antennas is compared to a wired nanolink formed by a waveguide, the study showed that the wireless link may work better than a plasmon waveguide in sending optical signal in nanoscale from one point to another, from a certain distance. In [15], a nanolink Yagi-Uda chip directives are proposed, the results show that the use of directional antennas increases the energy transfer (power ratio) and link efficiency, minimizing interference with other parts of the circuit. In [16], it is presented another wireless nanolink application formed by a transmitting nanoantenna Vivaldi and another receiver, to be used in chip, with that nanolink a high gain and bandwidth covering the entire spectrum of the C band of telecommunications. In [17], broadband nanolinks were analyzed using horn and dipole type optical nanoantennas, where the horn antenna had better performance, because better energy transfer at the nanolink and greater bandwidth were obtained in relation to the dipole link. These studies used identical transmitting and receiving antennas, and showed the feasibility of using wireless communication in the nanophotonics.

In this work, we present a comparative analysis of nanolinks formed by equal and different transmitting and receiving nanoantennas. The antennas used are

Yagi-Uda and dipole. The numerical analysis is performed by the method of moments (MoM) [18] and the finite element method (FEM) through the software COMSOL Multiphysics [19]. In this analysis, the transmission power and the near electric field are investigated for three nanolinks: Yagi-Uda/dipole, Yagi-Uda/Yagi-Uda and dipole/dipole. This work is organized as follow: Section 1 is the introduction, Section 2 presents the description of nanolinks, Section 3 presents the numerical model used in the analysis, Section 4 contain the numerical results, and Section 5 are the conclusions.

2. Description of nanolinks

In this work, three models of nanolinks are proposed and analyzed. The first is a nanolink formed by dipole/dipole antennas (**Figure 2**, without reflector and directors), the second by Yagi-Uda/dipole antennas (**Figure 2**) and the third by Yagi-Uda/Yagi-Uda antennas (**Figure 2**, with the receiving antenna equal to the transmitting antenna).

The geometry of the Yagi-Uda/dipole nanolink is presented in **Figure 2**, where a voltage source V_S excites the left nanoantenna, which functions as a transmitter (Yagi-Uda) and the right nanoantenna that acts as a receptor (dipole), connected to load impedance Z_C. The nanolink is located in the free space and is formed by cylindrical conductors of gold. The complex permittivity of this material is represented by the Lorentz-Drude model of Au [11]. The Yagi-Uda transmitting nanoantenna is composed of a dipole, a reflector and three directors (**Figure 1** left). The dipole of the transmitter, located in the $z = 0$ plane along the x-axis and centered at the origin, has total length $2h_{dT} + d_{dT}$, radius a_{dT} and voltage gap d_{dT}. The reflector has h_r length and a_r radius. The directors have the same length h_d and radius a_d. The parameters d_{hr} and d_{hd} are the distances between reflector and directors element to Yagi-Uda antenna, respectively (**Figure 1** left). The receiver antenna is a dipole (**Figure 1** right), located in the $z = 0$ plane and displaced at a d_{TR} relative to the dipole axis of the transmitting antenna, with total length $2h_{dR} + d_{dR}$, radius a_{dR}, gap length d_{dR} and load Z_C connected to its gap.

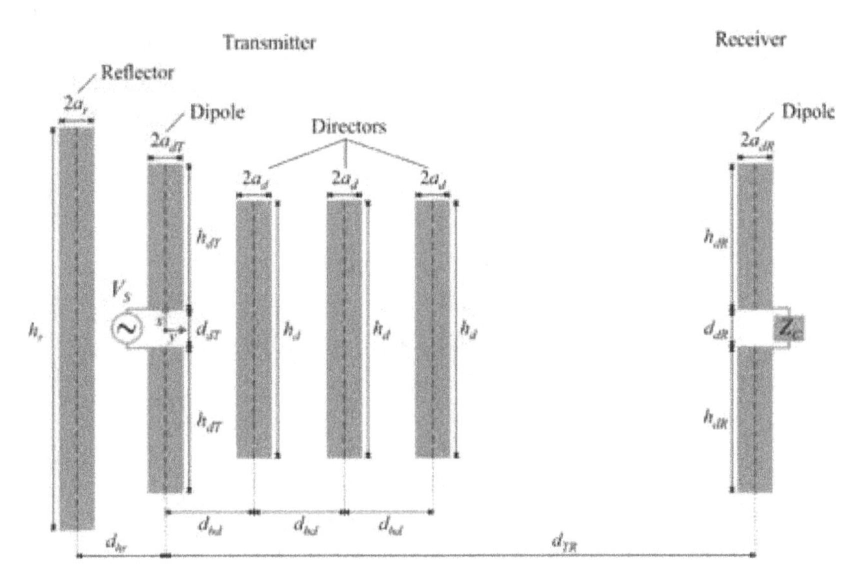

Figure 2.
Geometry of the nanolink composed by a Yagi-Uda antenna (transmitter) and a nanodipole (receiver).

3. Numerical model

In this section, we present the numerical methods used in the analysis of the wireless optical nanolinks described in the last section. The methods used here are MoM [18] and FEM [19].

3.1 Method of moments

The linear MoM presented here is based in the linear current approximation with an equivalent surface impedance model of the cylindrical conductors, with sinusoidal test and base functions [18]. The method will be explained for the particular example of a single dipole radiating in a free space, composed by plasmonic cylindrical elements made of gold.

Figure 3 shows the geometry of the original problem, the equivalent MoM and circuit models of the nanodipole. In this figure, L is the length of the arms, d is the nanodipole gap and a is the dipole radius. The total length of this antenna is $L_t = 2L + d$. In present analysis, we do not take into account the capacitance generated by the air gap (C_{gap}) of the nanodipole. In this case, our input impedance is equivalent to that Z_a presented in [20] without C_{gap}.

In the radiation problem of **Figure 3**, the gold material of the antenna is represented by the Lorentz-Drude model for complex permittivity $\varepsilon_{\text{Au}} = \varepsilon_0 \varepsilon_{r\text{Au}}$:

$$\varepsilon_{r\text{Au}} = \varepsilon_\infty - \frac{\omega_{p1}^2}{\omega^2 - j\Gamma\omega} + \frac{\omega_{p2}^2}{\omega_0^2 - \omega^2 + j\gamma\omega}, \tag{1}$$

where the parameters in this equation are as follows [1]: $\varepsilon_\infty = 8$, $\omega_{p1} = 13.8 \times 10^{15}\text{s}^{-1}$, $\Gamma = 1.075 \times 10^{14}\text{s}^{-1}$, $\omega_0 = 2\pi c/\lambda_0$, $c = 3 \times 10^8$ m/s, $\lambda_0 = 450$ nm, $\omega_{p2} = 45 \times 10^{14}\text{s}^{-1}$, $\gamma = 9 \times 10^{14}\text{s}^{-1}$, and ω is the angular frequency in rad/s. **Figure 4** presents the real and imaginary part of (1) versus wavelength (λ). This figure also shows the experimental data of [21]. We observe a good agreement between the results of the Lorentz-Drude model of (1) and the experimental data for $\lambda > 500$ nm.

The losses in metal are described by the surface impedance Z_s. This impedance can be obtained approximately by considering cylindrical waveguide with the mode TM_{01}. In this case, the surface impedance is given by [22].

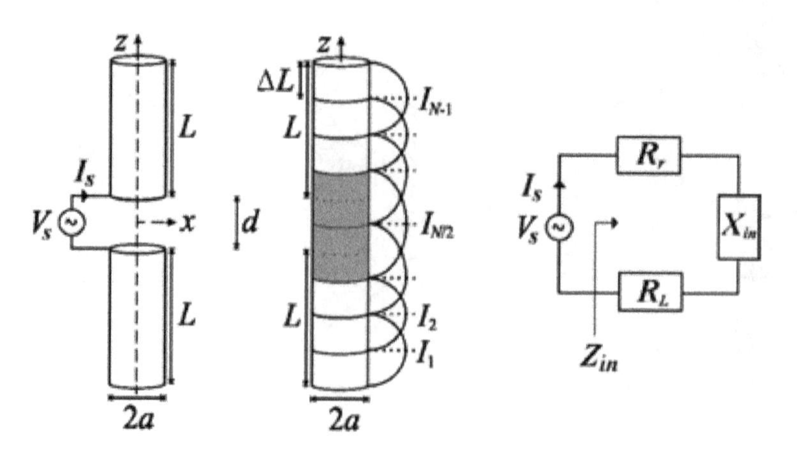

Figure 3.
Geometry of nanodipole: original problem (left), MoM model (middle), and equivalent circuit model (right).

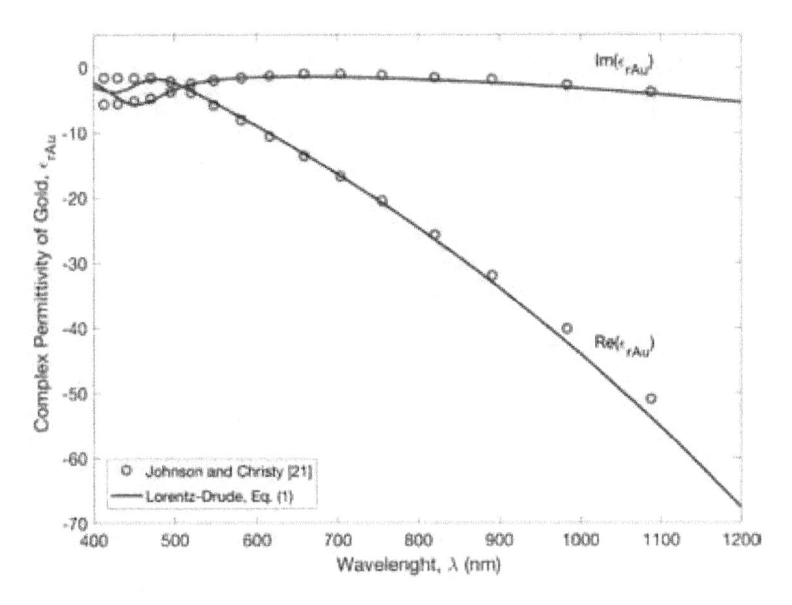

Figure 4.
Complex permittivity of gold obtained by Lorentz-Drude model of (1) and experimental data of Johnson-Christy [21].

$$Z_s = \frac{TJ_0(Ta)}{2\pi a j\omega \varepsilon_{Au} J_1(Ta)}, \quad (2)$$

being $T = k_0\sqrt{\varepsilon_{rAu}}$ and $k_0 = \omega\sqrt{\mu_0\varepsilon_0}$. The boundary condition for the electric field at the conductor surface is $\left(\overline{E}_s + \overline{E}_i\right) \cdot \overline{a}_l = Z_s I$, where $\overline{a}_l$ is a unitary vector tangential to the surface of the metal, $\overline{E}_s$ is the scattered electric field due to the induced linear current I on the conductor, $\overline{E}_i$ the incident electric field from the voltage source (**Figure 3**), and I is the longitudinal current in a given point of the nanodipole.

The integral equation for the scattered field along the length l of the nanodipole is given by

$$\overline{E}_s(\overline{r}) = \frac{1}{j\omega\varepsilon_0}\left[k_0^2 \int_l \overline{I} g(R) dl' + \int_l \frac{dI}{dl'} \nabla g(R) dl'\right], \quad (3)$$

where $g(R) = e^{-jk_0R}/4\pi R$ is the free space Green's function, and $R = |\overline{r} - \overline{r}'|$ is the distance between source and observation points. The numerical solution of the problem formulated by (1)–(3) is performed by linear MoM as follows. Firstly, we divide the total length $L_t = 2L + d$ in $N = 2N_a + 2$ straight segments, where N_a is the number of segments in $L - 0.5d$ with the size $\Delta L = (L - 0.5d)/N_a$ (white segments in **Figure 3**), and two segments in the middle with the size $\Delta L = d$ (gray segments in **Figure 3**). Later, the current in each segment is approximated by sinusoidal basis functions. The expansion constants I_n are shown in **Figure 3** where each constant defines one triangular sinusoidal current. To calculate these constants, we use $N - 1$ rectangular pulse test functions with unitary amplitude and perform the conventional testing procedure. As a result, the following linear system of equations is obtained:

$$V_m = Z_s I_m \Delta_m - \sum_{n=1}^{N-1} Z_{mn} I_n, \; m = 1, 2, 3, \ldots, N-1, \quad (4)$$

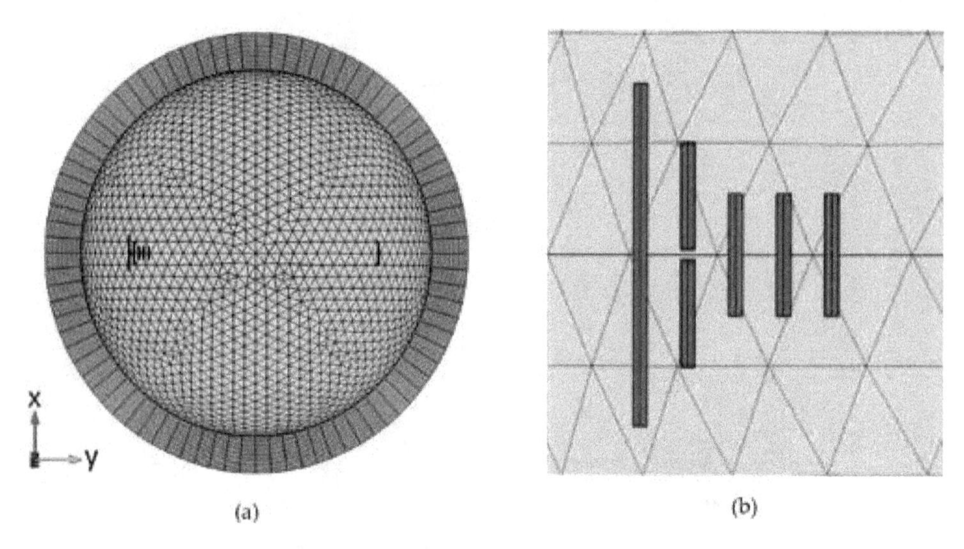

Figure 5.
*Mesh of the problem of **Figure 2** generated in COMSOL. (a) Nanolink mesh and its surroundings. (b) Enlarged Yagi-Uda antenna mesh.*

where Z_{mn} is the mutual impedance between sinusoidal current elements m and n, $\Delta m = 1/2[\Delta L_m + \Delta L_m + 1]$, and V_m is non-zero only in the middle of the antenna ($m = N/2$), where $V_N/2 = V_s$. The solution of (4) gives the current along the dipole and the input current I_s. For $V_s = 1$ V, the input impedance is $Z_{in} = 1/I_s = (R_r + R_L) + jX_{in} = R_{in} + jX_{in}$, where R_r, R_L, R_{in}, and X_{in} are the radiation resistance, loss resistance, input resistance, and input reactance, respectively. The total input power is calculated by $P_{in} = 0.5\text{Re}(V_s I_s^*) = 0.5(R_L + R_r)|I_s|^2 = 0.5R_{in}|I_s|^2 = P_r + P_L$, P_r is the radiated power, and P_L the loss power dissipated at the antenna's surface which is calculated by

$$P_L = 0.5\text{Re}(Z_s)\sum_{n=1}^{N-1}|I_n|^2\Delta_n, \tag{5}$$

The radiated power can be obtained by $P_r = P_{in} - P_L$, and the resistances by $R_r = 2P_r/|I_s|^2$ and $R_L = 2P_L/|I_s|^2$. The radiation efficiency is defined by $e_r = P_r/P_{in} = P_r/(P_r + P_L) = R_r/(R_r + R_L) = R_r/R_{in}$.

3.2 Finite element method

The nanolinks of **Figure 3** were also analyzed numerically by FEM. **Figure 5a** shows the mesh of the nanolink of **Figure 2** modeled in the COMSOL, where the antennas are in a spherical domain of air, with scattering absorbing condition (PLM) applied at their ends. **Figure 5b** shows an enlarged image of the Yagi-Uda nanoantenna mesh and its surroundings.

4. Numerical results

4.1 Isolated antennas in transmitting mode

In this section, the transmitting antennas Yagi-Uda and dipole (**Figure 2** without reflector and directors) are analyzed separately. For this analysis, the values of the antennas parameters are those shown in **Table 1**, where with these v alues the main

Variable	h_{dT}	d_{dT}	a	h_r	h_d	h_{dR}	d_{dr}	d	d_{TR}
Values	220	20	15	700	250	220	20	100	5000

Table 1.
Nanolink parameters used in simulations. All parameters are in nanometer (nm).

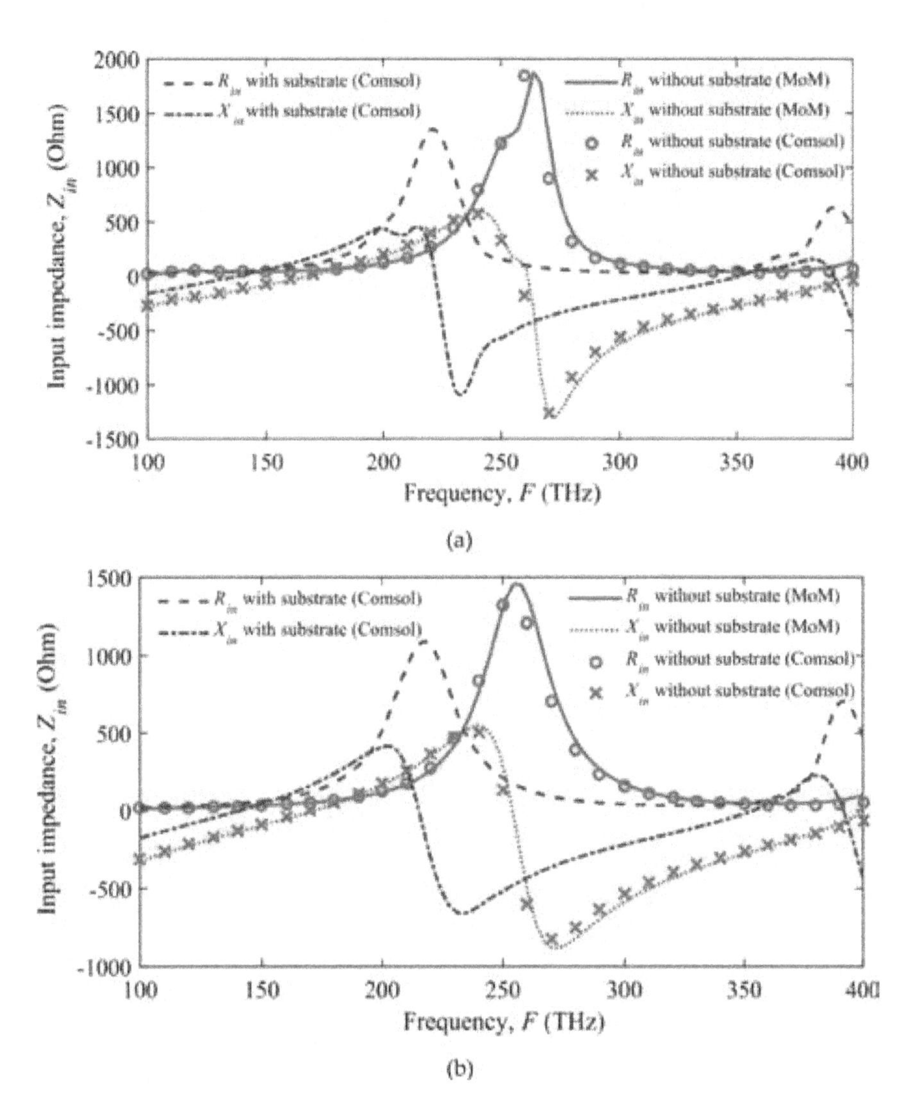

Figure 6.
Input impedance (Z_{in}) of the transmitting antennas Yagi-Uda and dipole. (a) Yagi-Uda without substrate, compared to MoM, and with SiO_2 substrate. (b) Dipole without substrate, compared to MoM, and with SiO_2 substrate.

resonances are in the frequency range of 100–400 THz considered. The parameters of the isolated dipole are based on [23] and those of the elements of the Yagi-Uda antenna were chosen so that the reflecting element was larger than the dipole and the smaller dipole directors. In **Table 1**, $a = a_r = a_{dT} = a_d = a_{dR}$ and $d = d_{hr} = d_{hd}$.

Figure 6 shows the input impedance (Z_{in}) for the Yagi-Uda antennas (**Figure 6a**) and dipole (**Figure 6b**). These input impedances were calculated by FEM with the COMSOL software, and by a linear MoM applied to cylindrical plasmonic nanoantennas [18], by coding the mathematical model in Matlab software [24]. We observe a good agreement between these two methods.

Also, the input impedances of the antennas are calculated for two situations, the first with the antennas in the free space (without substrate) and the second on a SiO_2 substrate with a permittivity of 2.15.

Comparing the input impedance result between the Yagi-Uda and free-space dipole antennas, it is noted that the first two resonant frequencies are close, which shows that the directors and reflectors do not significantly affect the original resonant frequencies of the isolated dipole. The main differences between these two transmitting antennas are observed near the frequencies of 175 and 260 THz, which correspond physically to the dipole resonances of the reflector and directors, respectively. These resonances can be observed in the distributions of currents in these frequencies, which are not shown here.

Figure 6 also shows the effect of the substrate in the input impedance and resonant properties of the antennas. It is observed that by placing the antennas on the substrate, their resonances are shifted to smaller frequencies in relation to the

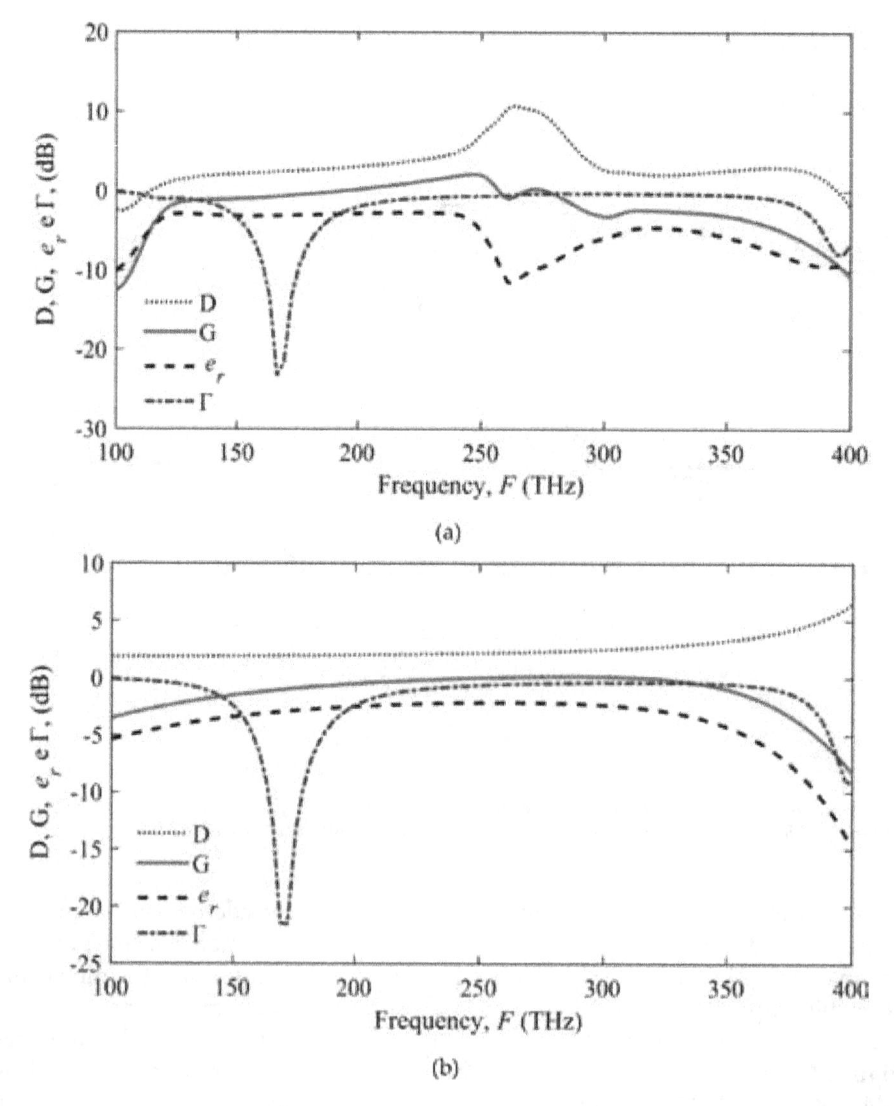

Figure 7.
Directivity (D), gain (G), radiation efficiency (e_r) and reflection coefficient (Γ) of antennas (a) Yagi-Uda and (b) dipole.

antennas in the free space. This effect of the substrate is similar to that observed in antennas in the microwave regime [25].

Figure 7 shows the results of directivity (D) and gain (G), radiation efficiency (e_r) and reflection coefficient (Γ), all in dB, versus frequency for the Yagi-Uda antennas (**Figure 7a**) and dipole (**Figure 7b**). The directivity and gain are calculated in the $+ y$ direction (**Figure 2**). In **Figure 7b** it is observed the conventional characteristic of the isolated dipole, where in a wide range of 150–300 THz one has approximately $D \approx 1.6$, $e_r \approx 0.6$ e $G \approx 1$ ($G = e_r D$). In the case of Yagi-Uda (**Figure 7a**), there is a peak of $D = 12$, near $F = 264$ THz, but at this frequency the radiation efficiency is minimal $e_r \approx 0.1$, and the gain is ($G = 0.86$). However, if greater gain and efficiency are desired rather than high directivity, the frequency near $F \approx 240$ THz is more adequate, where the maximum gain is approximately $G_{max} \approx 1.6$. The reflection coefficient of both antennas was calculated considering a transmission line with characteristic impedance of 50 Ω connected to antennas. With this result, it is observed that the best impedance matching for both antennas occurs around the first resonant frequency, however the maximum radiation efficiency occurs at higher frequencies.

Figure 8 shows the 3D far field gain radiation diagrams of the Yagi-Uda and dipole antennas, calculated at the frequency of 240 THz. It is observed that the maximum gain of the Yagi-Uda ($G_{max} \approx 1.6$) is approximately 60% greater than the maximum gain ($G_{max} \approx 1$) of the dipole. For the case of the Yagi-Uda antenna, the maximum gain occurs in the $+y$ direction with a small lobe in the $-y$ direction.

4.2 Nanolinks analysis

In this section, we present the results obtained in the analysis of the dipole/dipole nanolinks (**Figure 2**, without reflector and transmitter directors), Yagi-Uda/dipole (**Figure 2**) and Yagi-Uda/Yagi-Uda (**Figure 2**, with the receiving antenna equal to the transmitting antenna) for the frequency range of 100–400 THz. The parameters used for the receiving antennas are same as those of the transmitting antennas, with $Z_C = 50$ and 1250 Ω for each nanolink model.

Figure 9 shows the power transmission in dB (or power transfer function) for the three nanolinks, calculated by the ratio between the power delivered to the Z_C load and the power delivered by the source V_s at the transmitting antenna terminals. The results show that the Yagi-Uda/Yagi-Uda nanolink presents a small

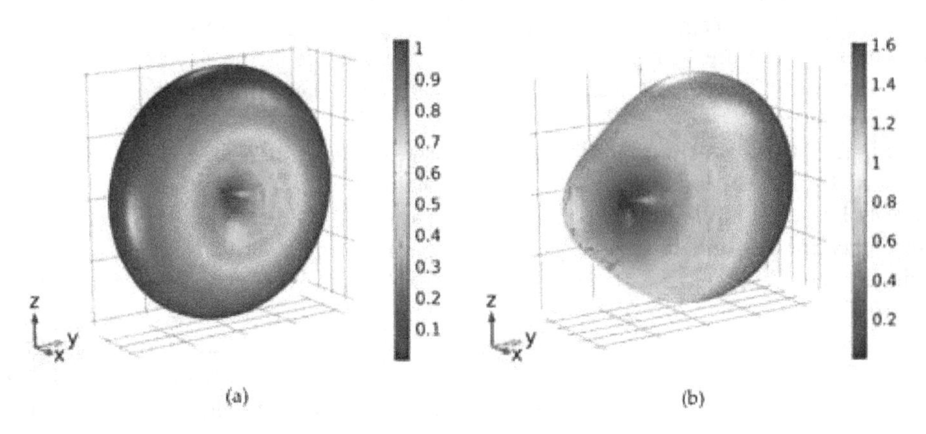

Figure 8.
3D far-field gain radiation diagram of (a) dipole antenna, and (b) Yagi-Uda (b), both in F = 240 THz.

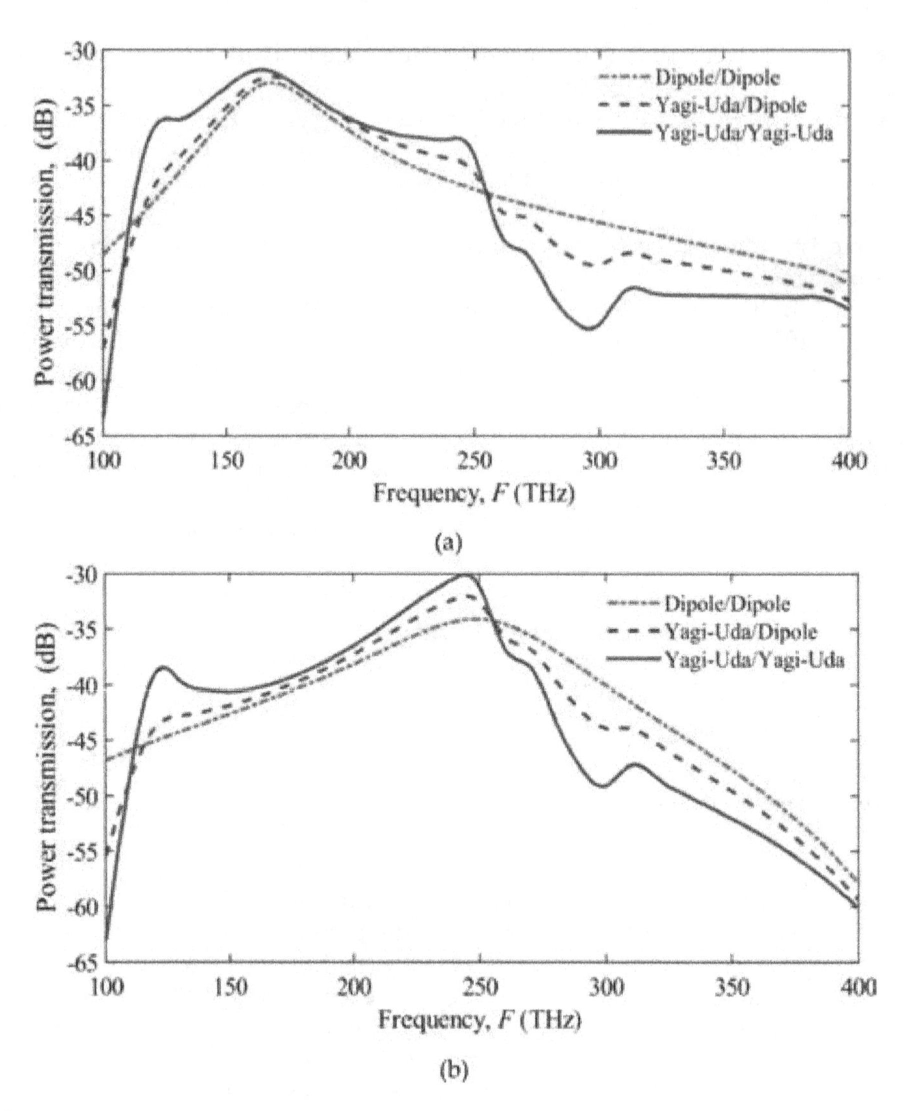

Figure 9.
Power transmission versus frequency for the dipole/dipole nanolinks, Yagi-Uda/dipole and Yagi-Uda/Yagi-Uda, for Z_C = 50 (a), e 1250 Ω (b).

improvement in power transmission, at some frequency points, in relation to the dipole/dipole and Yagi-Uda/dipole nanolinks. In addition, it can be observed that the links can operate with good transmission power at the frequency points 170 and 240 THz, for Z_C equal to 50 and 1250 Ω, respectively, where the power transmission are maximum.

Figure 10 shows the magnitude and phase of the electric near field, which is defined by $E = 20 \log_{10}(|\text{Re}\ (E_x)|)$, in the plane z = 25 nm, of the dipole/dipole nanolinks (a, b), Yagi-Uda/dipole (c, d) and Yagi-Uda/Yagi-Uda (e, f). The receiver antennas are positioned at a distance d_{TR} = 5 μm from the transmitting antennas, with F = 170 THz and Z_C = 50 Ω for the fields of figures (a), (c) and (e), and with F = 240 THz and Z_C = 1250 Ω for the cases of figures (b), (d) and (f). In all types of nanolinks shown in this figure, the radiated wave can be visualized by propagating from the transmitting antennas to the receiving antennas, with the appropriated wavelength. It is observed the amplitude decay of the electric field with the distance and the decrease of the wavelength with the increase of the frequency.

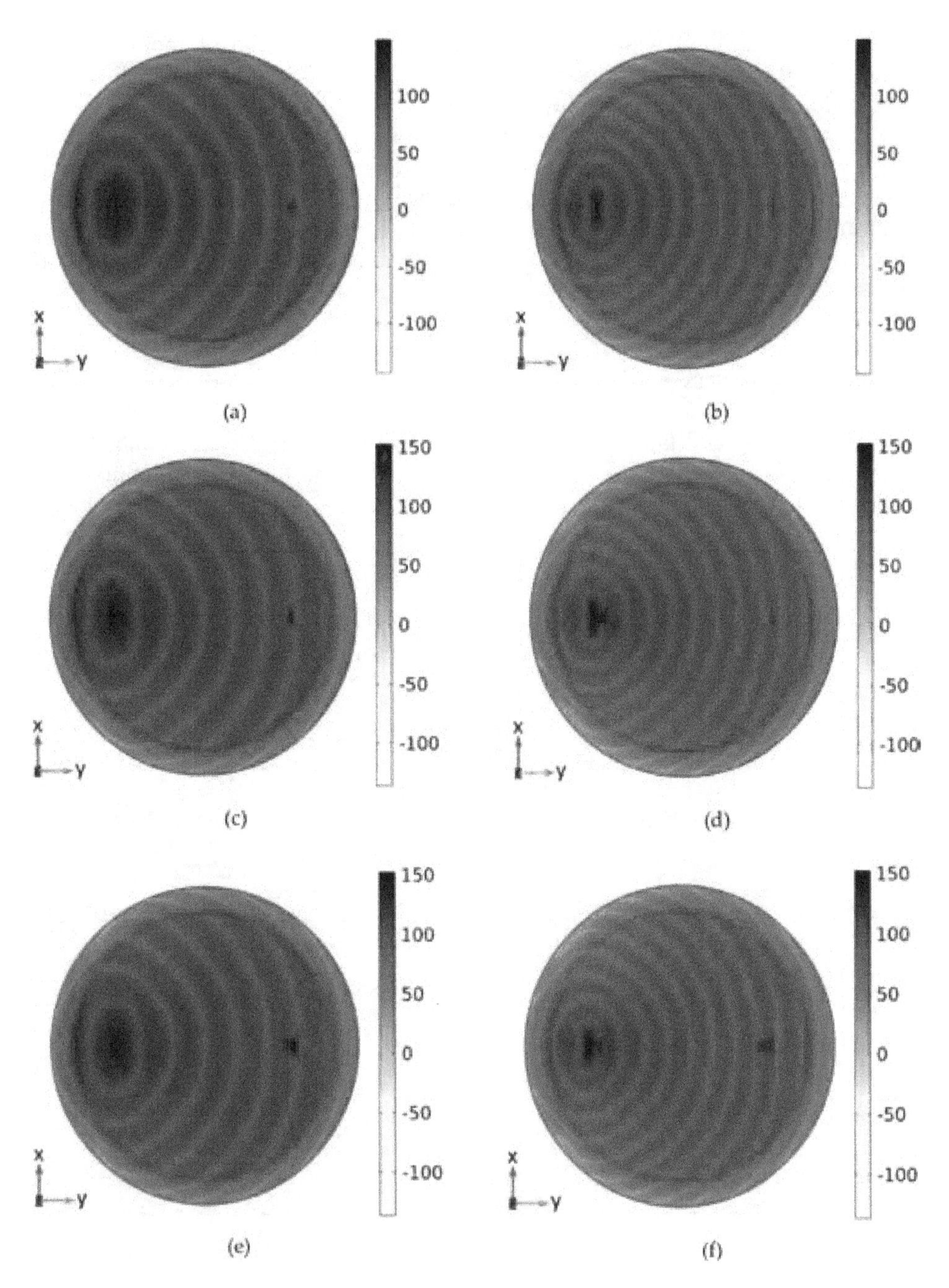

Figure 10.
Electric near field distribution of the magnitude and phase ($E = 20\ log_{10}(|Re\ (E_x)|)$), in the plane z = 25 nm, of the dipole/dipole nanolinks (a, b), Yagi-Uda/dipole (c, d) and Yagi-Uda/Yagi-Uda (e, f). The receiver antennas are positioned at 5 μm from the transmitting antennas, with F = 170 THz and Z_C = 50 Ω for figures (a), (c) and (e), and with F = 240 THz and Z_C = 1250 Ω for figures (b), (d) and (f).

5. Conclusions

It was presented in this work, a comparative analysis of nanolinks formed by Yagi-Uda and dipole plasmonic nanoantennas, where was investigated the power transmission for Yagi-Uda/Yagi-Uda, Yagi-Uda/dipole and dipole/dipole nanolinks types. These nanolinks were numerically analyzed by method of moments and finite element method. The results show that the Yagi-Uda/Yagi-Uda nanolink presents a small improvement in power transmission, at some frequency points, in relation to the other cases. In addition, the three links can operate with good power transmit at different frequency points, varying the load impedance of the receiving antenna, which is of great importance for future applications in nanoscale wireless communication. In future work, we intend to feed these nanolinks by more realistic sources, such as gaussian beams, to verify their influence on power transmission results.

Acknowledgements

The authors would like to thank the Mr. Mauro Roberto Collato Junior Chief Executive Officer of the Junto Telecom Company for the financial and emotional support to this project.

Author details

Karlo Queiroz da Costa[1*], Gleida Tayanna Conde de Sousa[1],
Paulo Rodrigues Amaral[1], Janilson Leão Souza[2], Tiago Dos Santos Garcia[1] and
Pitther Negrão dos Santos[1]

1 Department of Electrical Engineering, Federal University of Para, Belém-PA, Brazil

2 Federal Institute of Education, Science and Technology of Para, Tucuruí-PA, Brazil

*Address all correspondence to: karlo@ufpa.br

References

[1] Novotny L, Hecht B. Principles of Nano-Optics. New York: Cambridge; 2006

[2] Maier SA. Plasmonics: Fundamentals and Applications. New York: Springer; 2007

[3] Grober RD, Schoelkopf RJ, Prober DE. Optical antenna: Towards a unity efficiency near-field optical probe. Applied Physics Letters. 1997;**70**: 1354-1356

[4] Pohl DW. Near field optics as an antenna problem. Second Asia-Pacific Workshop on Near Field Optics. Beijing, China; 1999. pp. 9-21

[5] Novotny L. From near-field optics to optical antennas. Physics Today. 2011; **64**(N7):47-52

[6] Bharadwaj P, Deutsch B, Novotny L. Optical antennas. Advances in Optics and Photonics. 2009;**1**:438-483

[7] Novotny L, Hulst NV. Antennas for light. Nature Photonics. 2011;**5**:83-90

[8] Alù A, Engheta N. Theory, modeling and features of optical nanoantennas. IEEE Transactions on Antennas and Propagation. 2013;**61**(N4):1508-1517

[9] Giannini V, Dominguez AIF, Heck SC, Maier SA. Plasmonic nanoantennas: Fundamentals and their use in controlling the radiative properties of nanoemitters. Chemical Reviews. 2011;**111**: 3888-3912

[10] Krasnok AE. Optical nanoantennas. Physics-Uspekhi. 2013;**56**(N6):539-564

[11] Berkovitch N, Ginzburg P, Orenstein M. Nano-plasmonic antennas in the near infrared regime. Journal of Physics: Condensed Matter. 2012; **24** : 073202

[12] Atwater HA, Polman A. Plasmonics for improved photovoltaic devices. Nature Materials. 2010;**9**:205-213

[13] de Souza JL, da Costa KQ. Broadband wireless optical nanolink composed by dipole-loop nanoantennas. IEEE Photonics Journal. 2018;**10**:1-8

[14] Alù A, Engheta N. Wireless at the nanoscale: Optical interconnects using matched nanoantennas. Physical Review Letters. 2010;**104**:213902

[15] Solís D, Taboada J, Obelleiro F, Landesa L. Optimization of an optical wireless nanolink using directive nanoantennas. Optics Express. 2013;**21**: 2369-2377

[16] Bellanca G, Calò G, Kaplan A, Bassi P, Petruzzelli V. Integrated Vivaldi plasmonic antenna for wireless on-chip optical communications. Optics Express. 2017;**25**:16214-16227

[17] Yang Y, Li Q, Qiu M. Broadband nanophotonic wireless links and networks using on-chip integrated plasmônica antennas. Scientific Reports. 2016;**6**:19490

[18] da Costa KQ, Dmitriev V. Simple and efficient computational method to analyze cylindrical plasmonic nanoantennas. International Journal of Antennas and Propagation. 2014;**2014**. 8 pages

[19] COMSOL Multiphysic 4.4. COMSOL Inc. Available from: http://www.comsol.com

[20] Sabaawi AMA, Tsimenidis CC, Sharif BS. Analysis and modeling of infrared solar rectennas. IEEE Journal of Selected Topics in Quantum Electronics. 2013;**19**(N3): 9000208

[21] Johnson PB, Christy RW. Optical constants of the noble metals. Physical Review B. 1972;**6**:4370-4379

[22] Hanson GW. On the applicability of the surface impedance integral equation for optical and near infrared copper dipole antennas. IEEE Transactions on Antennas and Propagation. 2006;**54** (N12):3677-3685

[23] Da Costa KQ, Dmitriev V. Radiation and absorption properties of gold nanodipoles in transmitting mode. Microwave and Optical Technology Letters. 2015;**57**:1-6

[24] Matlab Software. Available from: https://www.mathworks.com/products/matlab.html

[25] Balanis CA. Antenna Theory-Analysis and Design. 3rd ed. New Jersey: John Wiley & Sons; 2005

6

Nanoplasmonic Arrays with High Spatial Resolutions, Quality and Throughput for Quantitative Detection of Molecular Analytes

Rishabh Rastogi, Matteo Beggiato, Pierre Michel Adam, Saulius Juodkazis and Sivashankar Krishnamoorthy

Abstract

Recent developments in nanoplasmonic sensors promise highly sensitive detection of chemical and biomolecular analytes with quick response times, affordable costs, and miniaturized device footprints. These include plasmonic sensors that transduce analyte-dependent changes to localized refractive index, vibrational Raman signatures, or fluorescence intensities at the sensor interface. One of the key challenges, however, remains in producing such sensors reliably, at low cost, using manufacturing compatible techniques. In this chapter, we demonstrate an approach based on molecular self-assembly to deliver wafer-level fabrication of nanoplas-monic interfaces, with spatial resolutions down to a few nanometers, assuring high quality and low costs. The approach permits systematic variation to different geometric variables independent of each other, allowing the significant opportunity for the rational design of nanoplasmonic sensors. The ability to detect small mol-ecules by SERS-based plasmonic sensing is compared across different types of metal nanostructures including arrays of nanoparticle clusters, nanopillars, and nanorod and nanodiscs of gold.

Keywords: nanofabrication, self-assembly, nanoplasmonic sensor, nanoarray, nanoclusters, nanopillar, nanoparticle, hot spots

1. Introduction

The optical properties of metal nanostructures and nanoarrays have been widely investigated in the context of exploiting particle plasmons to derive high performance within chemical and biosensing devices. These include sensing of range of analytes to diagnose the medical status of an individual [1]; detect the presence of chemical or biowarfare agents [2, 3], toxins, or adulterants in food [4]; or assess/monitor air, water, and soil quality in the environment [2, 5]. Plasmonic sensors offer a range of advantages over other analytical tools including high analytical and calibration sensitivity, quick response times, label-free detection opportunities, ease of integration within different sensor form factors, and the need for simple and portable instrumentation. Plasmonic sensors rely on metal nanostructures that act as nanoantennae to concentrate and enhance the electromagnetic field close to surface [6]. The enhanced EM field can be leveraged within different configurations, namely, localized surface plasmon resonance sensors (LSPR) that follow analyte-induced changes to local refractive index or report vibrational Raman or fluorescence intensities with high signal-to-noise ratios using surface-enhanced Raman spectroscopy (SERS) and metal-enhanced fluorescence (MEF), respectively. In all these cases, the performance of the sensor is critically linked to the optical properties of the plasmonic structures, which in turn correlates with their geometric attributes, namely, size, shape, aspect ratios, separation, distribution, and roughness. Control over geometries is the key to engineering profiles and intensities of the electromagnetic field around plasmonic nanostructures, which in turn determines the performance of the plasmonic transducer. The typical length scales for characteristic dimensions of plasmonic nanostructures correspond to spatial resolutions of few nanometers to few tens of nanometers. Geometric features such as pointed structures or nanoscale gaps have shown to concentrate and enhance EM field, thus acting as EM hot spots, as a function of decreasing radius of curvature or gap distances, with length scales down to sub-10 nm regime [7]. Enhanced EM fields at gap hot spots have shown to result in enhancement factors of the order of a million- to billion-fold in practice. A sizeable contribution to observe SERS signals (24% of the overall intensity) was observed to be rising out of <100 molecules per million when these molecules are positioned in EM hot spots with EM enhance-ment factors of the order of 10^9–10^{10} [8]. The EM hot spots help enhance the sensitivity of different plasmonic sensors, and especially those based on enhanced spectroscopies, namely, SERS and MEF. Gaining high sensitivity would thus be largely determined by the quality and number of EM hot spots [9]. However, addressing the production of EM hot spots with length scales of the order of only a few nanometers pose a profound and non-trivial challenge for nanofabrica-tion. Traditionally, such hot spots have been attained as a natural consequence of stochastic growth or deposition processes [10, 11], including electrochemical growth or roughening, de-wetting of nanoparticles or salts from solution phase [12–14], and de-wetting of thin films on the surface. The stochastic processes, by nature, result in a broad standard deviation in geometries, which reduces the number of most efficient hot spots and also leads to greater spot-to-spot variabil-ity in signal intensities [15, 16]. The randomness in geometry makes it particularly hard to predict response, to identify the source of issues, and to adopt a rational approach to optimize performance. Geometries with improved definition have remained forte of top-down lithography tools, for example, E-beam lithography, focused ion-beam milling, and X-ray interference lithography, which are all very time-consuming and also quite expensive [17–19]. The throughput and cost of fabrication is not only an issue for manufacturing but also reduces the efficacy of research due to the limitation in the number of samples available for investiga-tions [20] (**Figure 1**). From the manufacturing perspective, the throughput of the process would be a key determinant of the cost, which is driven to <$5 per sensor chip for point-of-care applications [28].

Techniques such as nanoimprint lithography and polymeric or colloidal self-assembly techniques allow enhancing throughput, albeit, at the cost of defects. The ability of self-assembly techniques in catering to the parallel fabrication of nanopatterns across arbitrarily large areas at low cost, as well as the several handles it offers toward tunability of structure dimensions down to molecular level, is unmatched by conventional lithography tools. However, self-assembly-based approaches carry limitations that demand careful attention to ensure their

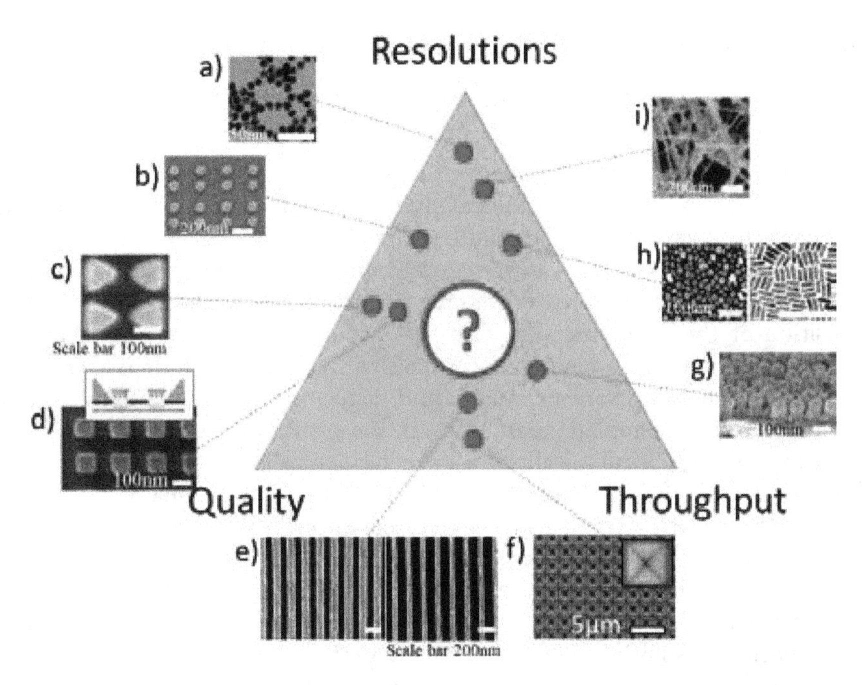

Figure 1.
Schematic representation of the trade-off between resolution, quality, and throughput in fabricating plasmonic nanoarrays with high spatial resolutions, showing the comparative advantages and limits of (a) nanoparticle assemblies [21] (b) top-down-bottom-up control over nanoparticle assemblies [19], (c) direct-write techniques [22], (d) nanostencil lithography [23], (e) nanoimprint lithography [24], (f) photolithography [25], (g) porous anodized alumina templates [10], and (h) stochastic, random structures, including island films [11, 26], and (i) stochastic chemical growth processes [27].

usefulness: (a) Poor uniformity, which is often a result of poor process optimization and in some cases due to the susceptibility of process parameters to environmental variables. This can be addressed by mapping the impact of environmental variables and factoring them within the process optimization. (b) Large standard deviations, either inherent to the primary templates produced by the technique or those that may creep in during different stages of processing. (c) Feature shape: Typically, self-assembled template patterns have a circular feature cross section. Rectangular, triangular, or other feature shapes with lower symmetry are uncommon. (d) In the absence of any external guidance, self-assembly techniques typically lead to a polycrystalline 2D hexagonal order. Long-range ordering, square, or other lattice types besides hexagonal symmetry remain uncommon and can be attained through guidance from a top-down lithographic tool [29–34]. The lack of long-range order and presence of point or line defects are better tolerated by plasmonic sensing applications, so long as the averaged properties are consistent and reproducible with low standard deviations. The chapter will present our approach to plasmonic nanoarrays with high spatial resolutions relying on hierarchical self-assembly of amphiphilic di-block copolymers into soft colloids and their subsequent quasipe-riodic organization when deposited on a planar surface [35]. The approach results in well-defined organic templates on the surface with nanometric control over the width, topography, and pitch, realized by control over parameters of molecular self-assembly. By understanding the impact of the different process parameters on the resulting geometric outcome, it is possible to deliver templates with high reproducibility and uniformity on full wafers, with a yield >90% [36]. These templates are translated into highly sensitive SERS-based plasmonic sensors, with control over metal nanogaps down to sub-10 nm regime.

2. Spatially controlled fabrication of nanoscale templates

2.1 Fabrication of self-assembled templates

Amphiphilic di-block copolymers can self-assemble into reverse micelles when dissolved in a solvent that selectively dissolves only the apolar block of the copolymer. These reverse micelles can be obtained or be induced into attaining a spherical morphology, with a feature size determined by their aggregation number. The micelle aggregation number is not only a function of the molecular weight and composition of the copolymer but also the quality of the solvent used and the presence of additives. As a consequence, the size can be varied independent of the molecular weight, using solvent quality and concentration of additives as useful handles to fine-tune template geometries and eventually those of resulting plasmonic arrays. The spherical micelles are solvent-laden colloidal structures which can be readily organized on a variety of surfaces to yield a 2D hexagonally ordered dot-array templates. These templates are subjected to physical or chemical means of pattern transfer to produce nanopatterns of desired materials. Pattern formation on surfaces using copolymer reverse micelle approach is fundamentally different from that of microphase separation in block copolymer thin films [37–39]. The size of reverse micelles in solution, which eventually determines the feature sizes on the surface, is independently variable by engineering the solvent quality or the use of additives. The standard deviation of the templates on the surface is determined by that of the micelles in solution phase before deposition, which in turn is governed by the intermicellar exchange process. The exchange process is slow due to the slower diffusion for larger molecular weight polymers or when solvent with high selectivity to the corona-forming blocks is employed [40]. During the spin-coating process, the solvent-laden micelles in solution deform on the surface to assume an ellipsoidal shape, with partial fusion of corona from adjacent reverse micelles resulting in the globally continuous organic film presenting periodic contrast in topography with an ultrathin film (<5 nm) in the background. The pitch of the ensuing pattern on the surface can be varied in steps <5% of its mean value, through control over evaporation rate or the concentration of the micelles in solution. The evaporation rate can be controlled using spin or dip coating speeds. The ability to vary the lattice periodicity within a certain window is attributed to a range of distances for which the PS blocks from the corona of adjacent micelles can still meet upon deformation and film formation. When this condition is not met, namely, at low solution concentrations or high spin speeds, the reverse micelles are spaced too far apart, resulting in patchy coverage. On the other hand, at high solution concen-trations or at low spin speeds, the excess concentration beyond what is necessary toward a monolayer appears as multilayers [36]. The topography of the reverse micelle film is a variable that can be determined by the relative humidity in the ambient environment during the coating process. This is attributed to the change in moisture that is likely to concentrate at the polar core-forming PVP and conse-quently increasing the interfacial tension resulting in resistance to collapse and as a result a higher topography. Under optimal conditions of coating, micelle arrays can be produced with a standard deviation of <10% in geometric attributes across a complete wafer.

In a specific example shown in **Figure 2**, reverse micelles of polystyrene-block-poly (2-vinylpyridine) (PS-b-P2VP) are obtained from dilute solutions of m-xylene. Here, a copolymer with a molecular weight of 81.5 kDa and a PDI of 1.10, at a concentration of 0.5% w/w in m-xylene, is spin-coated on a clean silicon surface at 5500 rpm resulting in a hexagonally ordered array with feature heights of 20 nm and pitch of 66 nm with typical standard deviations <15% in all geometric

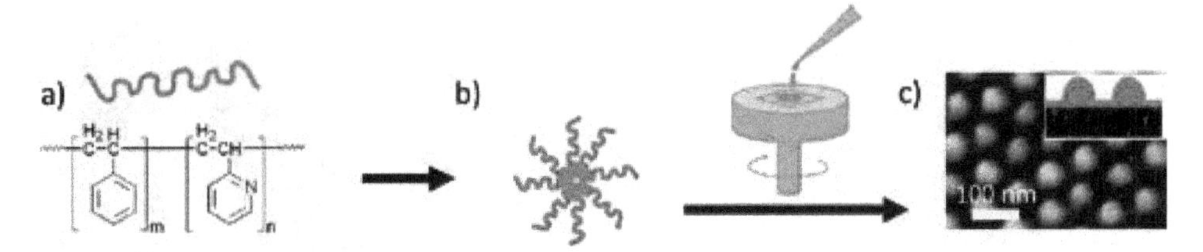

Figure 2.
Schematic representation of the self-assembly of amphiphilic di-block copolymer into reverse micelles and their subsequent assembly to form two-dimensional arrays on the surface.

variables. During the spin-coating process, the solvent-laden micelles in solution deform on the surface to assume an ellipsoidal shape, with the corona from adjacent reverse micelles coming together resulting in the globally continuous thin film presenting periodic topography and ultrathin film (<5 nm) in the background. The center-to-center distance or the pitch of the template arrays could be systemati-cally decreased in steps of <5 nm between 45 and 60 nm by increasing the solution concentrations from 0.6 to 1% at a fixed spin-speed of 5000 rpm or decreasing spin-speeds from 9000 to 2000 rpm at a fixed concentration of 0.7%. The ranges were found to constitute an optimal window of conditions where a continuous uniform film was obtained.

2.2 Reproducibility and scalability

Among key limitations encountered by self-assembly-based approaches, in general, is the scaling up to practically large areas while ensuring high consistency and reproducibility. The issues of reproducibility arise mainly due to the sensitiv-ity of the process outcome to environmental parameters. Such sensitivity also limits process scalability, due to inconsistencies encountered when coating large areas like full wafers and to limited batch-to-batch reproducibility. These issues are true also for the case of self-assembly of amphiphilic copolymers, and this can be addressed only by adequate investigations directed at mapping the impact of different environmental variables on the process outcomes. Several sources of variability were identified and addressed for the self-assembly of amphiphilic copolymers, including the presence of moisture and contaminants in solution; history of preparation (agitation and incubation), temperature, and humidity; dif-ferences in surface roughness or surface energy (e.g., due to organic or particulate contaminants on the surface); changes to solution concentrations due to solvent evaporation during use; and inadequate mixing of polymer. Under optimal condi-tions, the assemblies of reverse micelle feature exhibit standard deviations lower than 15% across full wafers.

A specific outcome of optimization of the templates represented in **Figure 2c** on a 100 mm silicon wafer can be seen in **Figure 3**, which shows the distribution for height, diameter, pitch, and nearest neighbors, at different regions of the wafer. Typical characterization involves AFM topography for heights and diameters (within errors of AFM tip-convolution) and SEM top view and image analysis using ImageJ or MATLAB. A representative AFM image recorded in the tapping mode is shown in **Figure 3b**. The characterization is critical for each batch of samples, and it is possible to scale the process to several batches of wafers [41]. Points A, B, C, and D corre-spond to four positions representing systematically increasing radial distances from the center to the edge of the wafer. AFM measurements at these points show standard deviation <15% for geometric variables and <10% variation of their mean values across the full wafer. Voronoi analysis of the AFM images shows a predominance of

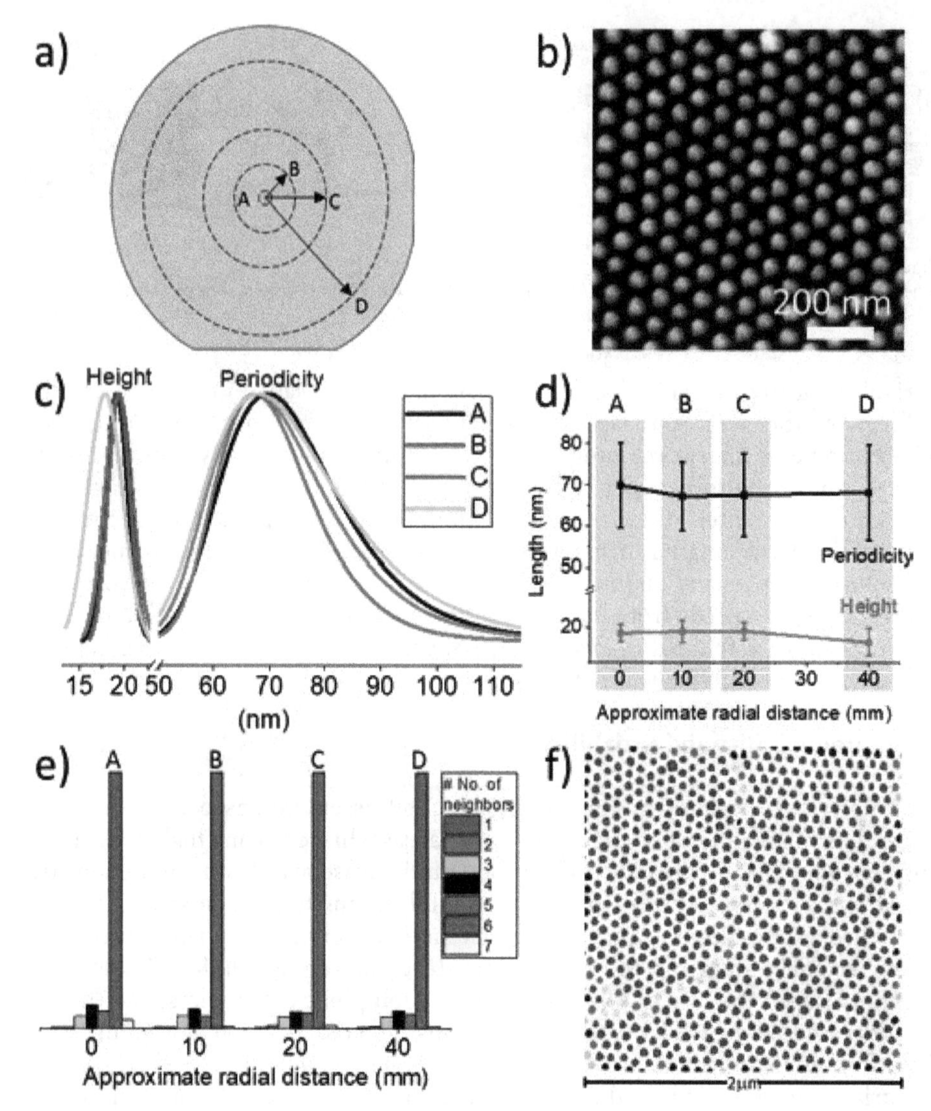

Figure 3.
Demonstration of scalability combined with uniformity on full wafers of optimized coatings: (a) Schematic of radially separated points from center to edge of 100 mm wafer, (b) where tapping mode AFM measurements are performed (image indicated for point a), (c) the distribution of height and pitch across full wafer, (d) with mean values plotted as function of radial distance from center, and error bars showing standard deviation in corresponding feature dimensions at a single point, (e) histogram of nearest neighbors showing predominantly six nearest neighbors as expected for hexagonal assembly, maintained across the wafer, and (f) representation of nearest neighbors using Voronoi analysis, with the features colored corresponding to the number of nearest neighbors as indicated in (e).

six nearest neighbors as expected for hexagonal packing, which is uniformly maintained across the wafer (**Figure 3e, f**). The outcome clearly demonstrates the feasibility for reliable scaling up of the technique to cater to nanostructures over large areas.

3. Nanoplasmonic arrays by pattern transfer

Fabrication of plasmonic nanoarrays starting from organic templates relies on pattern transfer approaches, for example, template-guided growth, deposition, or etching, to define noble metal nanoarrays with the conservation of pitch from

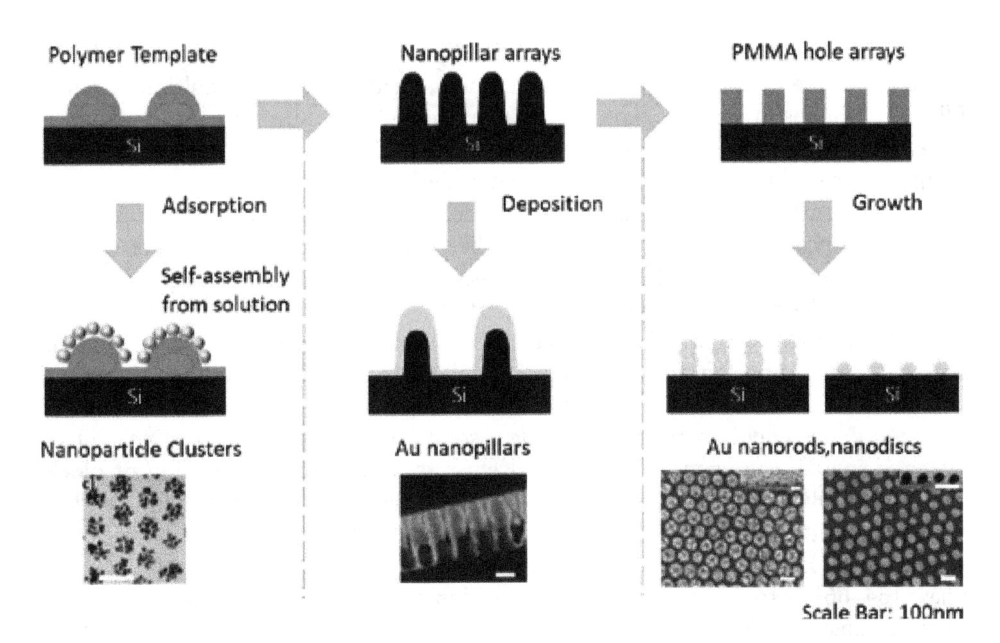

Figure 4.
Fabrication of plasmonic nanoarrays with different profiles, size, and distributions of metal nanostructures can be attained by control over pattern transfer processes. In all cases, the advantages of the original templates including the uniformity and scalability are preserved [42–44].

the original template. The pattern transfer approaches offer large flexibility in the geometry of the individual features, thus making it possible to fabricate plasmonic nanoarrays of different types, for example, nanoparticle cluster arrays, nanopillars arrays, nanorods, or nanodiscs. The pattern transfer parameters provide indepen-dent control over the size, shape, and aspect ratio of the features and should be optimized to ensure that they do not affect the spatial arrangement, uniformity, and reproducibility from the original template. Although the pattern transfer approaches are common in semiconductor fabrication, extending them to work at the scale of few nanometers requires rigorous optimization and quality assurance to ensure low standard deviations and reproducibility in geometries. Further in this section different pattern transfer methods to reach such three different plasmonic nanoarrays have been discussed in detail (**Figure 4**).

3.1 Nanoparticle cluster arrays

Clusters (used interchangeably with "aggregates" in this report) of metal nanoparticles behave differently from their isolated counterparts due to the collective optical behavior arising out of plasmonic coupling between the constituent nanoparticles [45–50]. Clusters of nanoparticles are known to behave as hot particles, with significantly enhanced electromagnetic fields at the inter-particle junctions [51–54]. Consequently, clusters exhibit higher extinction cross sections, with hot spots that can be excited at lower energies than the isolated particles. The aggregation-induced color change of gold nanoparticle suspensions caused by the analyte of interest has been the basis of several biological assays [55]. Such random aggregation typi-cally results in a large distribution in the number of particles per cluster, with a distinct lack of control over those numbers. To achieve clusters with desired optical properties, it is essential to be able to produce them with a narrow distribution in the size, shape, and spatial arrangement between nanoparticles within the cluster and between clusters in an array. Such clusters were demonstrated using templates fabricated by electron- beam lithography (EBL) [45–47, 53, 56–58] and controlling the size of the template to obtain dimers, trimers, quadrumers, and multimers. Since EBL is time-consuming and expensive to achieve high-resolution patterns spanning large areas, other low-cost means to achieve controlled nanoparticle aggregates using template-assisted means have been reported in the literature. These include the use of DNA, [59] surfactants, [60] block copolymers [60–62], carbon nanotubes [61], cylindrical micelles [62], or microorganisms like bacteria or viruses [63] as templates to attach nanoparticles. While the template-assisted cluster formation allows creating clusters with desired size (or a number of particles per cluster) and shape, they often fall short of abilities to control inter-cluster arrangement within an array. Such an arrangement is important to ensure reproducible inter-cluster plasmonic interactions as well as an ability to engineer them to achieve desired optical properties. In this direction, block copolymers are an excellent solution, as they allow effective control over the spatial arrangement of eventual clusters. Microphase separation of block copolymers (BCP) in thin films, as well as BCP reverse micelles, has been utilized as templates to attain clusters of metal nanoparticles on surfaces. These techniques have exploited the BCP templates either to organize one or more particles from solution phase [64–66] or to drive selective complexation of metal ions and in situ reduction to form clusters [67]. However, these fall short of opportunities for the rational design of the cluster properties. The approach based on copolymer reverse micelle template allows preparing clusters based on electrostatic self-assembly of preformed gold nanoparticles from the solution phase on to the features on the surface. The cluster dimensions (the number of nanoparticles per cluster) and the inter-cluster separation were controlled by control over the template size and their separation. The low standard deviation of the template enables low standard deviation in the geometric attributes of the clusters as well (**Figure 5**).

In a typical experiment, the reverse micelle arrays prepared from PS-b-PVP with a molecular weight of 114 kDa and PDI of 1.1 were spin coated from m-xylene solutions and subsequently exposed to an aqueous suspension of gold nanoparticles. Due to the presence of the pyridyl groups of PVP in the core of the features, the reverse micelle features are positively charged and attract negatively charged citrate-stabilized gold nanoparticles from the suspension. The electrostatic attraction between the nanoparticles occurs locally on the features and not in between them, therefore resulting in a patterned array of nanoparticle clusters. The number of nanoparticles in each feature is a function of its size. The inter-cluster separa-tions are controlled by the control over the pitch of the template. Extinction spectra of the nanoparticle clusters showed a peak at 620 nm, which was over 100 nm red-shifted from the plasmon resonance band of isolated Au nanoparticles. This is attributed to the strong plasmonic coupling between the nanoparticles within the cluster. Further, the absence of a strong contribution at 520 nm would confirm the geometric observation of the absence of isolated nanoparticles (**Figure 6**). Finite-difference time-domain (FDTD) simulations show inter-particle as well as inter-cluster hot spots. The 3D shape of the clusters was found to be necessary for the inter-cluster plasmonic coupling, and an increase in EM field at inter-particle and inter-cluster hot spots was found to correlate with decreasing inter-cluster separa-tions. The expectations from the geometry and optical properties of the nanoparti-cle cluster arrays were validated by SERS measurements of a probe molecule. Three different probe molecules were tested, namely, crystal violet, naphthalene thiol, and 2, 2′-bipyridine. In all three cases, the SERS intensities were found to be highest for the largest clusters with the smallest separations. The SERS intensities were higher for crystal violet than for smaller molecules due to resonance Raman effects, enabling SERS enhancement factors over 10^8. SERS quantitative assays showed the

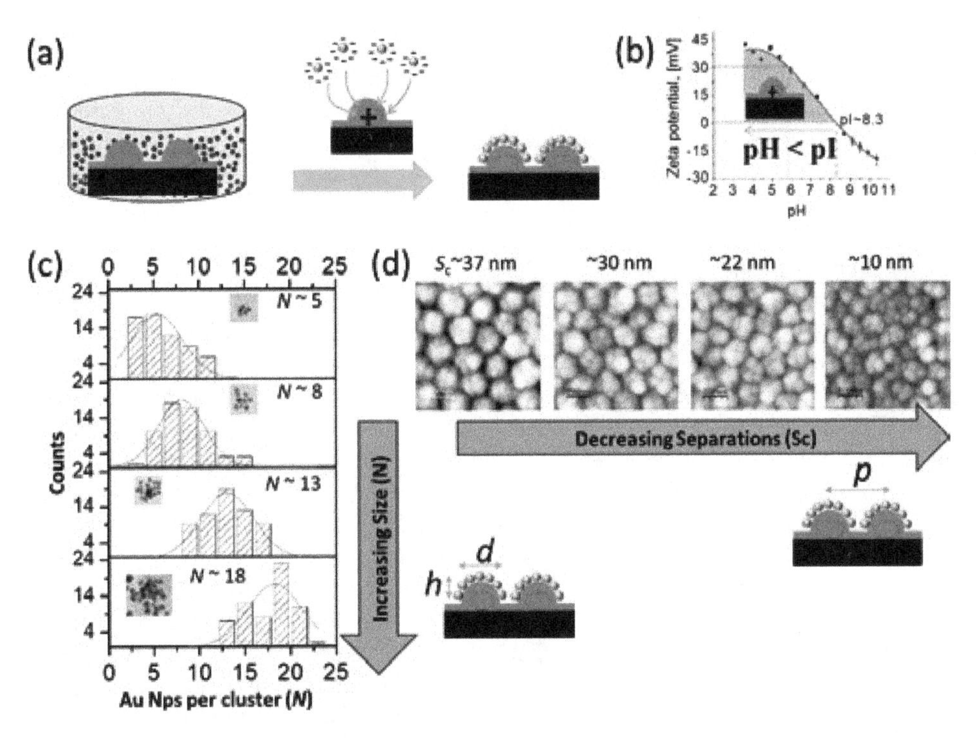

Figure 5.
(a) Schematic of the fabrication of gold nanoparticle cluster arrays by electrostatic attraction of negatively charged Au nanoparticles to positively charged templates on surface, (b) zeta potential titration showing the template acquiring positive potential for pH below 8.3, (c) systematic control over the size (pitch kept a constant), and (d) separation (S_c) between the nanoparticle clusters (size kept a constant at N~18) which is shown by the histograms and AFM measurements, respectively [42].

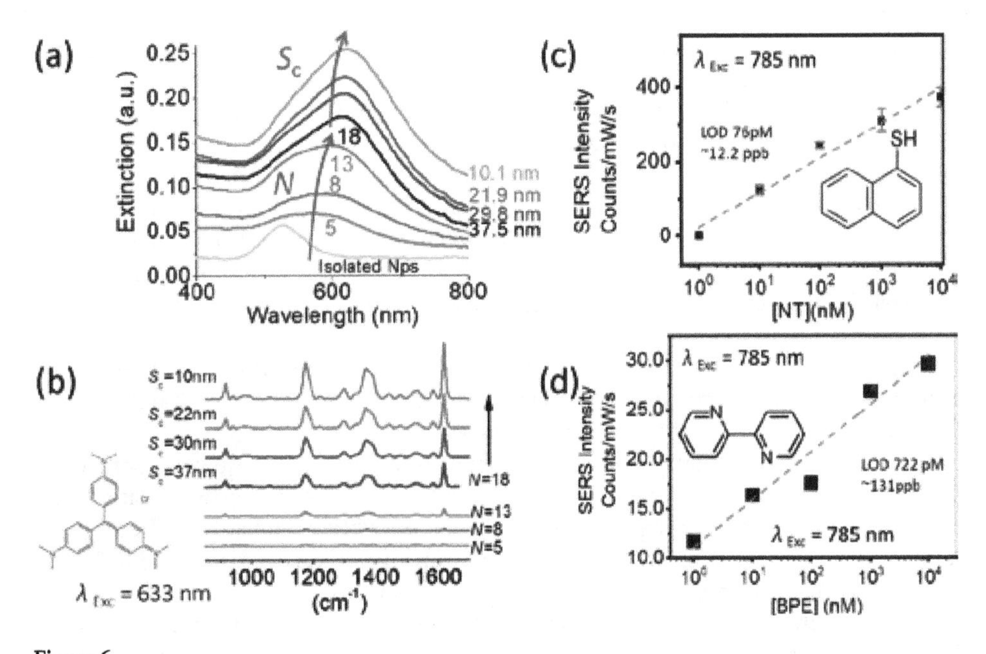

Figure 6.
(a) Graph shows the shift in the extinction peak of the nanoparticle clusters with change in separation, Sc, and change in no. of nanoparticles/cluster, N. (b) SERS spectra of CV shows the increase in the peak intensity at 1625 cm^{-1} with change in the separations, Sc, and no. of nanoparticles/cluster, N. (c) and (d) SERS intensity vs. concentration showing linear relationship for 1-NT and BPE, respectively [42].

lowest detection limits for naphthalene thiol to be an order of magnitude lower than 2,2′- bipyridine. This is attributed to the higher density of molecules for naphthalene thiol on the surface due to the covalent bonding of the thiols to the gold surface at a given concentration. The low standard deviations of the cluster arrays enabled signal intensity variations below 10% on several square millimeters on the surface.

3.2 Metal nanopillars arrays

Nanopillar shapes offer a unique opportunity to control and enhance EM field profiles and enhancements [68]. Nanopillar arrays for plasmonic sensing have been reported previously in the literature, for detection of disease markers and identification of bacteria as well as environmental pollutants [43, 69–71]. Such configurations can be obtained by metal deposition of optimal thickness on top of high aspect ratio dielectric pillars that can be produced by different approaches, namely, VLS growth [72], pulsed laser deposition [73], black Si production by RIE with gas plasmas [74], and patterned wet or dry etching with templates produced by other means, for example, photolithography, soft-lithography, EBL, NIL, NSL, and molecular self-assembly [75–80]. Nanopattern of pillars offers an advantage over stochastically arranged counterparts due to a better definition of the spatial relationship between the pillars that help rational enhancement to sensing performance. The stochastic arrangements, despite promising results, rely largely on empirical optimization and suffer from difficulty in identifying issues when they arise. Another aspect to consider is the stability of the pillar arrays, which is especially a concern for high aspect ratio structures [81–84]. High aspect ratio pillars can capture analyte between the pillars when they collapse by drying that may offer interesting means to trap analyte potentially at EM hot spots [85, 86]; the approach, however, gives less control over the collapse nor the concentration of analyte that is trapped, thus offering limited opportunity for rational design. The irreversible nature of such collapse would make it difficult to subject the pillars to solvent-based washing and drying steps, limiting them to a single time or single-step usage. Pillars of smaller aspect ratios would solve many of these issues,; however, it will need to be produced with high spatial resolutions to ensure a large number of EM hot spots and preferably with good geometric definition allowing rational optimiza-tion toward both enhancing EM hot spots and analyte capture on the surface. An approach based on colloids of amphiphilic copolymers allows the possibility to generate nanopillars that are stable and reusable while allowing significant oppor-tunities to rationally engineer the optical properties toward high plasmonic sensing performance.

Organic reverse micelle templates can be transferred into the underlying Si by adopting nanolithography processes based on reactive ion etching with halogen gas plasmas. Given the low template thickness (typically less than 30 nm) and small widths (sub-100 nm), nanolithography using these templates requires signifi-cant process optimization to ensure high selectivity to the underlying substrate, minimum undercut, and high anisotropy. It is possible to substitute the organic templates with a harder inorganic template to improve the selectivity and durability. This is achieved by transferring the organic template pattern into a thin dielectric film to generate harder dielectric masks for the next step of pattern transfer into Si. We had earlier shown a high degree of control over the incorporation of the organometallic precursor by using atomic layer deposition, with the resulting oxide nanostructures enabling nanolithography down to sub-10 nm regime [35]. This was followed by several other investigations in literature focusing on vapor phase incorporation of metal-organic precursors within different block copolymer domains using ALD processes [87–91]. It is important that the nanolithography process is optimized to not widen standard deviations between the features or across the wafer. The optimization processes across the full wafer level should take into account the "loading effects," which impacts the outcome of the reactive ion etching process depending on the proportion of exposed surface area available for etching.

Nanopillar arrays that are shorter and closer allows all advantages of the other reported approaches, while in addition also conserving the metal to be depos-ited. Nanopillar arrays of silicon obtained by molecular self-assembly approach (**Figure 8**) can subsequently be coated with gold or silver to prepare plasmonic nanopillar arrays. The conditions of the coating, including the choice of evaporation versus sputtering, the respective deposition parameters are critical to the structure and property of the resulting plasmonic arrays. The thickness of the films deposited in relation to the geometry of the underlying silicon pillars would be a key process parameter that determines the aspect ratio and the separation between the metal pillars. Both aspect ratio and the feature diameters increase, while the feature separations decrease systematically as a function of the thickness of metal depos-ited. The observed film growth is strongly anisotropic with slow growth laterally as compared with vertically. The growth on top of the silicon pillars correlates well with the thickness of the metal deposited, while the increase in diameter occurs much slower. The challenge in the processing includes the conformal deposition of metal, with good step edge coverage, especially when the deposition is made on top of closely separated high aspect ratio nanopillars.

In a specific example, organic templates prepared using PS-b-PVP with a molecular weight of 80.5 kDa was spin-coated on a 100mm Si wafer consisting of a 25 nm film of thermally grown SiO_2 layer (**Figure 7a**). The coating is exposed to brief O_2 plasma to remove the thin residual layer between the template features and expose the substrate beneath. The resulting organic template is then transferred into an underlying SiO_2 thin film using C_4F_8/CH_4 gas plasma. The resulting SiO_2 islands provide high selectivity in etching underlying Si using SF_6/C_4F_8 plasma to yield Si nanopillars. The RIE conditions employed resulted in silicon pillars with a positively tapered profile with a feature size of 40 nm (at half-width) and pitch of 78 nm and height of 120 nm, as measured by SEM (**Figure 7b**). The RIE process conditions can be varied to obtain pillars with other shapes as well (**Figure 7c**). The pillar arrays were found to be uniform throughout the wafer, as qualified by uniform color, with a variation of <10% as measured by reflectance spectroscopy [35]. **Figure 8** shows another case of similarly obtained nanopillars arrays prepared using

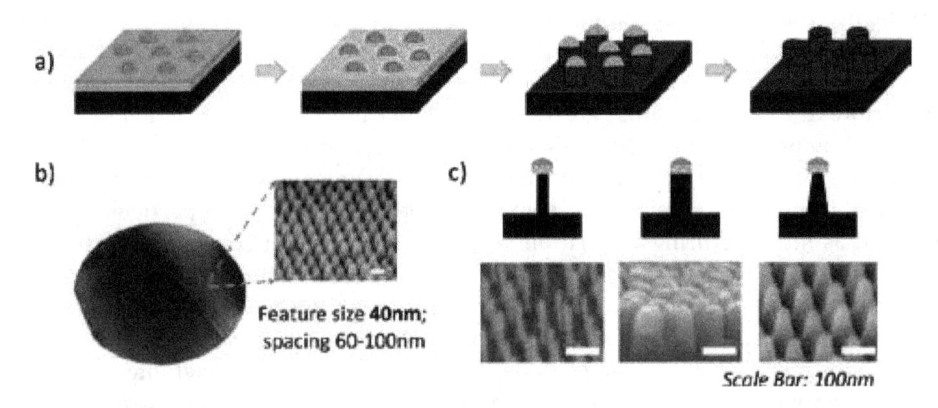

Figure 7.
(a) Fabrication process involved in pattern transfer from polymer template to fabricate silicon nanopillar arrays. (b) SEM image taken from a random position on the wafer with nanopillar arrays. (c) SEM images of different shapes of nanopillars obtained by varying etching conditions [35].

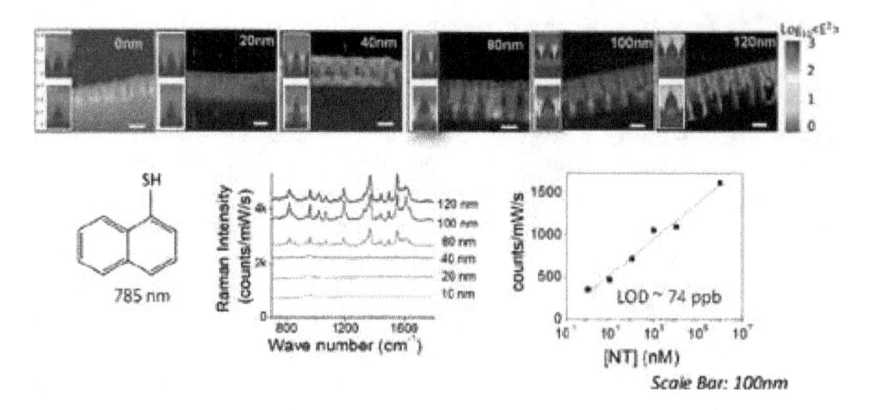

Figure 8.
(Top) Silicon nanopillar arrays with increasing thickness of gold (top, insets) show evolution in electromagnetic field profiles (bottom, left) and SERS spectra of 1-naphthalene thiol as function of metal pillar geometry (bottom, right) concentration dependence of intensity of the peak at 1371 cm^{-1} showing quantitative detection with low detection limits [43].

a different PS-b-P2VP (114 kDa, PDI - 1.1, fPS ~0.5) system. The Si pillar arrays are converted to plasmonic nanoarrays bycoating with thin layer of Cr, followed by systematically increasing thicknesses of gold by electron beam evaporation. The nanopillars coated with 5 nm Cr/120 nm of Au were found to result in gold pillars with separations below 10 nm. The optical modeling of the nanopillars proved EM hot spots with increasing intensity and with increasing metal thickness, as expected. This was correlated well with SERS experiments of 1-naphthalene thiol, where the evolution in SERS intensity was found to saturate at 100 nm. SERS-based plasmonic assays of naphthalene thiol show analytical sensitivity down to 74 ppb, low standard deviations in SERS intensities for different concentrations, and feasi-bility for quantification with large dynamic range. The approach established clear control over structure, property, and function to optimize the final performance of the plasmonic sensor.

3.3 Nanorods and nanodisc arrays from nanoimprint lithography from self-assembly derived high-resolution masters

The uniform templates obtained using reverse micelle approach is well-suited as masks for nanolithography to produce Si nanopillar arrays. These nanopillar arrays are highly interesting for exploitation as high-resolution molds for nanoimprint lithography (NIL) [44]. NIL is a convenient top-down patterning tool that allows replication of surface relief structures down to sub-10 nm feature sizes present in a mold into the polymer substrate [24, 92]. Replication using the NIL process is achieved by pressing the mold against a molten polymer film, followed by solidifying the polymer either by cooling below its Tg or by cross-linking, before removing the mold. Among the several top-down techniques known, NIL has particularly recognized as manufacturing compatible, scalable and low-cost solution for fabricating nanoscale templates of high resolution. The use of the silicon nanopillar arrays as NIL molds offers a distinct advantage of producing multiple copies of templates with identical pitch and width as the original pillars. Since these templates are used to produce metal arrays in the next step, it is possible to attain asymmetric metal features, with reproducibility in optical properties within and between samples. In a specific example, full wafers consisting of Si nanopillar arrays with a height of 120nm were fabricated as described in Section 3.1. They were subsequently diced into smaller pieces and used as NIL molds to replicate the pillar arrays into a thin film of PMMA (100 nm thick) coated on Si substrate. Prior to imprinting, the pillar arrays are functionalized with a perfluorosilane layer to enable anti-stiction property. The NIL process results in a nanoporous PMMA film with geometric characteristics that match with the Si pillar array mold (**Figure 9**). The pores were found to be ~120 nm deep, with a residual layer of ~10 nm thickness present beneath the pores. The residual layer was removed using a controlled O_2 plasma exposure and treated with HF in order to expose the bare Si substrate beneath. The porous template with through-holes was subsequently employed to guide the growth of metal to achieve gold nanorods from the surface through an electroless deposition process. The electroless deposition of gold was performed by galvanic displacement reac-tion where the oxidation of silicon substrate by HF provides electrons to reduce the Au(III) ions to Au(0), in a process earlier shown by Aizawa et al. [93, 94]. Exposure of the nanoporous template to an electroless plating bath consisting of 0.9% HF and 2.3 mM of $HAuCl_4$ for a duration of 1 min was sufficient to grow the nanorods within the pores. The selectivity of the process to the Si substrate ensured the absence of any non-specific metal deposition in unintended areas. The nanoporous PMMA template was then removed using an O_2 plasma exposure. The Au nanorod arrays were found to be 80 nm in height with a pitch identical with that of the NIL mold used. Upon annealing the nanorod array at 200°C for 2 h, a transformation in morphology into nanodiscs was observed. Such transformation that was observed to occur significantly below the melting point of bulk gold metal is likely due to compacting of nanoparticulate and porous gold features obtained upon electroless deposition. The nanoparticulate nature of nanorods is evident from the SEM measurements. The nanodiscs exhibit an ellipsoidal shape, with a diameter of 55.2 (± 4) nm and height of 35.3(± 5.4) nm, as measured using SEM and AFM. TEM cross section of the nanodisc arrays shows that the discs were present within depressions on the surface with a depth of ~5 nm. The depression below the nanodisc is presumably formed by substrate etching, due to the presence of HF in the electroless chemical bath. The transformation of the rod to disc morphology was found necessary in order to ensure mechanical stability of the metal arrays when exposed to solvents. It was found that the nanorod arrays disintegrated when dipped in aqueous solutions,

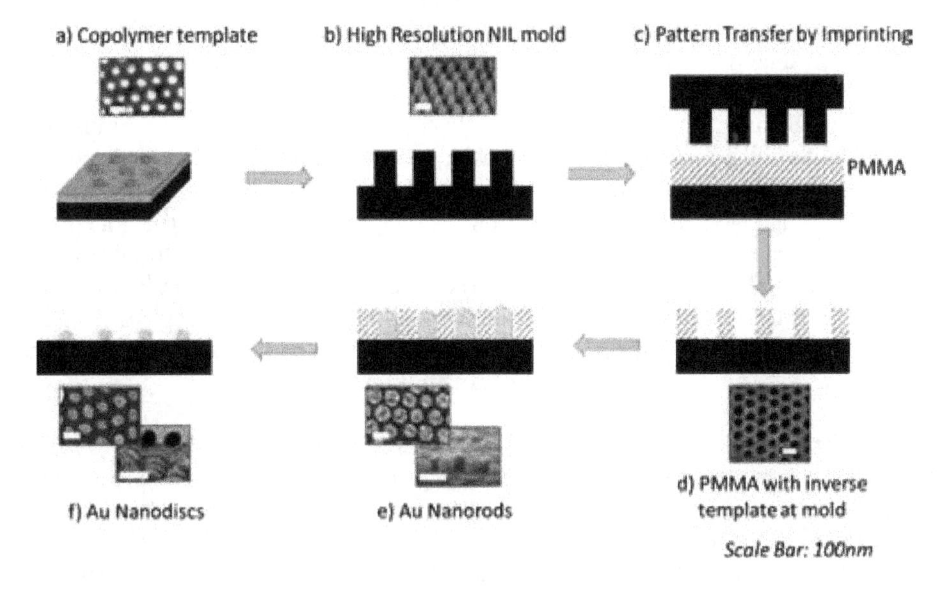

Figure 9.
Process steps involved in production of (a, b) high-resolution molds for NIL produced from self-assembled polymeric templates, (c, d) followed by replication onto PMMA thin film to form nanopore arrays and (e, f) selected area growth of gold nanorods within the pores and their subsequent transformation into nanodiscs [44].

while the nanodiscs remained undisturbed. Since our study primarily intended application of these arrays within a SERS-based sensor in liquid media, the nanorod arrays could not be considered further. Despite this instability, the nanorods could still find a use for plasmonic sensors functioning in the gas phase, or alternatively, if their mechanical stability can be improved through deposition of an overcoating of a dielectric, for example, alumina as shown earlier in literature [95]. The nanodisc arrays were evaluated for SERS performance against three different molecules, namely, naphthalene thiol that represented a covalently bound analyte, 2,2′-bipyridyl that represented non-covalently bound analyte, and crystal violet which repre-sented the possibility of resonance Raman effect. The SERS performance was benchmarked against commercially available Klarite substrates and was proved better for all three molecules. The approach demonstrated an inherently reproducible approach to the fabrication of plasmonic arrays, as several copies can be derived starting from the same nanopillars NIL stamp. The performance of these arrays nevertheless has scope for improvements by further EM enhancements by reducing feature separations. The inability to deliver low feature separations is a limitation of the approach, which can be overcome by post-processing of the porous templates obtained after NIL, for instance, via pore-widening approaches.

4. Conclusions and outlook

Molecular self-assembly using amphiphilic copolymers and colloids derived thereof can deliver nanoplasmonic interfaces with high spatial resolutions, with control over geometric variables in steps of only a few nanometers. The approach enables metal nanoarrays with spatial coherence between features, orthogonal control over the different geometric attributes, and standard deviations below 10%. These characteristics can be leveraged to better understand and predict the optical properties of these arrays, allowing rational routes to maximize plasmonic sensing performance. The self-assembly parameters at both the template production and pattern transfer stages could be rigorously controlled to ensure high uniformity, reproducibility, and scalability of the resulting plasmonic arrays on full wafers. The correlation of the geometry ⇔ optical ⇔ SERS performance was demonstrated with a combination of experiments and numerical simulations. Plasmonic nanoarrays presenting a large number of gap hot spots, with gap distances down to sub-10 nm length scale, are possible to obtain in case of nanoparticle cluster arrays and nano-pillar arrays. The homogeneously distributed hot spots over large areas present an opportunity to not only detect but also quantify the concentration of analytes, with large dynamic range with promisingly low limits of detection. Among the key challenges for future developments is to identify configurations that naturally drive the co-localization of analytes with EM hot spots to achieve maximize plasmonic signal enhancements. Further, the efforts to enhance EM fields solve only a part of the sensing challenge. In addition to maximizing the EM enhancements, the surface needs to be tailored to maximize analyte interactions and their concentrations on the surface.

Yet another challenge is the application of plasmonic arrays for biosensing. The high spatial resolutions sought for maximizing EM enhancements at gap or curvature hot spots are not compatible with the spatial requirements to accommodate large biomolecules like proteins. Further, the sensitivity of the plasmonic sensor extends typically to only a few nanometers from the surface. This is a challenge considering that the size of biomolecular interactions can already be a few tens of nanometers, for example, for an immunosandwich assay. Further, the plasmonic sensor needs to be adapted to work in complex media, for which the surface functionalization to avoid non-specific binding would also consume part of the sensitive space above the plasmonic interface. While the plasmonic sensors have held high promise for highly sensitive, fast responding, and portable configurations, they need significant transversal development cutting across topics beyond fabrication, physics, and photonics, to ensure reliable devices that address emerging analytical challenges.

Acknowledgements

Funding received from National Research Fund of Luxembourg (FNR) via the project PLASENS (C15/MS/10459961) and FNR-PRIDE (FNR PRIDE/15/10935 404) is gratefully acknowledged.

Author details

Rishabh Rastogi[1], Matteo Beggiato[1], Pierre Michel Adam[2], Saulius Juodkazis[3] and Sivashankar Krishnamoorthy[1*]

1 Materials Research and Technology Department, Luxembourg Institute of Science and Technology, Belvaux, Luxembourg

2 Institute Charles Delaunay CNRS, Physics, Mechanics, Materials and Nanotechnology, Department (PM2N), Nanotechnologies, Light, Nanomaterials and Nanotechnology team (L2N), University of Technology of Troyes (UTT), Troyes Cedex, France

3 Faculty of Science, Engineering and Technology, Swinburne University of Technology, Hawthorn, VIC, Australia

*Address all correspondence to: sivashankar.krishnamoorthy@list.lu

References

[1] Soler M, Huertas CS, Lechuga LM. Label-free plasmonic biosensors for point-of-care diagnostics: A review. Expert Review of Molecular Diagnostics. 2019;**19**(1):71-81

[2] Wabuyele MB et al. Portable Raman integrated tunable sensor (RAMiTs) for environmental field monitoring. Advanced Environmental, Chemical, and Biological Sensing Technologies II. 2004;**5586**:60

[3] Spencer KM, Sylvia JM, Marren PJ, Bertone JF, Christesen SD. Surface-enhanced Raman spectroscopy for homeland defense. Chemical and Biological Point Sensors for Homeland Defense. 2004;**5269**:1

[4] Bauch M, Toma K, Toma M, Zhang Q , Dostalek J. Plasmon-enhanced fluorescence biosensors: A review. Plasmonics. 2014;**9**(4):781-799

[5] Hammond JL, Bhalla N, Rafiee SD, Estrela P. Localized surface plasmon resonance as a biosensing platform for developing countries. Biosensors. 2014;**4**(2):172-188

[6] Maier SA, Atwater HA. Plasmonics: Localization and guiding of electromagnetic energy in metal/dielectric structures. Journal of Applied Physics. 2005;**98**(1):1-10

[7] Zhu W et al. Quantum mechanical effects in plasmonic structures with subnanometre gaps. Nature Communications. 2016;7:1-14

[8] Fang Y, Seong NH, Dlott DD. Measurement of the distribution of site enhancements in surface-enhanced raman scattering. Science. 2008;**321**(5887):388-392

[9] Balčytis A et al. From fundamental toward applied SERS: Shared principles and divergent approaches. Advanced Optical Materials. 2018;**6**(16):1-29

[10] Liu L et al. A high-performance and low cost SERS substrate of plasmonic nanopillars on plastic film fabricated by nanoimprint lithography with AAO template. AIP Advances. 2017;**7**(6):1-12

[11] Ye X et al. Improved size-tunable synthesis of monodisperse gold nanorods through the use of aromatic additives. ACS Nano. 2012;**6**(3):2804-2817

[12] Demirel MC et al. Bio-organism sensing via surface enhanced Raman spectroscopy on controlled metal/polymer nanostructured substrates. Biointerphases. 2009;**4**(2):35-41

[13] Chan S, Kwon S, Koo TW, Lee LP, Berlin AA. Surface-enhanced Raman scattering of small molecules from silver-coated silicon nanopores. Advanced Materials. 2003;**15**(19):1595-1598

[14] Kudelski A. Raman studies of rhodamine 6G and crystal violet sub-monolayers on electrochemically roughened silver substrates: Do dye molecules adsorb preferentially on highly SERS-active sites? Chemical Physics Letters. 2005;**414**(4-6):271-275

[15] Zeman EJ, Schatz GC. An accurate electromagnetic study of surface enhancement factors for Ag, Au, Cu, Li, Na, AI, Ga, In, Zn, and Cd. The Journal of Physical Chemistry. 1987;**91**(3):634-643

[16] Hildebrandt P, Stockhurger M. Surface-enhanced resonance Raman spectroscopy of Rhodamine 6G adsorbed on colloidal silver. The Journal of Physical Chemistry. 1984;**88**(24):5935-5944

[17] Huebner U, Boucher R, Schneidewind H, Cialla D, Popp J.

Microfabricated SERS-arrays with sharp-edged metallic nanostructures. Microelectronic Engineering. 2008;**85** (8):1792-1794

[18] Grand J et al. Optimization of SERS-active substrates for near-field Raman spectroscopy. Synthetic Metals. 2003;**139**(3):621-624

[19] Walsh GF, Forestiere C, Dal Negro L. Plasmon-enhanced depolarization of reflected light from arrays of nanoparticle dimers. Optics Express. 2011;**19**(21):21081

[20] Huang W, Yu X, Liu Y, Qiao W, Chen L. A review of the scalable nano-manufacturing technology for flexible devices. Frontiers of Mechanical Engineering. 2017;**12**(1): 99-109

[21] Lim IIS et al. Adsorption of cyanine dyes on gold nanoparticles and formation of J-aggregates in the nanoparticle assembly. The Journal of Physical Chemistry. B. 2006;**110** (13):6673-6682

[22] Gaddis AL. Geometrical Effects on Electromagnetic Enhancement to SERS from Metal Nanoparticle Dimer Arrays. [Masters Theses]. Knoxville, USA: University of Tennessee; 2009:1-78

[23] Vazquez-Mena O et al. High-resolution resistless nanopatterning on polymer and flexible substrates for plasmonic biosensing using stencil masks. ACS Nano. 2012;**6**(6):5474-5481

[24] Chou SY, Krauss PR, Renstrom PJ. Imprint lithography with 25-nanometer resolution. Science. 1996;**272**(5258):85-87

[25] McNay G, Eustace D, Smith WE, Faulds K, Graham D. Surface-enhanced Raman scattering (SERS) and surface-enhanced resonance Raman scattering (SERRS): A review of applications. Applied Spectroscopy. 2011;**65**(8): 825-837

[26] Sánchez-Iglesias A et al. Chemical seeded growth of Ag nanoparticle arrays and their application as reproducible SERS substrates. Nano Today. 2010;**5**(1):21-27

[27] Rekha CR, Sameera S, Nayar VU, Gopchandran KG. Simultaneous detection of different probe molecules using silver nanowires as SERS substrates. Spectrochimica Acta, Part A: Molecular and Biomolecular Spectroscopy. 2019; **213**(2018):150-158

[28] Krishnamoorthy S. Nanostructured sensors for biomedical applications-a current perspective. Current Opinion in Biotechnology. 2015;**34**:118-124

[29] Park S et al. Macroscopic 10 terabit/in 2 arrays from block copolymers with lateral order supplemental information. Science. 2009;**323**(February):1030

[30] Cheng JY, Ross CA, Thomas EL, Smith HI, Vancso GJ. Fabrication of nanostructures with long-range order using block copolymer lithography. Applied Physics Letters. 2002;**81**(19): 3657-3659

[31] Stoykovich MP et al. Directed self-assembly of block copolymers for nano-lithography: Fabrication of isolated features and essential integrated circuit geometries. ACSNano. 2007;**1**(3):168-175

[32] Yang JKW et al. Complex self-assembled patterns using sparse commensurate templates with locally varying motifs. Nature Nanotechnology. 2010;**5**(4):256-260

[33] Cetin AE et al. Handheld high-throughput plasmonic biosensor using computational on-chip imaging. Light: Science & Applications. 2014;**3**(1)

[34] Aksu S, Yanik AA, Adato R, Artar A, Huang M, Altug H. High-throughput nanofabrication of infrared plasmonic nanoantenna arrays for

vibrational nanospectroscopy. Nano Letters. 2010;**10**(7):2511-2518

[35] Krishnamoorthy S, Manipaddy KK, Yap FL. Wafer-level self-organized copolymer templates for nanolithography with sub-50 nm feature and spatial resolutions. Advanced Functional Materials. 2011; **21**(6):1102-1112

[36] Krishnamoorthy S, Pugin R, Brugger J, Heinzelmann H, Hinderling C. Tuning the dimensions and periodicities of nanostructures starting from the same polystyrene-block-poly(2-vinylpyridine) diblock copolymer. Advanced Functional Materials. 2006;**16**(11):1469-1475

[37] Mansky P, Haikin P, Thomas EL. Monolayer films of diblock copolymer microdomains for nanolithographic applications. Journal of Materials Science. 1995;**30**(8):1987-1992

[38] Lu J, Chamberlin D, Rider DA, Liu M, Manners I, Russell TP. Using a ferrocenylsilane-based block copolymer as a template to produce nanotextured Ag surfaces: Uniformly enhanced surface enhanced Raman scattering active substrates. Nanotechnology. 2006;**17**(23):5792-5797

[39] Spatz JP, Sheiko S, Möller M. Substrate-induced lateral microphase separation of a diblock copolymer. Advanced Materials. 1996; **8**(6):513-517

[40] Letchford K, Burt H. A review of the formation and classification of amphiphilic block copolymer nanoparticulate structures: Micelles, nanospheres, nanocapsules and polymersomes. European Journal of Pharmaceutics and Biopharmaceutics. 2007;**65**(3):259-269

[41] Pedrosa CR et al. Controlled nanoscale topographies for osteogenic differentiation of mesenchymal stem cells. ACS Applied Materials & Interfaces. 2019;**11**(9):8858-8866

[42] Yap FL, Thoniyot P, Krishnan S, Krishnamoorthy S. Nanoparticle cluster arrays for high-performance SERS through directed self-assembly on flat substrates and on optical fibers. ACS Nano. 2012;**6**(3):2056-2070

[43] Dinda S, Suresh V, Thoniyot P, Balčytis A, Juodkazis S, Krishnamoorthy S. Engineering 3D nanoplasmonic assemblies for high performance spectroscopic sensing. ACS Applied Materials & Interfaces. 2015;**7**(50):27661-27666

[44] Krishnamoorthy S, Krishnan S, Thoniyot P, Low HY. Inherently reproducible fabrication of plasmonic nanoparticle arrays for SERS by combining nanoimprint and copolymer lithography. ACS Applied Materials & Interfaces. 2011;**3**(4):1033-1040

[45] Nordlander P, Oubre C, Prodan E, Li K, Stockman MI. Plasmon hybridization in nanoparticle dimers. Nano Letters. 2004;**4**(5):899-903

[46] Brandl DW, Mirin NA, Nordlander P. Plasmon modes of nanosphere trimers and quadrumers. The Journal of Physical Chemistry. B. 2006;**110**(25):12302-12310

[47] Hentschel M, Saliba M, Vogelgesang R, Giessen H, Alivisatos AP, Liu N. Transition from isolated to collective modes in plasmonic oligomers. Nano Letters. 2010;**10**(7): 2721-2726

[48] Quinten M, Kreibig U. Optical properties of aggregates of small metal particles. Surface Science. 1986;**172**(3): 557-577

[49] Mirin NA, Bao K, Nordlander P. Fano resonances in plasmonic nanoparticle aggregates. The

Journal of Physical Chemistry. A. 2009;**113**(16):4028-4034

[50] Natan MJ. Concluding remarks: Surface enhanced Raman scattering. Faraday Discussions. 2006;**132**:321

[51] Ueno K, Juodkazis S, Mizeikis V, Sasaki K, Misawa H. Clusters of closely spaced gold nanoparticles as a source of two-photon photoluminescence at visible wavelengths. Advanced Materials. 2008;**20**(1):26-30

[52] Fan JA et al. Fano-like interference in self-assembled plasmonic quadrumer clusters. Nano Letters. 2010;**10**(11): 4680-4685

[53] Jin R. Nanoparticle clusters light up in SERS. Angewandte Chemie, International Edition. 2010;**49**(16): 2826-2829

[54] Le F et al. Metallic nanoparticle arrays: A common substrate for both surface-enhanced Raman scattering and surface-enhanced infrared absorption. ACS Nano. 2008;**2**(4):707-718

[55] Lee JS, Ulmann PA, Han MS, Mirkin CA. A DNA - gold nanoparticle-based colorimetric competition assay for the detection of cysteine. Nano Letters. 2008;**8**(2):529-533

[56] Talley CE et al. Surface-enhanced Raman scattering from individual Au nanoparticles and nanoparticle dimer substrates. Nano Letters. 2005;**5**(8): 1569-1574

[57] Camden JP et al. Probing the structure of single-molecule surface-enhanced Raman scattering hot spots. Journal of the American Chemical Society. 2008;**130**(38):12616-12617

[58] Lassiter JB et al. Fano resonances in plasmonic nanoclusters: Geometrical and chemical tunability. Nano Letters. 2010;**10**(8):3184-3189

[59] Ap A et al. Organization of nanocrystal molecules using DNA. Nature. 1996;**382**(6592):609-611

[60] Qiu P, Jensen C, Charity N, Towner R, Mao C. Oil phase evaporation-induced self-assembly of hydrophobic nanoparticles into spherical clusters with controlled surface chemistry in an oil-in-water dispersion and comparison of behaviors of individual and clustered iron oxide nanoparticles. Journal of the American Chemical Society. 2010;**132**(50):17724-17732

[61] Correa-Duarte MA, Liz-Marzán LM. Carbon nanotubes as templates for one-dimensional nanoparticle assemblies. Journal of Materials Chemistry. 2006;**16**(1):22-25

[62] Wang H, Lin W, Fritz KP, Scholes GD, Winnik MA, Manners I. Cylindrical block co-micelles with spatially selective functionalization by nanoparticles. Journal of the American Chemical Society. 2007;**129**(43): 12924-12925

[63] Blum AS et al. Cowpea mosaic virus as a scaffold for 3-D patterning of gold nanoparticles. Nano Letters. 2004;**4**(5): 867-870

[64] Lee W, Lee SY, Briber RM, Rabin O. Self-assembled SERS substrates with tunable surface plasmon resonances. Advanced Functional Materials. 2011;**21**(18):3424-3429

[65] Banwell CN, Sheppard N, Turner JJ. Directed deposition of nanoparticles using diblock copolymer templates. Advances in Molecular Spectroscopy. 2013;**15**(3):1183-1184

[66] Wang L, Montagne F, Hoffmann P, Heinzelmann H, Pugin R. Hierarchical positioning of gold nanoparticles into periodic arrays using block copolymer nanoring templates. Journal

of Colloid and Interface Science. 2011;**356**(2):496-504

[67] Förster S, Antonietti M. Amphiphilic block copolymers in structure-controlled nanomaterial hybrids. Advanced Materials. 1998;**10**(3):195-217

[68] Kugel V, Ji H-F. Nanopillars for sensing. Journal of Nanoscience and Nanotechnology. 2014;**14**(9):6469-6477

[69] Oh YJ, Kang M, Park M, Jeong KH. Engineering hot spots on plasmonic nanopillar arrays for SERS: A review. BioChip Journal. 2016; **10**(4):297-309

[70] Karadan P, Aggarwal S, Anappara AA,Narayana C, Barshilia HC. Tailored periodic Si nanopillar based architectures as highly sensitive universal SERS biosensing platform. Sensors Actuators, B Chemical. 2018;**254**:264-271

[71] Gudur A, Ji H-F. Bio-applications of nanopillars. Frontiers in Nanoscience and Nanotechnology. 2017;**2**(6):1-10

[72] Chattopadhyay S, Huang YF, Jen YJ, Ganguly A, Chen KH, Chen LC. Anti-reflecting and photonic nanostructures. Materials Science & Engineering R: Reports.2010;**69**(1-3): 1-35

[73] Zavaliche F et al. Electrically assisted magnetic recording in multiferroic nanostructures. Nano Letters. 2007;**7**(6):1586-1590

[74] Schneider L, Feidenhans'L NA, Telecka A, Taboryski RJ. One-step maskless fabrication and optical characterization of silicon surfaces with antireflective properties and a white color appearance. Scientific Reports. 2016;**6**(September):1-6

[75] Fujikawa S, Takaki R, Kunitake T. Fabrication of arrays of sub-20-nm silica walls via photolithography and solution-based molecular coating. Langmuir. 2006;**22**(21):9057-9061

[76] Cheung CL, Nikolić RJ, Reinhardt CE, Wang TF. Fabrication of nanopillars by nanosphere lithography. Nanotechnology. 2006;**17**(5):1339-1343

[77] Wang Y, Lee K, Irudayaraj J. Silver nanosphere SERS probes for sensitive identification of pathogens. Journal of Physical Chemistry C.2010; **114** (39):16122-16128

[78] Dieringer JA et al. Surface enhanced Raman spectroscopy: New materials, concepts, characterization tools, and applications. Faraday Discussions. 2006;**132**:9-26

[79] Shiohara A, Wang Y, Liz-Marzán LM. Recent approaches toward creation of hot spots for SERS detection. Journal of Photochemistry and Photobiology C Photochemistry Reviews. 2014;**21**:2-25

[80] Kim JH, Kang T, Yoo SM, Lee SY, Kim B, Choi YK. A well-ordered flower-like gold nanostructure for integrated sensors via surface-enhanced Raman scattering. Nanotechnology. 2009;**20**(23):235302

[81] Duan H, Yang JKW, Berggren KK. Controlled collapse of high-aspect-ratio nanostructures. Small. 2011;**7**(18): 2661-2668

[82] Lee SJ, Morrill AR, Moskovits M. Hot spots in silver nanowire bundles for surface-enhanced Raman spectroscopy. Journal of the American Chemical Society. 2006;**128**(7):2200-2201

[83] Chandra D, Yang S. Capillary-force-induced clustering of micropillar arrays: Is it caused by isolated capillary bridges or by the lateral capillary meniscus interaction force? Langmuir. 2009;**25**(18):10430-10434

[84] Gates BD, Xu Q, Thalladi VR, Cao T, Knickerbocker T, Whitesides GM. Shear patterning of microdominos: A new class of procedures for

making micro- and nanostructures. Angewandte Chemie, International Edition. 2004;**43**(21):2780-2783

[85] Hu M et al. Gold nanofingers for molecule trapping and detection. Journal of the American Chemical Society. 2010;**132**(37):12820-12822

[86] Schmidt MS, Hübner J, Boisen A. Large area fabrication of leaning silicon nanopillars for surface enhanced Raman spectroscopy. Advanced Materials. 2012;**24**(10):11-18

[87] Singh A, Knaepen W, Sayan S, El Otell Z, Chan BT, Maes JW, et al. Impact of sequential infiltration synthesis on pattern fidelity of DSA lines. In: Advances in Patterning Materials and Processes XXXII, Vol. 9425. International Society for Optics and Photonics; 2015. p. 94250N

[88] Ishchenko OM et al. Investigating sequential vapor infiltration synthesis on block-copolymer-templated titania nanoarrays. Journal of Physical Chemistry C. 2016;**120**(13):7067-7076

[89] Yin J, Xu Q, Wang Z, Yao X, Wang Y. Highly ordered TiO_2 nanostructures by sequential vapour infiltration of block copolymer micellar films in an atomic layer deposition reactor. Journal of Materials Chemistry C. 2013;**1**(5): 1029-1036

[90] Ku SJ et al. Highly ordered free-standing titanium oxide nanotube arrays using Si-containing block copolymer lithography and atomic layer deposition. Nanotechnology. 2013;**24**(8):085301

[91] Peng Q , Tseng YC, Darling SB, Elam JW. Nanoscopic patterned materials with tunable dimensions via atomic layer deposition on block copolymers. Advanced Materials. 2010;**22**(45):5129-5133

[92] Guo LJ. Nanoimprint lithography: Methods and material requirements. Advanced Materials. 2007;**19**(4):495-513

[93] Aizawa M, Buriak JM. Block copolymer templated chemistry for the formation of metallic nanoparticle arrays on semiconductor surfaces. Chemistry of Materials. 2007;**19**(21): 5090-5101

[94] Aizawa M, Buriak JM. Block copolymer-templated chemistry on Si, Ge, InP, and GaAs surfaces. Journal of the American Chemical Society. 2005;**127**(25):8932-8933

[95] Zhang X, Zhao J, Whitney AV, Elam JW, Van Duyne RP. Ultrastable substrates for surface-enhanced Raman spectroscopy: Al_2O_3 overlayers fabricated by atomic layer deposition yield improved anthrax biomarker detection. Journal of the American Chemical Society. 2006;**128**(31):10304-10309

7

Surface Plasmon Enhanced Chemical Reactions on Metal Nanostructures

Rajkumar Devasenathipathy, De-Yin Wu and Zhong-Qun Tian

Abstract

Noble metal nanomaterials as plasmonic photocatalysts can strongly absorb visible light and generate localized surface plasmon resonance (SPR), which in turn depends on the size, shape, and surrounding of the plasmonic metal nanomaterials (PMNMs). Remarkably, the high-efficiency conversion of solar energy into chemical energy was expected to be achieved by PMNMs. Therefore, researchers have chosen PMNMs to improve the photocatalytic activity toward targeted molecules. This enhancement can be achieved by the effective separation of photogenerated electrons and holes of the PMNMs in the presence of light. Surface-enhanced Raman spectroscopy (SERS) has been performed for obtaining information about the photochemically transformed surface species at molecular levels. A profound understanding of kinetic mechanisms is needed for the development of novel plasmonic catalysts toward various chemical transformations of targeted molecules. In this chapter, based on the above discussions, the participation of SPR excitation in PMNMs and photocatalysis toward chemical transformations of SERS-active organic molecules such as aromatic amino and nitro compounds based on PMNMs have been discussed in detail through theoretical and experimental studies. Eventually, a summary and the future directions of this study are discussed.

Keywords: plasmonic metal nanomaterials, surface plasmon resonance, surface-enhanced Raman scattering, chemical reaction, aromatic amino and nitro compounds

1. Introduction

Surface plasmon resonance (SPR) is the phenomenon resulting from the resonant oscillation of conduction electrons present at the interface between negative and positive permittivity of plasmonic metal nanomaterials (PMNMs) that can be induced by incident light. The electric field of the incident light and the free electrons present in PMNMs results in SPR, which yields strong intraband transition [1]. Thus, the above physical process permits the PMNMs to collect the incident light and focus the collected energy near the surface of PMNMs, which in turn transforms the light energy into energy associated with excited charge carriers. The features of PMNMs such as size, shape, and aggregation of nanoparticles as well as the permittivity of surrounding medium decide the intensity of SPR and resonant energy [2]. Several novel surface reactions, namely radiative and non-radiative relaxations, are initiated by the effect of SPR. The influence of SPR leads to the generation of hot electrons and holes, which are usually termed as hot carriers [3]. These hot carriers can stimulate chemical reactions to the molecules which are present vicinal to the PMNMs [4]. The exceptional characteristics of PMNMs have been extensively employed in numerous applications such as photovoltaic cells, plasmonic sensors, fuel cell fluorescence enhancement as well as local spectroscopies including surface-enhanced Raman spectroscopy (SERS) [5].

Among other analytical techniques, SERS has been deliberated as a commanding analytical technique for the study of surfaces and interfaces. The phenomenon of SERS was initially witnessed in pyridine adsorbed on a rough silver electrode surface in 1974 [6]. Recently, SERS has been utilized as an in situ potential powerful technique that offers unusual surface sensitivity and information regarding molecular fingerprint, to investigate the transient surface species as well as reaction mechanisms. Based on the unique features, SERS has been widely used in medical diagnosis, single molecule detection, pesticide analysis, safety inspections, and identification [7]. Notably, plasmonic metal nanomaterials (such as Ag, Au, and Cu), rough metal films, and gaps between metal surfaces and metal nanomaterials are the widely used SERS platforms for promoting photochemical reactions [8]. The molecules comprising of mostly aromatic or unsaturated bonds are generally targeted for SERS measurements. The Raman signal of the surface species present vicinal to the plasmonic PMNMs can be considerably promoted by the enhanced electromagnetic field, when excited with the appropriate light. Later, the electrochemical SERS had been also employed extensively to investigate the electrochemical processes (interfacial structures, surface reactions, and adsorption of species on electrodes at different potentials) [9]. This measurement has been performed in noble transition nanometals as well as single crystal surfaces for the chemical transformation of aromatic molecules in recent years [10].

From the results of former studies, substituted and unsubstituted aromatic amines as well as nitro compounds have been employed as demonstrative models for aromatic compounds. It is also evident from the earlier studies that an azo species p,p-dimercaptoazobenzene (DMAB) can be selectively resulted from the adsorbed substituted or unsubstituted p-aminothiophenol (PATP) and p-nitrothiophenol (PNTP) on the surface of PMNMs [11]. The PMNMs, pH of the solution, substrate, wavelength, and power of irradiation as well as the environmental conditions can have a strong impact on the above-mentioned conversion [12]. The study of this reaction mechanism is still a challenging issue. The complete thermodynamic and kinetic data are also required for the above system. Therefore, the interpretation of reaction mechanisms, which assist in the optimization of experimental conditions and increasing yield as well as selectivity, can be given by theoretical studies on model reactions.

In this chapter, we have given a brief explanation of SPR and its rapid relaxation of nanometals toward surface hot carriers in Section 2. Then, charge transfer mechanism (metal-to-molecule and molecule-to-metal) and the surface catalytic coupling/condensation of aromatic amines and aromatic nitro compounds on noble metal nanomaterials have been discussed through theoretical and experimental results. Since the experimental conditions are highly dependent on the effect of pH and adsorption of molecule, one of the aromatic amines (PATP) has been taken as a model compound to discuss clearly the effect of pH and adsorption of molecule on the nanometals in Section 3. As DMAB can be selectively resulted from the adsorbed PATP and PNTP on the surface of plasmonic nanostructures during the photochemical reaction, the reported SERS results from chemically synthesized DMAB with experimentally derived DMAB from PATP and PNTP have also been discussed in Section 3. Aerobic oxidation-assisted aromatic amine based on nanoplasmonic photocatalysts will be discussed in Section 4. Finally, the reaction mechanisms and future research prospective have been given Sections 5.

2. Surface plasmon resonance effect on PMNMs and its fast relaxations

The phenomenon of collective excitation of free electrons present in the noble metals when subjected to an external field generates plasmon resonance. For bulk metals, the density of the free electrons decides the characteristic frequency (ω_p) of plasmon resonance, that is,$\omega_p = (4\pi_e n/m\ )\ \ e;^{1/2}$ here n = density of conducting electrons, m_e = effective mass, and e = charge unit of electrons [13]. The plasmon resonance frequencies of metals such as gold, silver, and copper in bulk were found to be 9.0, 9.0, and 7.9 eV, respectively. Noticeably, the energy of interband transition is lower compared to the transition energies. This leads to the retardation of intra-band transition, which in turn results in a fairly large damping constant. The fairly accurate relaxation time has been calculated as about 10 fs [14].

Compared to gold, silver, and copper bulk metals, the frequency of SPR gradually shifts to the longer wavelength as the size of metals gets decreased under the surface effect. The lower excitation energy of silver nanostructures hinders the interband pathway, but considerably unusual optical characteristics are displayed by the intraband transition [15]. **Figure 1A** shows the SPR of a spherical PMNMs excited by visible light and the absorption and scattering of light on the noble PMNMs are primarily decided by the effect of SPR. The SPR effect is vital for the formation of sub-wavelength area (hot spot), which results from the conversion of far-field light irradiation into near-field photonic energy [16]. The probing molecules can display the effect of SERS/actuate the reactions of surface photochemistry, as they are adsorbed on the sub-wavelength area. The existence of SPR effect can be observed on transition metals unlike noble metals (copper, gold, and silver). The shorter lifetime of SPR is however due to the following reasons: (i) the radiative relaxation based on the photon emission or (ii) non-radiative relaxation via producing hot carriers. The property of metals, size of nanostructures, energy and the polarization of lasers can also determine the process of relaxation [17].

The size, shape, and dielectric constant of the medium of environment in which the single metal nanoparticles are present determine the frequency of SPR. The SPR lifetime is around 10 fs as the 2.2-eV incident photonic energy falls on gold nanospheres (**Figure 1B**). This results in higher energy photogeneration (around 2.0 eV) of hot electrons compared to the Fermi level, but the energy of the hot hole (lower than the Fermi level) was estimated to be around 1.0 eV [19]. The extinction spectrum of a spherical silver nanoparticle (diameter = 15 nm) shows 380 nm as its SPR frequency. A considerable decrease in the lifetime of hot carriers was observed as the size of silver nanoparticles increased [20]. For instance, in the excitation light, the major distribution of hot carriers was seen surrounding the Fermi level for

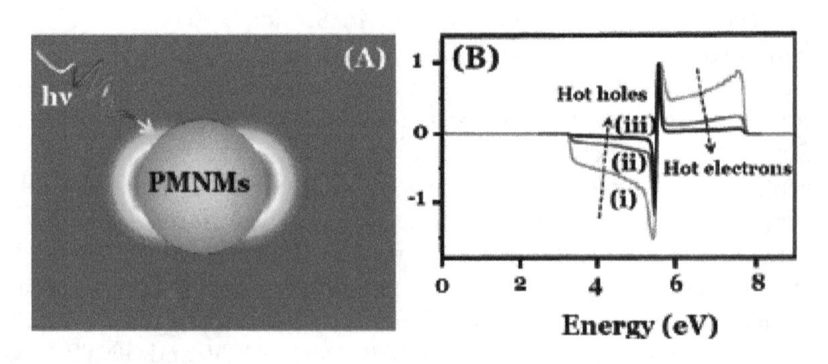

Figure 1.
(A) SPR of a spherical PMNM excited by visible light. (B) Normalized distribution of hot electrons and hot holes on Au slab with thicknesses: (i) 10 nm, (ii) 20 nm, and (iii) 40 nm with incident photonic energy of 2.22 eV.

the spherical silver nanoparticle possessing a diameter of 25 nm. When the size of metal nanoparticles increases, a considerable increase in the probability of radiation procedure was witnessed unlike nanoparticles with smaller size. In the case of PMNMs, as the lifetime of SPR gets longer, a larger probability in the distribution of hot carriers (high energy) was observed. The lifetime of SPR is longer for spherical silver nanoparticles relative to gold nanoparticles based on the respective interband transition energy values (3.2 and 2.3 eV) [19, 20].

3. Photoinduced charge transfer of aromatic nitro and amino molecules on PMNMs

To illustrate the reaction mechanism, the electroreduction of p-nitrobenzene to aminobenzene derivatives on the surface of noble metals was monitored by SERS. The essential uses of p-substituted aniline as intermediates can be seen in pharmaceuticals and dyes industries. The formation of p-substituted aniline as a result of electrochemical reduction of p-substituted nitrobenzene was revealed from the vanished Raman peak at about 1338 cm^{-1} (corresponding to the symmetric stretching vibration of nitro group in p-substituted nitroben-zene) and appearance of new peaks. Whereas, the mechanistic pathway for the electrochemical polymerization of aminobenzene or aniline was also studied by SERS [21]. Gold colloidal nanoparticles, silver nanofilms, and copper nanoparticles have improved the adsorption and photochemical processes of substituted amino- and nitrobenzenes [22–24]. The reaction mechanisms and the equivalent kinetics are however essential to comprehend the photochemical reactions. All the vibrational frequencies witnessed by our DFT calculations are not highly sensitive to the functional groups such as —OH, —COOH, —SH, —CN, and —NO_2, which are substituted at para positions of nitrobenzene and aminobenzene. All these substituted aminobenzenes and nitrobenzenes on nanoscale noble metals were chemically transformed to azobenzene derivatives after the photochemical reaction (**Figure 2A**). However, the SERS related to the N=N stretching modes of para-substituted azobenzene exposes an appar-ently high sensitivity [25]. In addition, the electron density of para functional groups decides the relative intensities of the peaks. The peak appeared at around 1430 cm^{-1} is weaker compared to the peak at around 1390 cm^{-1} in the case of electron-donating functional groups (hydroxyl, amine, and thiol). For electron-withdrawing functional groups (nitro, nitrile, and carboxylic acid), the peak detected at about 1390 cm^{-1} is however weaker than the peak at about 1430 cm^{-1} (**Figure 2B**).

We have given the possible photochemical reaction mechanism for the reduction and oxidation of substituted aromatic nitro and amino compounds into corresponding azobenzene derivatives (**Figure 3A** and **B**). The substituted amino molecules transformation into azobenzene derivatives in the presence of light involves two steps: 1. Oxidation of amino molecules, and 2. Coupling of two oxidized amino molecules into disubstituted azobenzenes, whereas, the reduction of amino-substituted nitro molecules transformation into azobenzenes in the presence of light involves three steps: 1. Reduction of nitro molecules, 2. Condensation of reduced nitromolecules to oxadia-ziridine, 2,3-bis (4-mercaptophenyl), and 3. Further oxidation of Oxadia-ziridine can convert 2,3-bis (4-substituted phenyl) into disubstituted azobenzene. The experimental results from the oxidation/reduction of different para-substituted aminobenzene/nitrobenzene into para-substituted azobenzenes nevertheless depends on the experimental conditions along with the laser powers

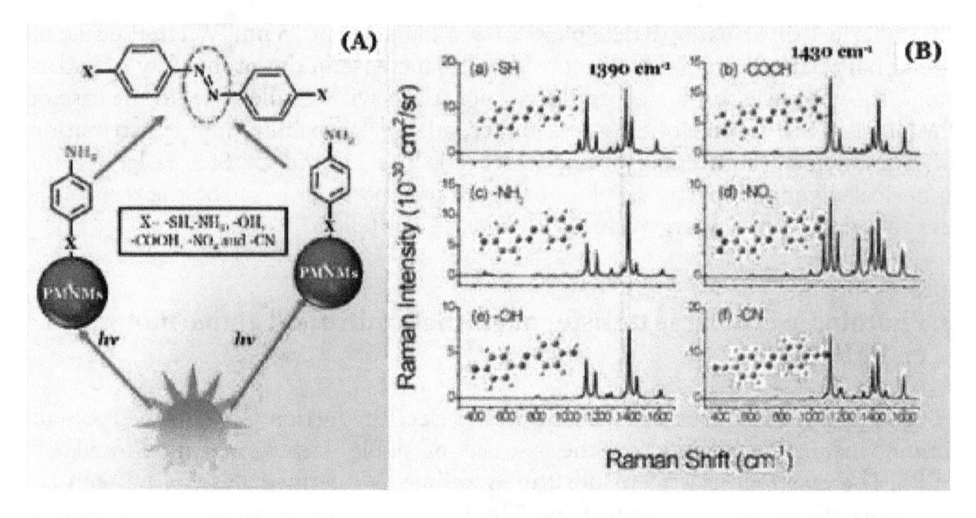

Figure 2.
(A) Schematic representation for photochemical transformation of substituted amino and nitrobenzenes to azobenzene derivatives. (B) Raman spectra of azobenzene derivatives with various para functional groups simulated by PW91PW91/6-311 + G(d, p): (a) —SH, (b) —COOH, (c) —NH₂, (d) —NO₂, (e) —OH, and (f) —CN.

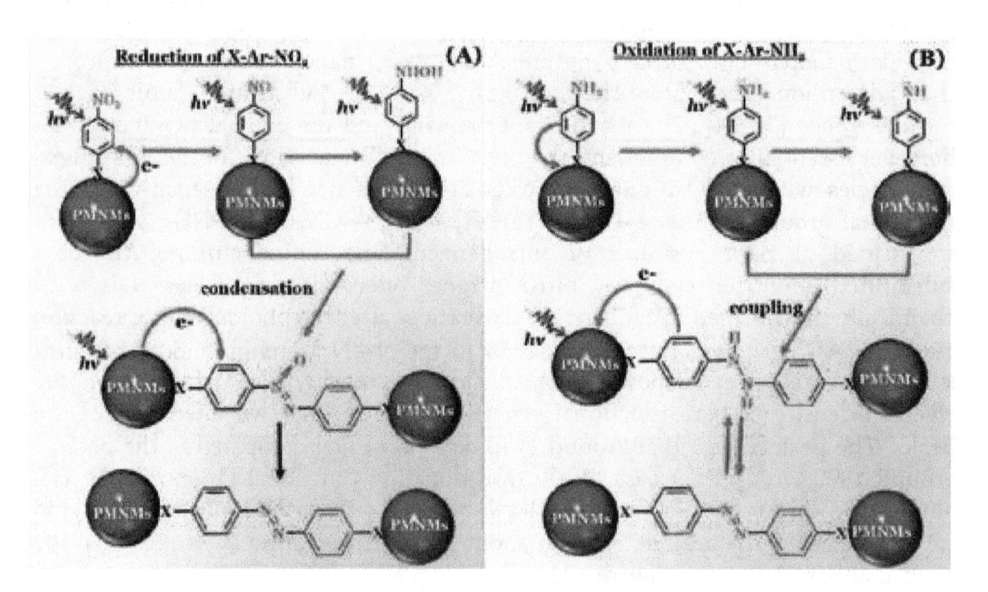

Figure 3.
Schematic representation of reaction mechanism for the photochemical reduction of X-ArNO₂ (A) and oxidation of X-ArNH₂ (B) (X = —OH, —COOH, —SH, —CN, and —NO₂).

employed for Raman excitation, electrolyte pH as well as adsorption that promotes the electroreduction of nitro groups into aromatic amines or azo compounds. Thus, numerous studies have been developed for studying the surface reactions of the above mentioned SERS molecules.

Among other substituted aromatic nitro compounds, SERS has been employed widely to study the photochemical as well as electrochemical reduction of p-nitro-benzoic acid (PNBA) and p-nitro-thiophenol (PNTP) [26]. The decrease in the intensities of several original peaks and appearance of several new peaks take place in the course of cathode polarization/laser irradiation. Nonetheless, the detailed study for the above spectral alterations is still contentious. The appeared new peaks are in concordance with the observed SERS results of aromatic amines adsorbed on silver nanostructures [27]. Thus, some studies equate the new peaks to aromatic amines as reduction products. For example, Sun et al. described that, the product formed can be both aniline species and azo compound or individual aniline species/ azo compound. They proposed the charge transfer mechanism where the transfer of charge between silver island films and p-nitrobenzoic acid arose from the laser excitation [28]. The high-power laser irradiation also resulted in the formation of p, p-azodibenzoic acid through reductive coupling reaction [29].

p-aminothiophenol (PATP) is a molecule possessing thiol group at the para position of aniline. It can form self-assembled monolayer (SAM) on the surface of nanometals. Additionally, PATP has been targeted as a significant surface probe molecule in the areas of SERS and nanoscience. An exceptional and sturdy SERS signal has been displayed by PATP. However, the study of enhancement mechanism is still challenging since/in 1990s. An intense potential-dependent peak of SERS for three excitation wavelengths (488, 514.5, and 633 nm) was seen at 1430 cm^{-1} [30]. As the excitation wavelength increases, the potential analogous to the maximum Raman peak intensity proceeds toward the direction of more negative potential [30]. This reveals that these light irradiations can perform the charge transfer from the PMNMs to adsorbed surface species. From DFT calculations of metallic cluster models, we have proposed a clear theoretical pathway for the charge transfer mechanism from PATP to PMNMs and PMNMs to PNTP in (**Figure 4**). In the course of incident light excitation on the surface of rough/colloidal silver and gold, the participation of sur-face plasmon and charge transfer processes have been described. At this juncture, for oxidation and reduction reactions, the metal nanoparticles act as sink of electrons and electron source. The Fermi level of PMNMs which exists in between HOMO and LUMO of PNTP or PATP (adsorbate) can be altered by the applied potential. The resonant charge transfer can occur as the energy difference between the ground state ψ_g and the photon-driven charge transfer excited state ψ_{CT} equates the excit-ing radiation energy. Thus, the charge transfer takes place from the molecule to the

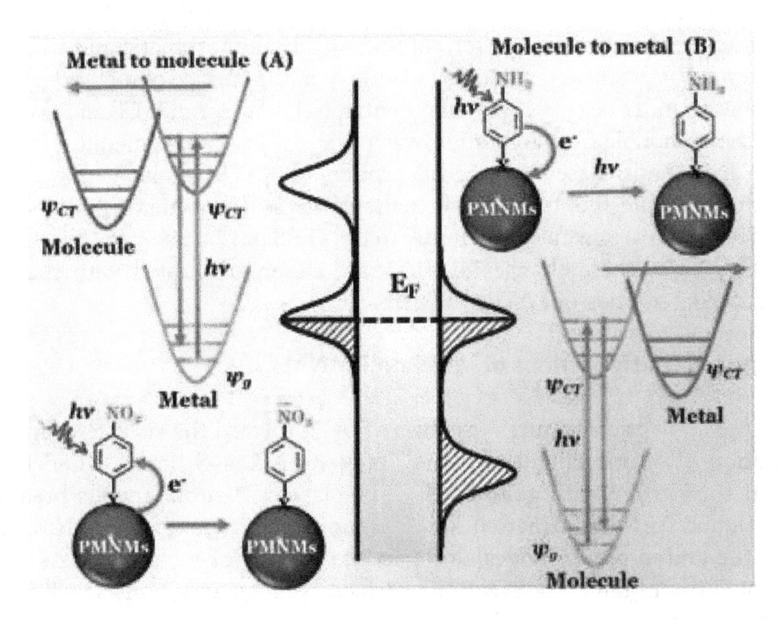

Figure 4.
(A) Schematic representation of charge transfer from metal to molecule: an electron excitation from the Fermi level of PMNMs to the LUMO of aromatic nitro compounds. (B) Charge transfer from molecule to metal: an electron excitation from the HOMO of aromatic amine to the Fermi level of PMNMs.

surface of silver for X–$ArNH_2$ and from silver surface to molecule for X–$ArNO_2$. In both cases, X symbolizes the thiol functional group at the para position. Later, the excited surface complex possibly will go through one of the two dissimilar deexcita-tions as follows: (i) based on the reverse charge transfer back to ground state which is followed by a radiative process, where, this deexcitaion (purely physical) comprises of either Raman scattering or fluorescence emission and (ii) photochemical reaction. Here, a neutral $ArNO_2H$ radical can be formed by the excited nitro radical anion, which takes up proton from proton donors present in solution. Further reactions take place from the so formed neutral $ArNO_2H$ radical. The neutral ArNH radical formed from the excited amine radical cation by giving out a proton. This ArNH radical can undergo nitrogen-nitrogen coupling reaction (dimerization), indicating the molecules are strongly adsorbed on the electrode surface (**Figure 2**). Thus, the charge transfers from molecule (PATP) to metal nanoparticles and metal nanopar-ticles to molecule (PNTP) occur. In fact, the experimentally observed charge transfer should be responsible to a process from silver electrode to adsorbed DMAB species, as proposed from previous DFT calculations [25].

3.1 Effect of pH toward photochemical reaction on PATP-adsorbed PMNMs

In order to give a clear explanation of the effect of pH, adsorption of molecules for surface catalytic coupling reactions, and oxygen-assisted photocatalytic reactions on PMNMs, we have chosen only para-aminothiophenols (PATPs) as representative compounds. The SERS signals are highly sensitive to the pH of the electrolyte and applied potential. It is evident from the previous SERS measurements of the adsorp-tion of PATP on the surface of rough silver and gold electrodes, reported by Hill and Wehling [31]. The property of SERS significantly varies from acidic to alkaline medium. For the case of anodic and cathodic polarization, SERS spectra of PATP show substantial variations in alkaline solution. In contrast to alkaline medium, a good reversibility behavior (in intensity of SERS) was seen in acidic medium based on changing potentials. As the applied potentials move toward the negative direction, the SERS peaks (1130, 1390, and 1440 cm^{-1}) retained virtually in the alkaline medium. In order to describe the above character, an isomerization of aromatic and quinonoidic configurations was postulated for PATP adsorbed on the surface of metal electrodes in acidic medium. Under negative applied potential (−1.4 V vs. Ag|AgCl) in pure sodium disulfide, the quinonoidic configuration was proposed in alkaline medium to support the above hypothesis. Based on the adsorption of PATP in the nanoscale cavity between the substrate (gold) and silver nanoparticles, the isomerization and charge transfer mechanism are given concurrently to the SERS mechanism by Zhou et al. [32]. Till now, we know these changes of SERS are closely associated with the surface catalytic coupling reaction of PATP to DMAB.

3.2 Adsorption configurations of PATP on PMNMs

The three adsorption features feasible as PATP toward the surface of metal are: (i) formation of strong chemical bonds (Au—S or Ag—S bond) when the easy binding of thiol group with gold/silver takes place, (ii) simultaneous breaking of S—H bond, and (iii) the formation of weak coordination bonds (Au—N or Ag—N bond) as the amino group moves closer to the surface of metal [33]. As a result, there is concurrent binding between the surface of metal and thiol as well as amino groups. On the metal surface, the top, bridge, and hollow sites possessing large adsorption energies can hold the sulfur atoms (**Figure 5A**). The adsorption con-figuration exposing a skewed angle with regard to the surface has been observed as the top/bridge sites hold the thiol group. In the case of hollow sites, the molecular

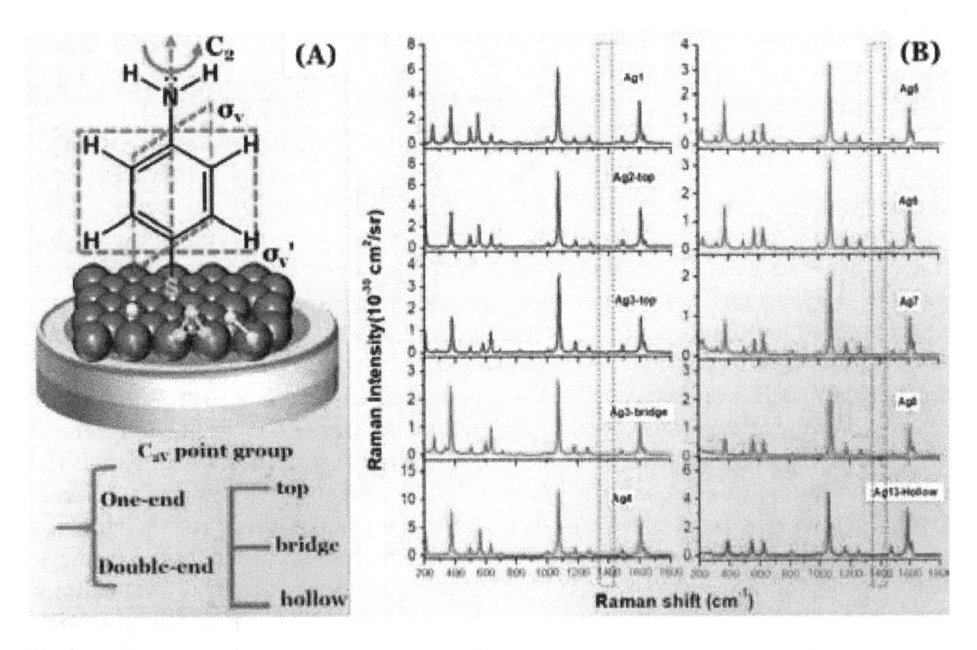

Figure 5.
*(A) One-end adsorption configuration at a top site, a bridge site, a hollow site, and double end configuration through the interaction between C_{2V} point group of amino nitrogen binding and silver. (B) Simulated Raman spectra of PATP with different silver clusters using DFT theoretical methods (B3LYP/6-311 + G**(C, N, S, and H)/LANL2DZ(Ag)).*

symmetric axis of PATP can be visualized as perpendicular with respect to the surface. A skewed angle (about 60°) was observed with regard to the line normal to the surface.

The adsorption configurations of PATP on PMNMs can be easily understood from our DFT calculations [33]. The simulated Raman spectra of PATP adsorbed on various silver clusters have been shown in **Figure 5B**. The frequencies of intense peaks observed at 379, 630, 1001, 1071, 1167, 1336, 1476, and 1596 cm^{-1} can be related with the peak frequencies at 379, 630, 1010, 1080, 1181, 1334, 1489, and 1588 cm^{-1}. Here, the latter peak frequency values represent the PATP adsorbed (in aqueous acidic solution) on a rough silver electrode [31]. The strongest Raman peaks appear at 1071 and 1596 cm^{-1}. These peaks symbolize the absolutely symmetric mixed vibrational modes of C—C and C—S stretching and the C—C bonds stretching parallel to C_2 axis when we assume PATP with a symmetric point group of C_{2v}. Here we assumed that the backbone of the adsorbed PATP has a local and approximate C_{2V} symmetry of point group, as shown in **Figure 4A**. The totally symmetric in-plane bending vibration of C—H bonds can be observed at 1181 as well as 1489 cm^{-1}. In the case of free as well as adsorption states, PATP has been anticipated to be in C_{2V} point group. The four vibrational modes (1125, 1286, 1322, and 1426 cm^{-1}) with low Raman intensity present in between 1100 and 1450 cm^{-1} possess b_2 symmetry, which correspond to the asymmetric stretching of C—C bonds and in-plane bending vibrations of C—H bonds. Our results along with the study of vibrational analysis of free and adsorbed PATP revealed the deficit vibrational fundamental frequencies at about 1390 cm^{-1}. This shows the significant dissimilarity between experimental as well as theoretical Raman spectra [30, 31]. In spite of the photon-driven charge transfer Herzberg-Teller vibronic coupling, a likely intense peak of SERS cannot be offered by the chemical enhancement mechanism of PATP. The spectra of PATP adsorbed on silver, gold, as well as copper were investigated in various configurations in order to comprehend the vibronic coupling enhancement in the SERS spectra of PATP. These conclusions lead to the initial hesitation for the strong SERS peaks only related to the earlier elucidated photo-driven charge transfer enhancement mechanism of adsorbed PATP.

In analysis of low-lying excited states, a photo-driven charge transfer reaction takes place from PATP toward the surface of metal. This has been also revealed from our density functional theory (DFT) studies [34]. Additionally, the energies of low-lying excited states have been evaluated by using a molecule-metal cluster modeling system. The charge transfer energies for PATP-to-silver clusters (~2.28 eV) and PATP-to-gold clusters (~2.08 eV) were estimated in the case of PATP-M_n clusters, where n = 13. The energies of transition from M_n clusters to PATP for charge transfer excited states were additionally examined to be higher than 3.0 eV [35]. This reveals that in common SERS measurements, the incident photonic energies are lower when compared to the charge transfer energies from metal toward PATP. Furthermore, it has been found that the interband transition energies of gold are lower than the energies of charge transfer and the interband transition energies of silver are nearer to the energies of charge transfer. Thus, under the irradiation of visible light, the photon-driven charge transfer should take place from PATP towards metal surfaces. This direction of charge transfer has been revealed from our early DFT calculations. When the wavelength of laser increases, the maximum potential in the potential-dependent SERS intensity profile should travel toward the positive direction. Our theoretical results were in concordance with the results of theoretical studies from other groups [36, 37]. In contrast, our results were incoherent with the earlier proposed SERS results of PATP adsorbed on surfaces of various metals [30]. This above deviation deliberates the additional uncertainty on the SERS signal appeared.

3.3 Surface catalytic coupling reactions on PMNMs

We observed that the mentioned charge transfer mechanism cannot be suitable for all the experimental studies. Here, two main key contradictions were observed as follows: (i) SERS peaks at 1140, 1390, and 1426 cm^{-1} appeared in Raman excitation wavelengths from 488 to 1064 nm for PATP-adsorbed silver and gold surfaces [32, 38]. The same SERS peaks were also acquired in the nanocavity between silver nanoparticle and smooth gold substrate that employs the wavelength of 1064 nm [32]. The UV-Visible absorption peak at 295 nm ($\pi \rightarrow \pi^*$ transition) of PATP in methanol solution resembles transition energy of about 4.20 eV, whereas, SERS peaks arising from photo-driven charge transfer are in the range of 1.16–2.54 eV (corresponding to the incident photonic energy), which would contradict the predicted charge transfer transition. Therefore, if the charge transfer enhancement mechanism was categorized as a resonance-like Raman scattering process, this is the inconsistent large energy gap between the intramolecular excited state and the charge transfer excited state [39–41]. (ii) Another important contradiction is the pH effect. In acidic solutions, the reversible behavior of intensity ratios of SERS peaks (from1440 and 1080 cm^{-1}) was described with respect to the potentials applied in the earlier studies. Some studies used isomerization to elucidate the reversibility nature with applied potentials. In the case of alkaline solutions, the elu-cidation of irreversible behavior is not possible. In addition, the correlation between the reversible and irreversible nature in both acidic as well as basic solutions was not simply explained on the basis of charge transfer mechanism.

Motivated by the experimental results, the probable surface species have been reassessed. We have proposed three different surface species for adsorption of PATP on the surface of PMNMs (**Figure 6A**). (i) PATP was oxidized to 4′-mercapto-4-aminodiphenylamine by increasing the potential anodically at PATP-adsorbed gold and platinum metal surfaces. Conversely, a quite different simulated Raman

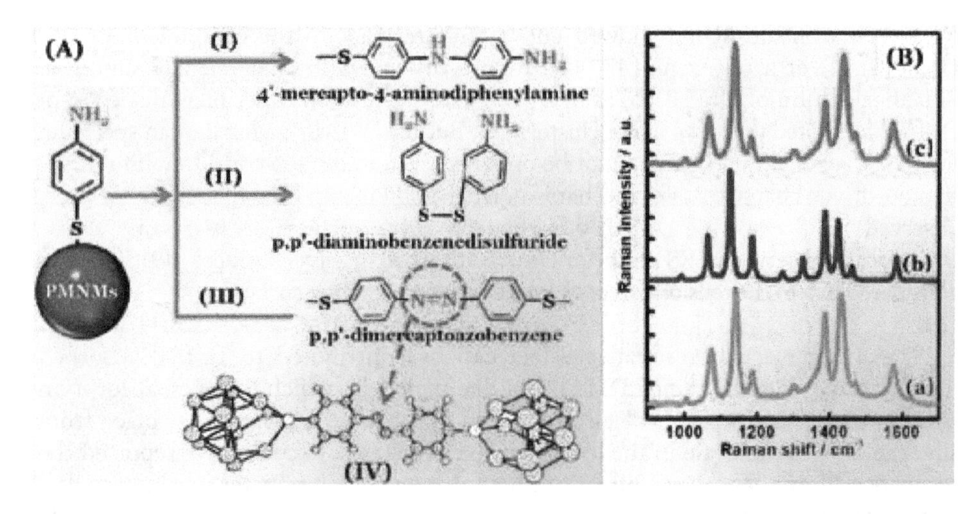

Figure 6.
*(A) The three possible reaction pathways of PATP adsorbed on nanostructured metal surfaces after the photochemical reactions (I), (II), and (III). A theoretical adsorption configuration of DMAB adsorbed on rough metal surfaces (IV). (B) SERS spectrum of synthesized DMAB (a), simulated Raman spectrum of DMAB on two silver clusters using DFT (PW91PW91/6-311 + G**(C, N, S, and H)/LANL2DZ(Ag) level (b)), and SERS spectrum from PATP adsorbed on silver nanoparticles measured with the excitation at 632.8 nm (c).*

spectrum was observed for the PATP adsorbed on silver electrodes [30]. (ii) The formation of p,p′-diaminobenzenedisulfuride was observed for PATP adsorbed on silver films, wherein the disulfide compound was believed to form from its corresponding azo compound [42]. However, the disulfide bond will be shattered because of the strong Ag-S bond on silver surfaces [43]. (iii) Chemical transforma-tion of PATP to DMAB (p,p′-dimercaptoazobenzene) was due to the surface cata-lytic coupling reaction on noble metal surfaces. We suggested that PATP adsorbed on noble metal surfaces can transform to DMAB under irradiation of visible laser based on our DFT and experimental results [33, 44].

DMAB complex was calculated on the basis of static polarizability derivatives. Moreover, DMAB complex and DMAB-Ag_n complexes were also obtained by a single-end configuration, but the observed Raman spectra were found to be very similar [33, 34]. In particular, the vibrations of azo (N═N)and benzene ring result in the strong Raman peaks. The theoretical and experimental results from Sun and Xu support the above interpretations [45, 46]. Considerable influence of Ag—S vibrational frequencies was also observed by the corresponding strong Raman peaks because of the localization interaction of sulfur and silver clusters. Furthermore, the active Raman modes (A_g and B_g) are the irreducible representations for symmet-ric center trans DMAB to $C_{2h.}$ The strong Raman peaks at 1130, 1390, and 1440 cm^{-1} thus mainly appeared from azo group and benzene ring symmetric vibrations of DMAB [(44)]. The peak at 1130 cm^{-1} resembles C—N symmetric stretching vibration and the peaks at 1390 and 1440 cm^{-1} show the mixed vibrations of the N═Nbond stretching and the C—H in-plane symmetric bending [33, 44]. By our DFT calculations, we concluded that proper functional is very important to predict the N═Nbond distance, since the positions of latter two strong peaks are crucial. Theoretical frequencies were in good agreement with the experimental frequencies when the PW91PW91 functional was combined with the triple-zeta Gaussian basis set 6-311 + G**, whereas, the theoretically and experimentally observed vibra-tional frequencies are pointedly overestimated when B3LYP functional was used. Two different N═N bond distances for DMAB were calculated to be 1.273 Å (by PW91PW91/6-311 + G**) and 1.256 Å (by B3LYP/6-311 + G**) [33]. Other reports by gas electron diffraction showed comparable result with the experimental value (1.260(8) Å) of azobenzene [47, 48]. **Figure 6B** shows the theoretically simulated Raman spectrum of DMAB (b) is in a good agreement with experimental spectra of DMAB interacted with two silver clusters (a). But for DMAB, such a Raman spectrum for PATP adsorbed on silver cannot be obtained even under the consid-eration of the photon-driven charge transfer mechanism. Duan and Luo reported that experimentally observed SERS peaks of DMAB adsorbed on silver surfaces are the same as the theoretically observed SERS peaks for DMAB adsorbed on silver surfaces [49]. This was considered due to the tension effect of the adsorption on silver surfaces.

The intramolecular resonance effect can be contributed to DMAB adsorbed metallic nanostructures since DMAB is a dye molecule, which owns its absorption band in the visible region. We also observed that charge transfer takes place from silver to DMAB molecule in the low-lying excited states. Recently, we reported the resonance-like enhancement effect of photo driven charge transfer mechanism and intramolecular electronic transition [33, 34]. As the molecular orbitals are mainly distributed in the DMAB azo group(>C—N=N—C<), azo Raman peaks at 1130, 1390, and 1440 cm^{-1} strongly match with the photonic energies to the intramolecular resonance energy, relative to other Raman peaks at 1078 and 1596 cm^{-1}. Because of surface plasmon resonance and intramolecular resonance effects, SERS peak of DMAB appeared very easily after the oxidation of PATP through the surface catalytic coupling reaction. Based on the deep studies, the SERS spectrum of DMAB follows the surface-enhanced resonance Raman spectroscopy in the visible region. This study also revealed that the enormous enhancement effect can be observed in the SERS spectrum of PATP adsorbed on silver or gold nanostructures. The key factors such as acidity and applied potentials influence the stability of DMAB at electrochemical interfaces [50]. Therefore, DMAB can be converted into PATP in acidic solution, due to the reversibility of applied potentials in acidic solution [51]. Even though, DMAB is more stable in the alkaline solution, it can be reduced into PATP at more cathodic or negative potentials.

The charge transfer mechanism was also engaged to study the Raman spectra of DMAB at silver/gold surfaces and PATP on silver or gold substrates. In the case of DMAB, the photon-driven charge transfer is from metal (silver or gold) to DMAB, whereas, photo-driven charge transfer occurs from PATP to metals under visible light for PATP adsorbed on silver or gold substrates. Therefore, charge transfer directions are opposite for low-lying excited states of PATP and DMAB. Moreover, the reversibility or irreversibility of DMAB is strongly dependent on acidity of aqueous solutions.

Kim and coworkers studied the abnormal SERS peaks arising from the view of CT mechanism in various conditions such as pH, rotation, temperature, and reducing agent as follows [52]. They observed SERS peaks at 1130, 1390, and 1430 cm^{-1} in an acidic solution with pH = 3. SERS signals of the—NO_2 symmetric stretching gradually disappeared when PNTP was adsorbed on a rotation platform with 3000 circles per minute modified with silver nanoparticles. They visualized the strong peaks at 1130, 1390, and 1430 cm^{-1} after 30 min. It was believed that the strong peaks appeared only from the PATP, and not from PNTP. They concluded that no photochemical reaction occurred for PATP at the boundary of ice and silver nanoparticles at liquid nitrogen temperature (77 K) on the basis of nonexistence of reaction for PNTP at the same boundary [52]. Even when strong reductants such as $NaBH_4$ exist, they still observed abnormally strong SERS peaks. It may be assumed that such a significant SERS band should be observed only in the SERS spectra of PATP and DMAB adsorbed on noble metals.

The reaction mechanisms and their dynamics of PATP and PNTP are very different. The activation energy of the rate determination steps for PATP and PNTP on silver surfaces was calculated to be 5 and 12 kcal/mol, respectively. It indicates that the PATP oxidation reaction rate is quite fast so that an early photochemical reaction cannot be identified. The lack of thermodynamic properties and kinetic informa-tion for these reactions are observed in certain studies, which deal with the reaction mechanisms. Our opinions regarding the above issue as well as the anticipated reaction mechanisms will be discussed in the next section.

4. Aerobic oxidation-assisted aromatic amine on PMNMs photocatalysts

The mechanism for the electrochemical reduction of aromatic nitro compounds into aromatic amines at electrode interface has been reported by Haber in 1898 [53]. The catalytic oxidation of aniline to azobenzene on TiO_2-reinforced gold nanopar-ticles (in the presence of oxygen atmosphere) was proposed by Grirrane et al. [54]. After that, several oxidation and reduction reactions of aminobenzene and nitrobenzene have been reported to form azobenzene. However, limited number of mechanisms (which strongly depend on experimental conditions) have been investigated as silver/gold nanomaterials based on photochemical and photoelec-trochemical processes under visible light [55]. The respective absorption bands for PATP and PNTP occur around 300 and 326 nm revealing the absence of absorption peaks of the above two molecules in visible light region [30]. Interestingly, many studies reported that the surface reactions of PATP and PNTP chemically trans-formed into DMAB in the presence of visible light regions using SERS. Therefore, the adsorbed and photochemical reacted PATP and PNTP molecules on PMNMs needed for the new pathways under SERS investigation.

The activation of O_2 is found to be one of the most significant steps in the case of catalytic oxidation reactions. The activation of O_2 into two weakly-bonding oxygen atoms has been achieved by employing silver anions and gold clusters [56]. The binding nature of negative metal clusters with oxygen is stronger, compared to neutral metal clusters. The surface plasmon-mediated injection of an excited electron from the optically excited silver nanoparticles toward adsorbate can activate O_2 into two weakly-bonding oxygen atoms as reported by Christopher et al. [17, 57]. Oxygen adsorption on the surface of plasmonic nanostructures may involve direct as well as indirect charge transfer reactions. On the other hand, no clear evidences are available to describe that either the direct or indirect charge transfer reactions played the significant part [58]. The study of activation mechanism of oxygen is the chal-lenging part, because of the catalytic oxidation reaction taking place on the surface of plasmonic metal nanoparticles, which involves the direct participation of O_2 in metal-oxygen battery systems.

The renowned surface plasmon resonance has been displayed in **Figure 7A**. After the optical excitation of surface plasmon, there are two probable channels of relaxation. The process of scattering (radiative decay into photons) or absorp-tion (non-radiative decay into electron-hole pairs) may result in surface plasmon damping (**Figure 7B**). When the volume of nanoparticle increases, the ratio of scattering to absorption gets increased on the basis of discrete dipole approximation (DDA) simulation and Mie theory calculation [59]. Aimed at the nanoparticles in very small size, the generation of electron-hole pair acts as the foremost pathway to decay. In order to demonstrate the dynamics of excited electrons and holes as well as energy levels, several theoretical models were developed. Based on first-principle TD-DFT calculations [60], we have recently studied the photoinduced excitation of small metal clusters. The evaluated TD-DFT results are in concordance with the dis-tribution results of plasmonic carriers in gold and silver nanoparticles, which were acquired from the quantum equation of motion for the density matrix (18). Because

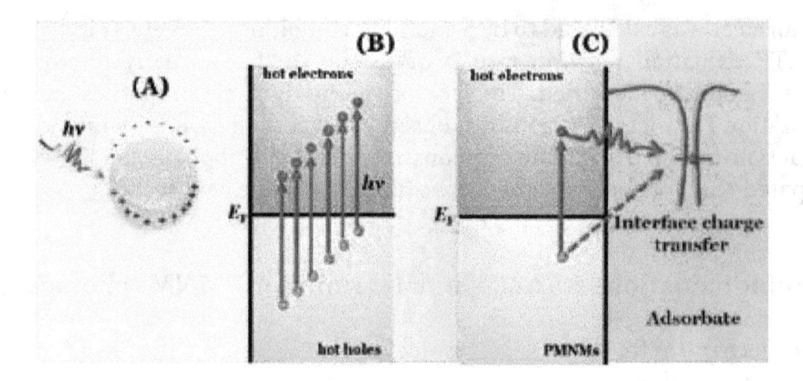

Figure 7.
(A) Schematic representation of coherent optical excitation of SPR on noble metal surfaces. (B) Hot electron-hole pairs created by non-radiative decay of SPR. (C) Surface plasmon-induced hot electron injection to adsorbed oxygen molecules.

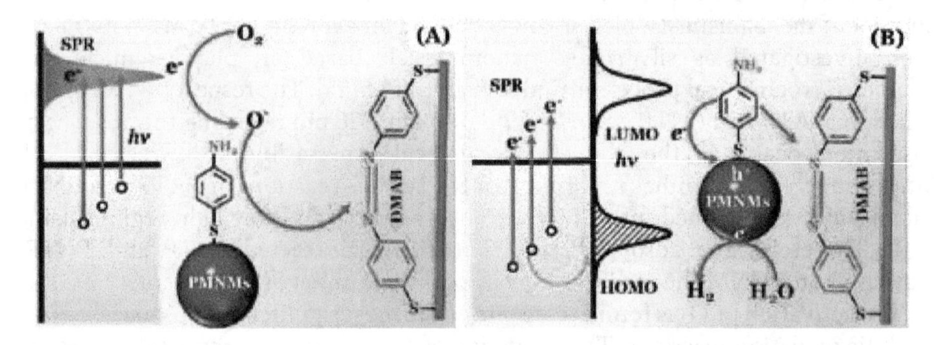

Figure 8.
(A) Schematic representation of surface oxidation coupling reaction for PATP-adsorbed PMNMs by surface oxygen species activated by hot electrons under the excitation by visible light. (B) The hot-hole oxidization mechanism for PATP adsorbed PMNMs in the solid/liquid interfaces.

of their higher energy, hot electrons will extend further away from the nanoparticle surface than an equilibrium electron distributed at the Fermi level. As shown in **Figure 7C,** the electron acceptor (oxygen adsorbed on a nanoparticle surface) located nearer can accept the hot electron tunneling to its unoccupied orbitals. This effective reaction is equivalent to the process occurring in the metal-semiconductor composites [20].

The broader absorption bands observed in the region of visible light are due to the influence of sturdy surface plasmon resonance. The generation of hot electron-hole pairs by the process of irradiation relaxation depends on the lifetime (10–20 fs) of localized SPR. The oxygen molecules can be reduced into active oxygen species (under oxygen atmosphere) by the hot electrons (reducing nature) produced at the surface of metal. Thus, adsorbed PATP on silver nanoparticles can be oxidized by the active oxygen atoms (**Figure 8A**). Here, the oxidation of amine group results in the formation of respective imine/free radical. Later, hydroazobenzene was formed due to dimerization. Further oxidation leads to the formation of adsorbed DMAB on the surface of silver nanoparticles [60].

In the absence of active oxygen species, the hot-hole oxidation mechanism for the conversion of PATP into DMAB has been displayed in **Figure 8B**. The presence of interfacial defects/small clusters lengthens the lifetimes of hot electrons or holes on the surface of PMNMs. The energy of the hot hole at interfaces and the energy position of the occupied orbital were alike for the adsorption of PATP on the surface of metals [25, 60]. Thus, PATP can form a cation free radical upon oxidation. The intermediate formed can give out a proton to neutral imine free radical in alkaline/neutral solution. This can be further dimerized and oxidized for the formation of DMAB, parallel to the oxidization reaction based on surface-active oxygen species [61]. Because of the SPR effect, a larger value of rate constant has been exposed by the hot-hole oxidization mechanism unlike the direct charge transfer process.

The experimental environments decide the hot-hole oxidation mechanism for the adsorption of PATP on the surfaces of noble PMNMs. Various investigations have been reported for the generation of electron-hole pairs, which prompts the chemical reactions on the surface of semiconductors. Once the electron-hole pairs got separated, the electrons (present in conduction band) involve in the process of reduction and the holes (present in valence band) are responsible for initiating oxidation. An apparent decrease in the lifetimes of hot electrons and holes was observed as the persistent distribution of energy band seen around the Fermi level on the PMNMs [62]. On the surface of noble PMNMs, the lifetime of hole is frequently found to be in femtoseconds [63]. The lifetime of excitation electrons (associated with its excitation energy) in the excited sp-band is lower than the lifetime of d-band holes [64]. The involvement of hole in the chemical processes has been deliberated in limited reports. The relationship of excitation wavelengths with the pH values has been presented by our results. The conversion of PATP into DMAB based on the oxidation coupling reaction is a multistep reaction comprising of electron/hole transfer as well as proton transfer steps. The values of applied potential, pH of solution, and incident photonic energy decide the chemical potentials of hot electrons, holes as well as protons in the electrochemical SERS measurements [60]. As the incident photonic energy increases, an increase in energies of hot electrons and holes was observed. In contrast, a decrease in their lifetimes with regard to the Fermi level was observed. When the pH value of the solution increased, a decrease in the chemical potential of protons present in the interfacial solution phase was observed. The increase in the energy of hot-hole pairs as well as the value of pH supports the oxidation of PATP into DMAB with loss of two electrons and two protons for each surface reactant. This is complicated but interesting.

5. Summary and prospects

As the molecules from liquid/gas phase proceed toward the surface of metal, considerable variations in the optical physical and chemical characteristics are observed. SERS as well as surface photocatalytic chemical reactions explain the above characteristic variations on PMNMs. Herein, the oxidation and reduction reactions of PATP and PNTP on the surface of PMNMs were described. In addition, the reaction mechanisms based on these aromatic compounds were investigated in the region of visible light, where the surface plasmon resonance appears on nanostructures of noble metals. The photo-driven charge transfer and the interfacial chemical reactions take place because of the hot electrons and holes generated from SPR relaxation. The hot electrons actuate the surface-activated oxygen species, which result in the surface catalytic coupling reaction in gas/solid interfaces. Relative to the activation barrier of gold nanostructures, the activation barrier of silver nanostructures is smaller. It is evident from the DFT calculations in simulated SERS and chemical reaction pathways. These calculations and SERS measurements describe the reason for the occurrence of highly selective and rapid oxidation coupling reaction of PATP Relative to photooxidation of PATP on PMNMs, the photoreduction of PNTP is apparently slower. A steady decrease in the SERS signal of the symmetric stretching vibration (nitro group) was observed in the photoreduction reaction. The emphatic evidences for the photooxidation of adsorbed PATP on metal nanomaterials is inadequate compared to the evidences for the conversion of PNTP to DMAB. The main reason is the former reaction is much faster than the latter one. The primary outcomes for the photoreduction and photooxidation processes have been given through theoretical as well as experimental investigations. In addition, the effect of pH, power of laser, incident wavelengths, and oxygen assisted on photoreduction/photooxidation were tried to be made clear.

The study of the effect of enhancement mechanism resulting from surface plasmon resonance on the thermodynamic and dynamics of chemical reactions of metals in nanoscale dimensions is an interesting area. The SPR can enhance the local surface optical electric field on the surface of PMNMs, which in turn increases the charge transfer probability between metal and molecules. The chemical reac-tions on the surface were actuated by the hot electrons and holes generated from the SPR relaxation. The processes result in surface transient active species inclusive of anions with negative charge and free radicals carrying positive charge. However, there is lack of evidence available for the initial reaction steps taking place on the surface of PMNMs.

The major concern for hot electrons can be observed only in the photochemical reactions on the surface of PMNMs. A contemporary consideration of hot electrons as well as the hot holes is essential for a complete system of chemical reaction. For the transformation of light energy in-to chemical energy, a stable and reliable system can be created by PMNMs. The hot holes were attracted to the surface of PMNMs after the completion of PATP photooxidation. But, the retention of hot electrons reduces the surface oxidation species as well as hydrated protons and water molecules into hydrogen gas. The hot electrons present on the surface of PMNMs moreover result in hydrated electrons, which reduces the oxygen species in solution phase. Here, the oxidation reactions are actuated by the hoarded hot holes on the surface of PMNMs. In addition, water molecules in aqueous solution can be oxidized into oxygen gas in the presence of hot holes. On the basis of SPR effect, these photochemical surface reactions actuate chemical processes for the transformation of light energy to chemical energy. The above mentioned surface plasmon-mediated chemical reactions, can find promising applications in the enhancement of light energy efficiency in photocatalytic fuel cells and dye-sensitized and new perovskite solar cells.

Acknowledgements

This work was financially supported by the National Science Foundation of China (21533006 and 21373172).

Author details

Rajkumar Devasenathipathy, De-Yin Wu* and Zhong-Qun Tian
State Key Laboratory of Physical Chemistry of Solid Surfaces, Department of Chemistry, College of Chemistry and Chemical Engineering, Xiamen University, Xiamen, China

*Address all correspondence to: dywu@xmu.edu.cn

References

[1] Schuller JA, Barnard ES, Cai W, Jun YC, White JS, Brongersma ML. Plasmonics for extreme light concentration and manipulation. Nature Materials. 2010; **9**(3):193

[2] Hartland GV. Optical studies of dynamics in noble metal nanostructures. Chemical Reviews. 2011;**111**(6): 3858-3887

[3] Zhan C, Wang Z-Y, Zhang X-G, Chen X-J, Huang Y-F, Hu S, et al. Interfacial construction of plasmonic nanostructures for the utilization of the plasmon-excited electrons and holes. Journal of the American Chemical Society. 2019;**141**(20):8053-8057

[4] Zhao L-B, Liu X-X, Zhang M, Liu Z-F, Wu D-Y, Tian Z-Q. Surface plasmon catalytic aerobic oxidation of aromatic amines in metal/molecule/ metal junctions. The Journal of Physical Chemistry C. 2016;**120**(2):94 4-955

[5] Yu H, Peng Y, Yang Y, Li Z-Y. Plasmon-enhanced light–matter interactions and applications. npj Computational Materials. 2019;**5**(1):45

[6] Fleischmann M, Hendra PJ, McQuillan AJ. Raman spectra of pyridine adsorbed at a silver electrode. Chemical Physics Letters.1974;**26**(2):163-166

[7] Liu Y, Yang D, Zhao Y, Yang Y, Wu S, Wang J, et al. Solvent-controlled plasmon-assisted surface catalysis reaction of 4-aminothiophenol dimerizing to p, p'-dimercaptoazobenzene on Ag nanoparticles. Heliyon. 2019;**5**(4):e01545

[8] Golubev AA, Khlebtsov BN, Rodriguez RD, Chen Y, Zahn DR. Plasmonic heating plays a dominant role in the plasmon-induced photocatalytic reduction of 4-nitrobenzenethiol. The Journal of Physical Chemistry C. 2018;**122**(10):5657-5663

[9] Tsai M-H, Lin Y-K, Luo S-C. Electrochemical SERS for in situ monitoring the redox states of PEDOT and its potential application in oxidant detection. ACS Applied Materials and Interfaces. 2018;**11**(1):1402-1410

[10] Wu D-Y, Li J-F, Ren B, Tian Z-Q. Electrochemical surface-enhanced Raman spectroscopy of nanostructures. Chemical Society Reviews. 2008; **37** (5):1025-1041

[11] Zhang Q , Wang H. Mechanistic insights on plasmon-driven photocatalytic oxidative coupling of thiophenol derivatives: Evidence for steady-state photoactivated oxygen. The Journal of Physical Chemistry C. 2018;**122**(10):5686-5697

[12] Kim M, Lin M, Son J, Xu H, Nam JM. Hot-electron-mediated photochemical reactions: Principles, recent advances, and challenges. Advanced Optical Materials. 2017;**5**(15):1700004

[13] Kreibig U, Vollmer M. Optical Properties of Metal Clusters. Berlin: Springer; 1995

[14] Watanabe K, Menzel D, Nilius N, Freund H-J. Photochemistry on metal nanoparticles. Chemical Reviews. 2006;**106**(10):4301-4320

[15] Moskovits M. Surface-enhanced spectroscopy. Reviews of Modern Physics. 1985;**57**(3):783

[16] Alvarez-Puebla R, Liz-Marzán LM, García de Abajo FJ. Light concentration at the nanometer scale. The Journal of Physical Chemistry Letters. 2010; **1**(16):2428-2434

[17] Christopher P, Xin H, Linic S. Visible-light-enhanced catalytic oxidation reactions on plasmonic silver nanostructures. Nature Chemistry. 2011;**3**(6):467

[18] Govorov AO, Zhang H, Demir HV, Gun'ko YK. Photogeneration of hot plasmonic electrons with metal nanocrystals: Quantum description and potential applications. Nano Today. 2014;**9**(1):85-101

[19] Manjavacas A, Liu JG, Kulkarni V, Nordlander P. Plasmon-induced hot carriers in metallic nanoparticles. ACS Nano. 2014;**8**(8):7630-7638

[20] Govorov AO, Zhang H, Gun'ko YK. Theory of photoinjection of hot plasmonic carriers from metal nanostructures into semiconductors and surface molecules. The Journal of Physical Chemistry C. 2013;**117**(32): 16616-16631

[21] Gao P, Weaver M, editors. Surface-enhanced Raman-spectroscopy as a vibrational probe of electrochemical reaction-mechanisms-the electroreduction of nitrobenzene. Journal of the Electrochemical Society. 1987;**134**(3): c132

[22] Bello JM, Narayanan VA, Vo-Dinh T. Surface-enhanced Raman scattering interaction of p-aminobenzoic acid on a silver-coated alumina substrate. Spectrochimica Acta Part A: Molecular Spectroscopy. 1992;**48**(4):563-567

[23] You T, Jiang L, Yin P, Shang Y, Zhang D, Guo L, et al. Direct observation of p, p′-dimercaptoazobenzene produced from p-aminothiophenol and p-nitrothiophenol on Cu_2O nanoparticles by surface-enhanced Raman spectroscopy. Journal of Raman Spectroscopy. 2014;**45**(1):7-14

[24] Baia M, Toderas F, Baia L, Popp J, Astilean S. Probing the enhancement mechanisms of SERS with p-aminothiophenol molecules adsorbed on self-assembled gold colloidal nanoparticles. Chemical Physics Letters. 2006;**422**(1-3):127-132

[25] Zhao L-B, Huang Y-F, Liu X-M, Anema JR, Wu D-Y, Ren B, et al. A DFT study on photoinduced surface catalytic coupling reactions on nanostructured silver: Selective formation of azobenzene derivatives from Para-substituted nitrobenzene and aniline. Physical Chemistry Chemical Physics. 2012;**14**(37):12919-12929

[26] Ling Y, Xie WC, Liu GK, Yan RW, Tang J. The discovery of the hydrogen bond from p-Nitrothiophenol by Raman spectroscopy: Guideline for the thioalcohol molecule recognition tool. Scientific Reports. 2016;**6**:31981

[27] Osawa M, Ikeda M. Surface-enhanced infrared absorption of p-nitrobenzoic acid deposited on silver island films: Contributions of electromagnetic and chemical mechanisms. The Journal of Physical Chemistry. 1991; **95**(24):9914-9919

[28] Sun S, Birke RL, Lombardi JR, Leung KP, Genack AZ. Photolysis of p-nitrobenzoic acid on roughened silver surfaces. The Journal of Physical Chemistry. 1988;**92**(21):5965-5972

[29] Roth P, Venkatachalam R, Boerio F. Surface-enhanced Raman scattering from p-nitrobenzoic acid. The Journal of Chemical Physics. 1986;**85**(2):1150-1155

[30] Osawa M, Matsuda N, Yoshii K, Uchida I. Charge transfer resonance Raman process in surface-enhanced Raman scattering from p-aminothiophenol adsorbed on silver: Herzberg-Teller contribution. The Journal of Physical Chemistry. 1994;**98**(48): 12702-12707

[31] Hill W, Wehling B. Potential-and pH-dependent surface-enhanced Raman scattering of p-mercapto aniline on silver and gold substrates. The Journal of Physical Chemistry. 1993;**97**(37):94 51-9455

[32] Zhou Q, Li XW, Fan Q, Zhang XX, Zheng JW. Charge transfer between

metal nanoparticles interconnected with a functionalized molecule probed by surface-enhanced Raman spectroscopy. Angewandte Chemie, International Edition. 2006;**45**(24):3970-3973

[33] Wu DY, Liu XM, Huang YF, Ren B, Xu X, Tian ZQ. Surface catalytic coupling reaction of p-Mercaptoaniline linking to silver nanostructures responsible for abnormal SERS enhancement: A DFT study. The Journal of Physical Chemistry C. 2009;**113**(42):18212-18222

[34] Wu D-Y, Zhao L-B, Liu X-M, Huang R, Huang Y-F, Ren B, et al. Photon-driven charge transfer and photocatalysis of p-aminothiophenol in metal nanogaps: A DFT study of SERS. Chemical Communications. 2011;**47**(9):2520-2522

[35] Zhao L-B, Huang R, Huang Y-F, Wu D-Y, Ren B, Tian Z-Q. Photon-driven charge transfer and Herzberg-Teller vibronic coupling mechanism in surface-enhanced Raman scattering of p-aminothiophenol adsorbed on coinage metal surfaces: A density functional theory study. The Journal of Chemical Physics. 2011;**135**(13):134707

[36] Gibson JW, Johnson BR. Density-matrix calculation of surface-enhanced Raman scattering for p-mercaptoaniline on silver nanoshells. The Journal of Chemical Physics. 2006;**124**(6):064701

[37] Sun M, Xu H. Direct visualization of the chemical mechanism in SERRS of 4-Aminothiophenol/metal complexes and metal/4-Aminothiophenol/metal junctions. ChemPhysChem. 2009;**10**(2):392-399

[38] Osawa M, Matsuda N, Yoshii K, Uchida I. Charge-transfer resonance process in surface-enhanced Raman-scattering from p-aminothiophenol adsorbed on silver - Herzberg-Teller contribution. The Journal of Physical Chemistry. 1994;**98**(48):12702-12707

[39] Lombardi JR, Birke RL, Lu T, Xu J. Charge-transfer theory of surface enhanced Raman spectroscopy: Herzberg-Teller contributions. The Journal of Chemical Physics. 1986;**84**: 4174-4180

[40] Albrecht AC. On the theory of Raman intensities. The Journal of Chemical Physics. 1961;**34**:1476-1484

[41] Kambhampati P, Child CM, Foster MC, Campion A. On the chemical mechanism of surface enhanced Raman scattering: Experiment and theory. The Journal of Chemical Physics. 1998;**108**(12):5013-5026

[42] Lu Y, Xue G. Study of surface catalytic photochemical reaction by using conventional and Fourier transform surface enhanced Raman scattering. Applied Surface Science. 1998;**125**(2):157-162

[43] Patrito EM, Cometto FP, Paredes-Olivera P. Quantum mechanical investigation of thiourea adsorption on Ag(111) considering electric field and solvent effects. The Journal of Physical Chemistry. B. 2004;**108**:15755-15769

[44] Huang Y-F, Zhu H-P, Liu G-K, Wu D-Y, Ren B, Tian Z-Q. When the signal is not from the original molecule to be detected: Chemical transformation of Para-aminothiophenol on ag during the SERS measurement. Journal of the American Chemical Society. 2010; **132**(27):9244-9246

[45] Fang Y, Li Y, Xu H, Sun M. Ascertaining p,p′-dimercaptoazobenzene produced from p-aminothiophenol by selective catalytic coupling reaction on silver nanoparticles. Langmuir. 2010;**26**(11):7737-7746

[46] Huang Y, Fang Y, Yang Z, Sun M. Can p,p′-Dimercaptoazobisbenzene Be produced from p-aminothiophenol by surface photochemistry reaction in the junctions of a Ag

nanoparticle-molecule-Ag (or Au) film? The Journal of Physical Chemistry C. 2010;**114**(42):18263-18269

[47] Tsuji T, Takashima H, Takeuchi H, Egawa T, Konaka S. Molecular structure and torsional potential of trans-azobenzene. A gas electron diffraction study. Journal of Physical Chemistry A. 2001;**105**(41):9347-9353

[48] Briquet L, Vercauteren DP, Perpete EA, Jacquemin D. Is solvated trans-azobenzene twisted or planar? Chemical Physics Letters. 2006;**417** (1-3):190-195

[49] Duan S, Ai Y-J, Hu W, Luo Y. Roles of Plasmonic excitation and protonation on photoreactions of p-Aminobenzenethiol on Ag nanoparticles. The Journal of Physical Chemistry C. 2014;**118**(13):6893-6902

[50] Hill W, Wehling B. Potential-dependent and pH-dependent surface-enhanced Raman-scattering of p-mercaptoaniline on silver and gold substrates. The Journal of Physical Chemistry. 1993;**97**(37):9451-9455

[51] Kim K, Kim KL, Shin KS. Photoreduction of 4,4'- Dimercaptoazobenzene on Ag revealed by Raman scattering spectroscopy. Langmuir. 2013;**29**(1):183-190

[52] Kim K, Choi J-Y, Shin KS. Surface-enhanced Raman scattering of 4-Nitrobenzenethiol and 4-Aminobenzenethiol on silver in icy environments at liquid nitrogen temperature. The Journal of Physical Chemistry C. 2014;**118** (21):11397-11403

[53] Lund H. Cathodic reduction of nitro and related compounds. Organic Electrochemistry. 2001;**4**:379-409

[54] Grirrane A, Corma A, García H. Gold-catalyzed synthesis of aromatic azo compounds from anilines and nitroaromatics. Science. 2008;**322** (5908):1661-1664

[55] Zhu H, Ke X, Yang X, Sarina S, Liu H. Reduction of nitroaromatic compounds on supported gold nanoparticles by visible and ultraviolet light. Angewandte Chemie, International Edition. 2010;**49**(50):9657-9661

[56] Huang W, Zhai H-J, Wang L-S. Probing the interactions of O_2 with small gold cluster anions (Au_n^-, n=1– 7): Chemisorption vs physisorption. Journal of the American Chemical Society. 2010;**132**(12):4344-4351

[57] Christopher P, Xin H, Marimuthu A, Linic S. Singular characteristics and unique chemical bond activation mechanisms of photocatalytic reactions on plasmonic nanostructures. Nature Materials. 2012;**11**(12):1044

[58] Linic S, Aslam U, Boerigter C, Morabito M. Photochemical transformations on plasmonic metal nanoparticles. Nature Materials. 2015;**14**(6):567

[59] Tcherniak A, Ha J, Dominguez-Medina S, Slaughter L, Link S. Probing a century old prediction one plasmonic particle at a time. Nano Letters. 2010;**10**(4):1398-1404

[60] Zhao L-B, Zhang M, Huang Y-F, Williams CT, Wu D-Y, Ren B, et al. Theoretical study of plasmon-enhanced surface catalytic coupling reactions of aromatic amines and nitro compounds. The Journal of Physical Chemistry Letters. 2014;**5**(7):1259-1266

[61] Huang Y-F, Zhang M, Zhao L-B, Feng J-M, Wu D-Y, Ren B, et al. Activation of oxygen on gold and silver nanoparticles assisted by surface plasmon resonances. Angewandte Chemie, International Edition. 2014;**53**(9):2353-2357

[62] Goldmann A, Matzdorf R, Theilmann F. Experimental hot-electron

and photohole lifetimes at metal surfaces—What do we know? Surface Science. 1998;**414**(1-2):L932-L9L7

[63] Chulkov E, Borisov A, Gauyacq J, Sánchez-Portal D, Silkin V, Zhukov V, et al. Electronic excitations in metals and at metal surfaces. Chemical Reviews. 2006;**106**(10):4160-4206

[64] Knoesel E, Hotzel A, Wolf M. Ultrafast dynamics of hot electrons and holes in copper: Excitation, energy relaxation, and transport effects. Physical Review B. 1998;**57**(20):12812

8

Plasmonic Nanoantenna Array Design

Tao Dong, Yue Xu and Jingwen He

Abstract

Recently, wireless optical communication system is developing toward the chip level. Optical nanoantenna array in optical communication system is the key component for radiating and receiving light. In this chapter, we propose a sub-wavelength plasmonic nanoantenna with high gain operating at the standard optical communication wavelength of 1550 nm. The designed plasmonic antenna has a good matching with the silicon waveguide in a wide band, and light is fed from the bottom of the nanoantenna via the silicon waveguide. Furthermore, we design two kinds of antenna arrays with the proposed plasmonic nanoantenna, including one- and two-dimensional arrays (1 × 8 and 8 × 8). The radiation characteristics of the antenna arrays are investigated and both arrays have high gains and wide beam steering range without grating lobes.

Keywords: plasmonic nanoantenna, localized surface plasmon, integrated optical antenna arrays, integrated photonic devices, radiation characteristics

1. Introduction

In recent years, silicon-based integrated photonic devices have been developing rapidly. In particular, integrated optical antenna arrays have broad application prospect in many fields, such as optical communication, light detection and ranging (LiDAR), vehicle autonomous driving, security monitoring, and display advertising [1–5]. Nanoantenna is a key part of the optical antenna array for converting guided light and free space light with specific directivity. Based on the light interference principle, beam steering is realized by controlling the phase of the light radiated by each nanoantenna in the optical antenna array. In order to realize optical phase control, the concept of optical phased array (OPA) is proposed [6–8]. In the field of silicon-based photonics, OPA is a highly integrated on-chip system, which consists of light division network, phase shifters, and optical antenna array [9–11]. In optical communication, OPA is required to have high gain, narrow beam, and wide steering range. However, at present, monolithic integrated OPAs suffer from low gain, small beam steering range, and wide beam width, which are mainly due to the low radiation efficiency of the nanoantenna, the large element spacing (the spacing between adjacent antenna in the antenna array), and the limited scale of the optical antenna array [12, 13].

The most commonly used silicon-based nanoantenna in the integrated opti-cal antenna array is dielectric grating antenna. Generally, dielectric grating antenna refers to the periodic micro/nanostructure etched on dielectric substrate.

The existing dielectric grating antenna suffers from large footprint and bidirectional radiation, which result in large element spacing and waste of radiation energy of the optical nanoantenna array [14–18]. In a uniform antenna array, the element spacing larger than the operating wavelength will lead to the appearance of grating lobes in the radiation pattern of the antenna array, which will limit the steering range of the optical nanoantenna array.

In order to obtain a miniaturized optical antenna with high radiation efficiency, plasmonic nanoantenna is proposed [19–22]. Plasmonic nanoantenna is composed of metal and dielectric. When electromagnetic wave impinges on the interface of metal and dielectric, it will couple and oscillate with the surface electrons of the metal, and surface plasmon polarization (SPP) is generated. When SPP is unable to transmit along the interface and is confined, the SPP is called localized surface plasmon (LSP). LSP can confine the electromagnetic wave into a space far less than a wavelength. Based on the LSP resonance effect, electromagnetic wave will be enhanced and radiated into free space by plasmonic antenna. Taking advantage of this character, plasmonic nanoantennas can realize a tiny size [22]. However, the traditional plasmonic nanoantennas [20, 22] are fed by plasmonic waveguides, in which the impedance matching band is narrow and high loss is introduced. In addition, the traditional plasmonic nanoantenna does not radiate light along the direction perpendicular to the plane where the antenna is located, which also limits the light steering range of the optical nanoantenna array [1, 2, 20, 22].

In this chapter, we propose a plasmonic nanoantenna with sub-wavelength footprint and high gain operating at the standard optical communication wavelength of 1550 nm, i.e., 193.5 THz [23]. The proposed plasmonic nanoantenna consists of a silver block and a silicon block, and its footprint is much smaller than that of dielectric grating antenna. Unlike recent studies on the plasmonic nanoantennas, an impedance matching between the proposed plasmonic nanoantenna and a silicon waveguide is achieved in a wide band. Light is fed from the bottom of the plasmonic nanoantennas by the silicon waveguide and is radiated vertically upward without bidirectional radiation. This kind of bottom fed plasmonic nanoantenna is suit-able for the expansion of the nanoantenna array. Based on the proposed plasmonic nanoantenna, two plasmonic nanoantenna arrays including 1×8 and 8×8 arrays are designed. The radiation characteristics of the plasmonic nanoantenna arrays are simulated and discussed in detail.

2. Plasmonic nanoantenna

2.1 Radiation characteristics of the designed plasmonic nanoantenna

Figure 1 illustrates the schematic diagram of the proposed plasmonic nanoantenna. The plasmonic nanoantenna is composed of silicon (Si) block, silver (Ag) block, and a silicon waveguide with a silicon dioxide (SiO_2) coating. As shown in **Figure 1**, a silicon waveguide through the silver block is connected with the silicon block for feeding light into the plasmonic nanoantenna. The lengths of the silicon waveguide along the x- and y-axis are 450 nm and 220 nm respectively, which are suitable for the light propagation with high efficiency. In each block, the length of the edge along the$x$- and y-axis is defined as width and length. The width, length, and height of the silicon block are represented by W_1, L_1, and H_{Si}, respectively.
Similarly, the parameters W_2, L_2, and H_{Ag} represent the width, length, and height of the silver block, respectively. The geometric parameters of the nanoantenna are as follows: W_1 = 625 nm, L_1 = 850 nm, H_{Si} = 300 nm, W_2 = 1100 nm, L_2 = 1100 nm, and H_{Ag} = 200 nm.

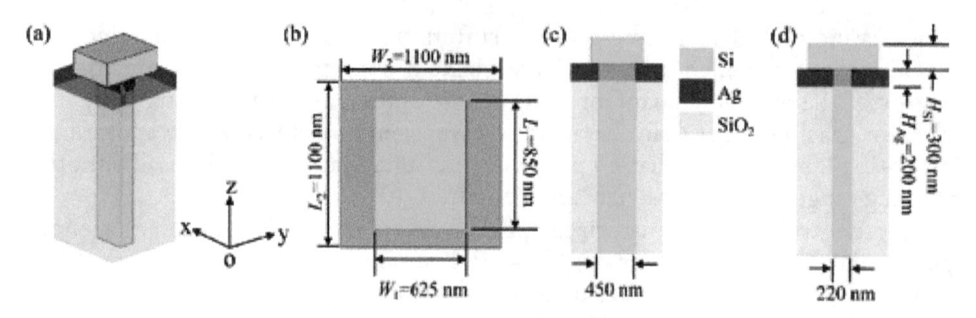

Figure 1.
The plasmonic nanoantenna in the (a) perspective, (b) overhead, (c) front, and (d) side views.

Some electromagnetic simulations for investigating the radiation characteristics of the plasmonic nanoantenna are performed with the commercial software of CST-Microwave Studio. For simulation, a beam of light with TE polarization (x-polarization) is fed into the plasmonic nanoantenna via the silicon waveguide. At the frequency of 193.5 THz, the relative dielectric constants of the materials of silicon and silicon dioxide are 12.11 and 2.1, respectively. The relative dielectric constant of silver is −129 + j3.28 at 193.5 THz [24]. Due to the small imaginary part and the large real part, silver has a very low ohmic loss that makes it an ideal metal to generate LSP.

Figure 2 shows the simulated optical field distributions in the cross-sections of the designed plasmonic nanoantenna at the frequency of 193.5 THz. **Figure 2(a)** and **(b)** illustrates the optical fields in the x-o-y cross section and x-o-z

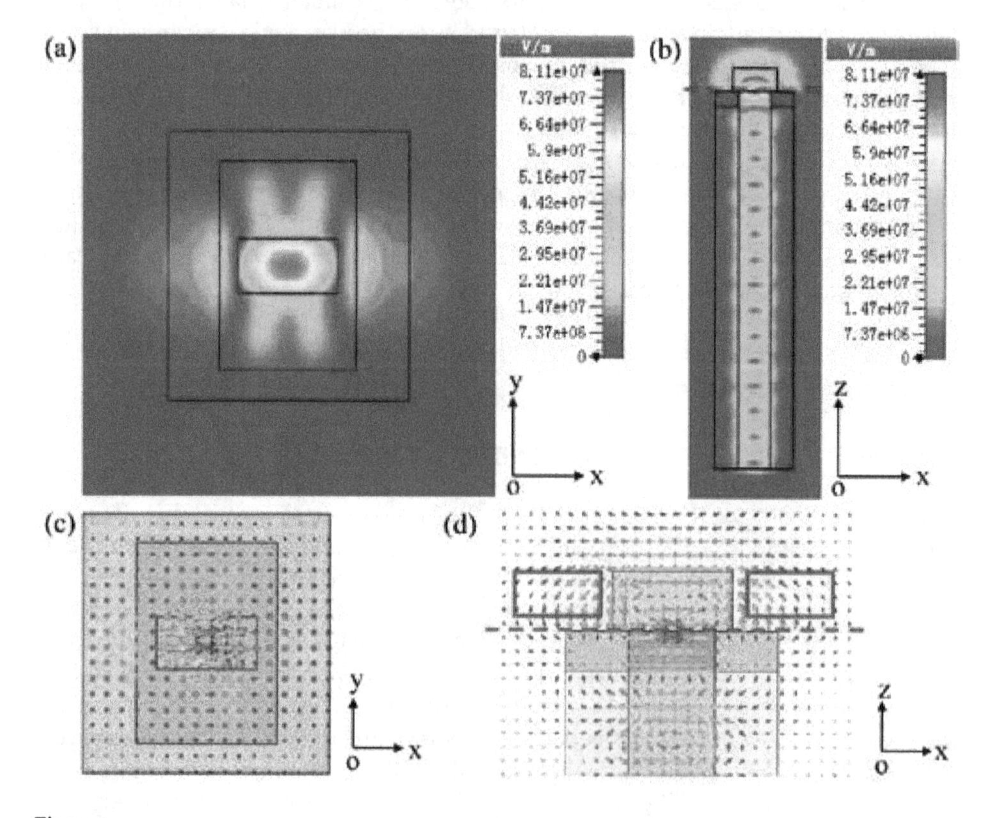

Figure 2.
Simulated optical field in the (a) x-o-y cross-section and (b) x-o-z cross-section of the designed plasmonic nanoantenna at 193.5 THz. Optical field vector distribution maps in the (c) x-o-y and (d) x-o-z cross-sections of the nanoantenna at 193.5 THz.

cross-section, respectively. **Figure 2(a)** represents the interface of the silicon and silver blocks marked by a red dashed line in **Figure 2(b)**. As shown in **Figure 2(b)**, the optical field changes periodically in the silicon waveguide and an optical enhancement happens at the interface of the silicon and silver blocks. Such a field distribution indicates that the light propagates along the silicon waveguide until it reaches the junction of the silver and silicon blocks. Due to the introduction of metallic material silver, SPPs are generated on the surface of the silver block and localized surface plasma resonance (LSPR) occurs at 193.5 THz. The LSPR results in the enhancement of the optical field and the radiation of the nanoantenna.

The optical vector distribution maps in the x-o-y and x-o-z cross-sections are displayed in **Figure 2(c)** and **(d)**, respectively. **Figure 2(c)** represents the interface of the silicon and silver blocks marked by a red dashed line in **Figure 2(b)**. As shown in the areas marked by two red boxes in **Figure 2(d)**, the optical vector fields in the regions on both sides of the silicon block point up and down, respectively. The optical vector fields in the opposite direction indicate that the phase difference of the optical fields on both sides of the silicon block is 180°. Therefore, the optical fields are superposed above the plasmonic nanoantenna.

Radiation characteristics of the plasmonic nanoantenna in **Figure 1** including far-field radiation pattern, gain, and return loss are analyzed by using electromagnetic simulation. **Figure 3** displays the far-field radiation pattern of the designed plasmonic nanoantenna at 193.5 THz. The parameters θ and φ represent the elevation and azimuth angles in free space, respectively. It is clearly seen that light is radiated vertically. The plasmonic nanoantenna has a smooth main lobe without bidirectional radiation. In order to observe clearly, the far-field radiation patterns in two orthogonal planes of $\varphi = 0°$ (x-o-z) and $\varphi = 90°$ (y-o-z) at 193.5 THz are extracted and displayed in **Figure 4(a)**. The black and blue curves indicate the far-field radiation pattern in the planes of $\varphi = 0°$ and $\varphi = 90°$, respectively. The two curves coincide at $\theta = 0°$ indicating a vertical radiation with a gain of 8.45 dB. On the two planes of $\varphi = 0°$ and $\varphi = 90°$, the half-power beam widths (HPBWs) of the main lobe are 51.3° and 43.7°, respectively. The return loss of the designed plasmonic nanoantenna is shown in **Figure 4(b)**. It can be found that the return loss is less than −10 dB in the frequency range from 176.7 to 248.5 THz, which means that the designed plasmonic nanoantenna has a good matching with the silicon wave-guide within a bandwidth of 71.8 THz.

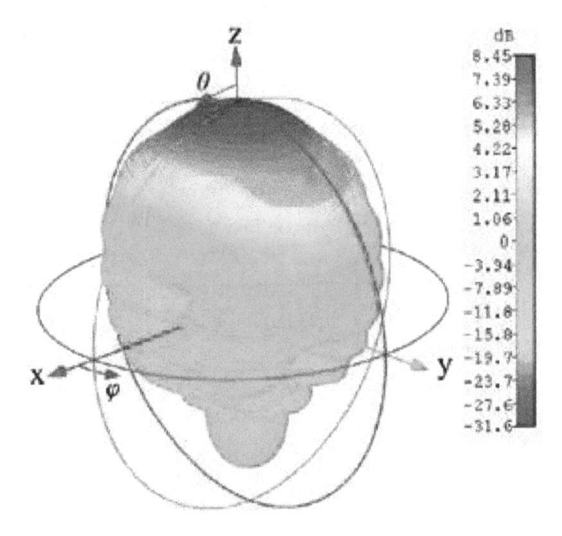

Figure 3.
Far-field radiation pattern of the plasmonic nanoantenna at 193.5 THz.

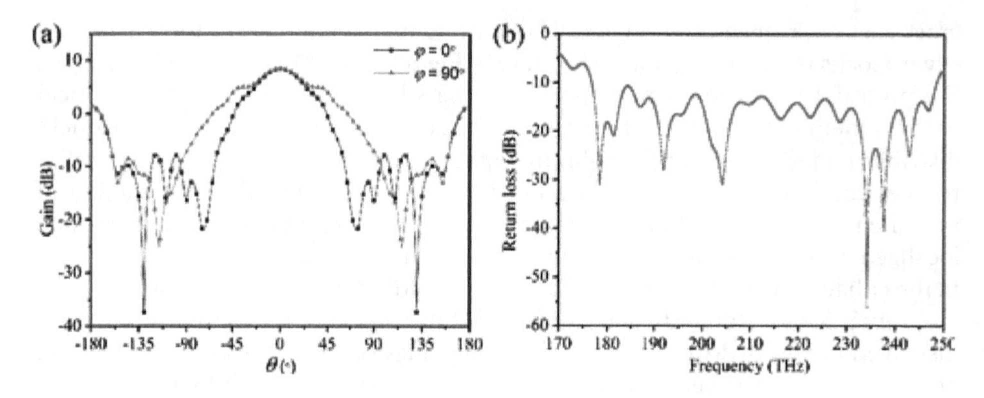

Figure 4.
(a) Far field radiation patterns in the planes of φ = 0° and φ = 90° and (b) return loss of the designed plasmonic nanoantenna at 193.5 THz.

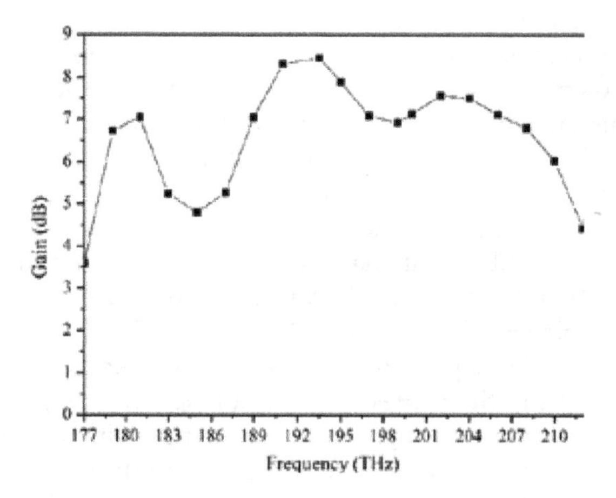

Figure 5.
Gain of the designed plasmonic nanoantenna at the frequency range of 177–212 THz.

Gain is an important radiation characteristic of the optical antennas, which represents the ability of an antenna to radiate optical power in a given direction. **Figure 5** shows the calculated gain of the designed plasmonic nanoantenna. It is clearly seen that at the center frequency of 193.5 THz, the gain of the antenna reaches its maximum value of 8.45 dB.

2.2 Parameter analysis of the plasmonic nanoantenna

In order to understand the impacts of the geometric parameters of the silicon and silver blocks on the radiation characteristics of the plasmonic nanoantenna, a series of electromagnetic simulations are performed and the simulation results are analyzed in detail. We choose the plasmonic nanoantenna mentioned above (W_1 = 625 nm, L_1 = 850 nm, H_{Si} = 300 nm, W_2 = 1100 nm, L_2 = 1100 nm, and H_{Ag} = 200 nm) as the simulation model.

Firstly, the influences of the width (W_1) and length (L_1) of the silicon block in the plasmonic nanoantenna on the gain are analyzed, and the results are given in **Figure 6**. **Figure 6(a)** shows the variation of the gain with the width of the silicon block (W_1) increasing from 545 to 705 nm. It can be found that the gain increases first and decreases with the increase of W_1, and the gain reaches the maximum value of 8.45 dB when W_1 = 625 nm. It also proves that the width of 625 nm satisfies the condition of electromagnetic resonance. Similarly, the variation of the gain with the

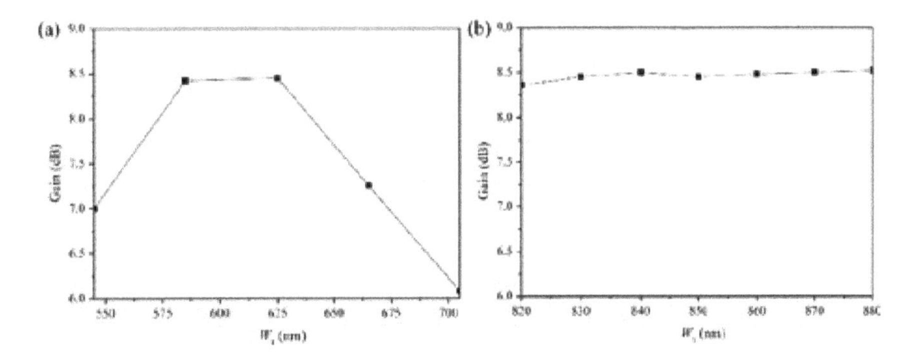

Figure 6.
The simulated gain of the plasmonic nanoantenna varies with (a) the width W_1 and (b) the length L_1 of the silicon block at the frequency of 193.5 THz, respectively.

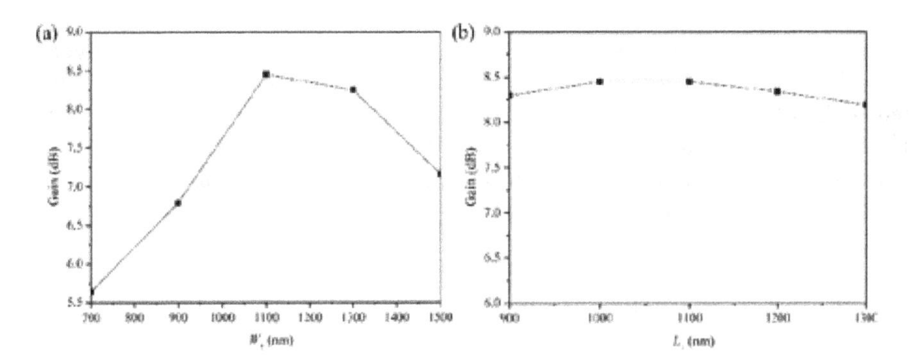

Figure 7.
The simulated gain of the plasmonic nanoantenna varies with (a) the width W_2, and (b) the length L_2 of the silver block at the frequency of 193.5 THz, respectively.

length of the silicon block (L_1) increasing from 820 to 880 nm is studied, as shown in **Figure 6(b)**. It can be seen that the length of the silicon block (L_1) has little effect on the gain. Since the polarization state of the incident light is TE mode (the polarization direction along to the x-axis), the electromagnetic resonance is determined by the geometric dimension along the x-axis. Therefore, the influence of the width of the silicon block on the gain is more obvious than that of the length of the silicon block.

Secondly, the influences of the width (W_2) and length (L_2) of the silver block in the plasmonic nanoantenna structure on the gain are analyzed, and the results are given in **Figure 7**. **Figure 7(a)** displays the variation of the gain with the width of the silver block (W_2) increasing from 700 to 1550 nm. The variation of the gain with the length of the silver block (L_2) increasing from 900 to 1300 nm is shown in **Figure 7(b)**. It is clearly seen that when W_2 is increased, the gain will increase first and then decrease. The increase of the width of the silver block means the increase of the antenna aperture, which will result in the increase of the gain. However, in the proposed nanoantenna, the gain will decrease when W_2 and L_2 are larger than 1100 nm because of high absorption loss of metal in optical wavelength. The length of the silver block has little effect on the gain, which depends on the polarization state of the incident light.

Finally, the influences of the heights of the silicon block (H_{Si}) and silver block (H_{Ag}) in the plasmonic nanoantenna structure on the gain are also analyzed, as shown in **Figure 8(a)** and **(b)**, respectively. When H_{Si} and H_{Ag} are increased, the gain of the antenna increases first and then decreases. Compared with the silver block, the height of the silicon block has a greater influence on the gain. The gain reaches the maximum when H_{Si} = 300 nm and H_{Ag} = 200 nm, respectively.

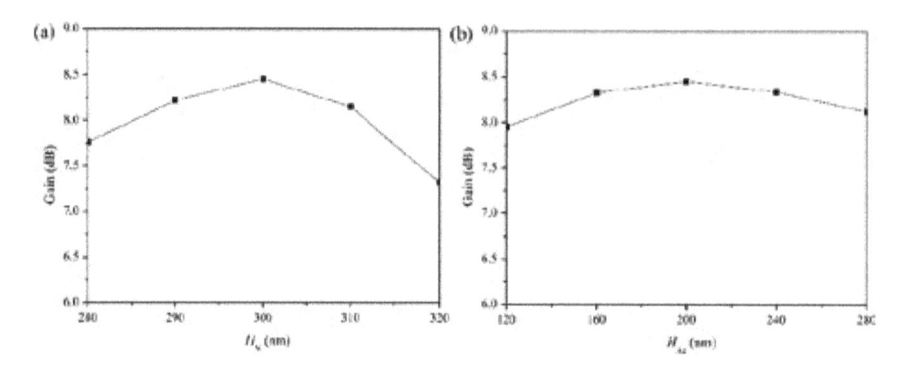

Figure 8.
The simulated gain of the plasmonic nanoantenna varies with the heights of the (a) silicon and (b) silver block at the frequency of 193.5 THz.

The simulation results show that the geometric parameters of the silver and silicon blocks will affect the radiation characteristics of the antenna. For TE polarized incident wave, the widths of the silicon and silver blocks have significant effect on the gain, and the lengths of the silicon and silver blocks have little effect on the gain. It is necessary to optimize these parameters to obtain a plasmonic nanoan-tenna with high gain.

3. Plasmonic nanoantenna arrays

In practical applications, the antenna array is used to realize beam steering on the basis of the optical field superposition principle. The beam deflection is realized by changing the phase of the light radiated by each antenna in the array. Usually, radiation characteristics of the antenna array, including steering angle, beam width, gain, return loss, and mutual coupling between each antenna [25] should be considered. Utilizing the proposed plasmonic nanoantenna, one-dimensional (1-D) and two-dimensional (2-D) arrays are designed, and their radiation characteristics are investigated in detail.

3.1 1 × 8 plasmonic nanoantenna array

In the research of 1-D array, a 1 × 8 array is designed using the proposed plasmonic nanoantenna, as shown in **Figure 9**. In the 1 × 8 array, the nanoantennas are arranged in a row along the x-axis with element spacing of d. In the simulation, optical signals are fed from the waveguide port of each antenna, i.e., Port1 to Port8, respectively. For the array with uniform element spacing, the element spacing larger than one wavelength will lead to the appearance of the grating lobes in the far-field radiation pattern of the array, which will limit the beam steering range. In our design, the element spacing is set to $0.7\lambda_0$ (i.e., d = 1085 nm), which is much smaller than the element spacing reported in Ref. [1]. Thus, the whole 1 × 8 array is 8680 nm in length and 1085 nm in width.

The return loss of each waveguide port in the designed 1 × 8 array is calculated and shown in **Figure 10**. The curves S11 to S88 represent the return losses of Port1 to Port8, respectively. The results in **Figure 10** show that the return losses of all ports are less than −22.5 dB in the frequency region 191.5–196.2 THz. It means that the reflectivity of each port is less than 0.6%. Such low return losses also prove that there is a good impedance match between the designed plasmonic nanoantenna and the silicon waveguide in a wide bandwidth. To research the coupling effect

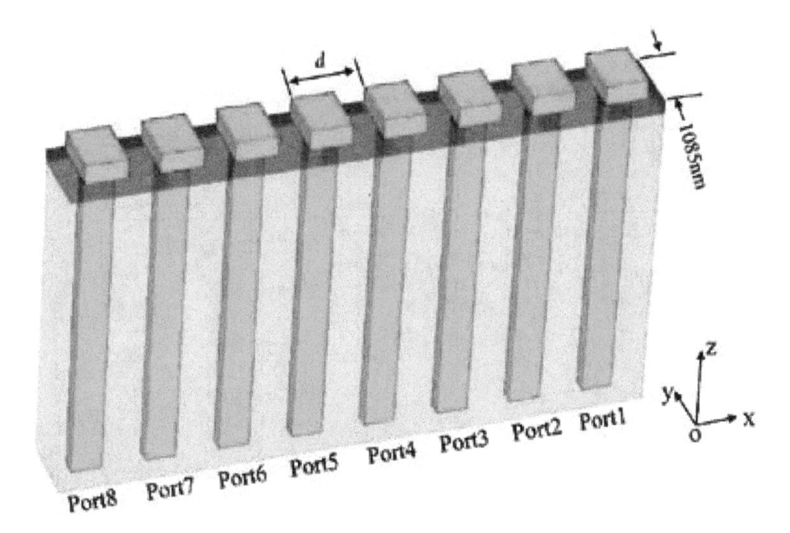

Figure 9.
Structure of 1 × 8 array.

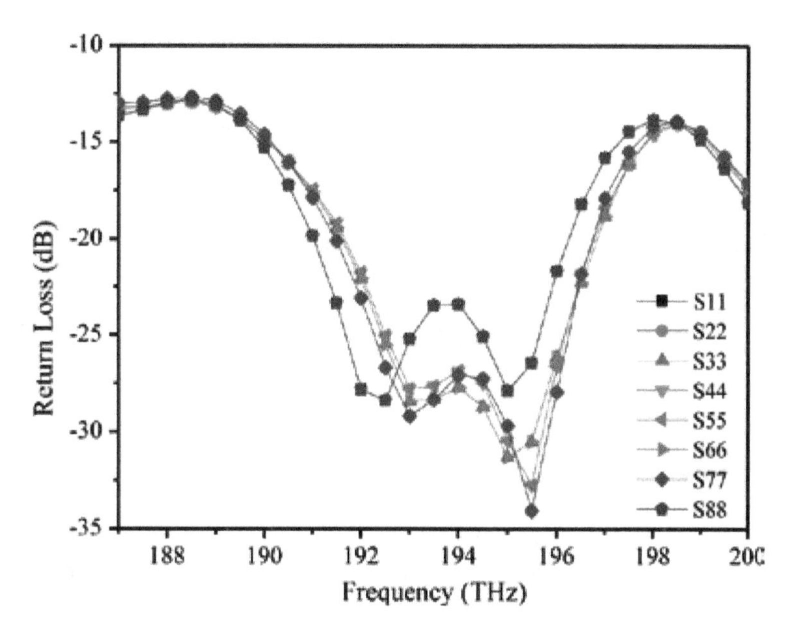

Figure 10.
Return losses of the ports from Port1 to Port8 in the 1 × 8 array with $d = 0.7\lambda_0$.

among the ports, the mutual couplings between other ports and Port1 are shown in **Figure 11**. The mutual coupling decreases as the distance between Port1 and other ports increases at 193.5 THz.

In the simulation, the radiation pattern of the 1 × 8 array is studied by feeding optical signals with the equal amplitude and the same phase into each port simultaneously. **Figure 12(a)** and **(b)** display the far-field radiation patterns of the 1 × 8 array in the planes of $\varphi = 0°$ and $\varphi = 90°$ at 193.5 THz, respectively. The simulation results show that the 1 × 8 array radiates light vertically upward with a gain of 14.5 dB and the HPBWs of the main lobe on the planes of $\varphi = 0°$ and $\varphi = 90°$ are 9.0° and 96.4°, respectively.

According to the filed superposition principle, beam steering can be realized by changing the phase of light radiated by each antenna in the antenna array. In theory,

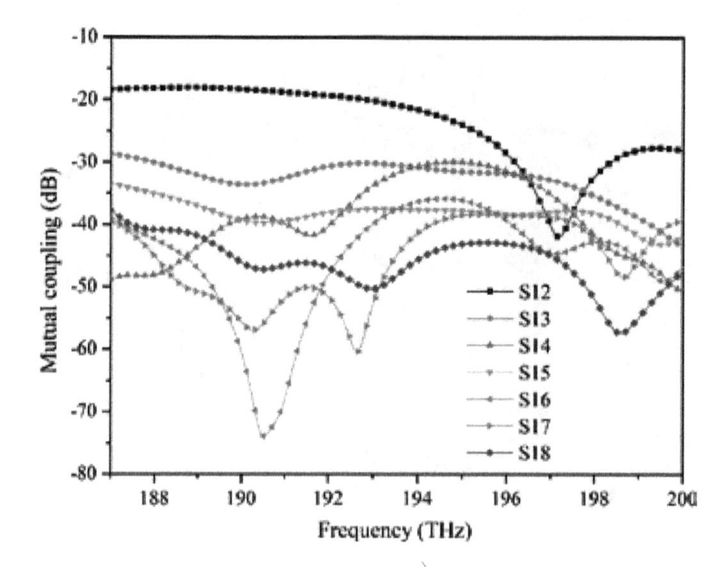

Figure 11.
The mutual coupling between other ports and port1.

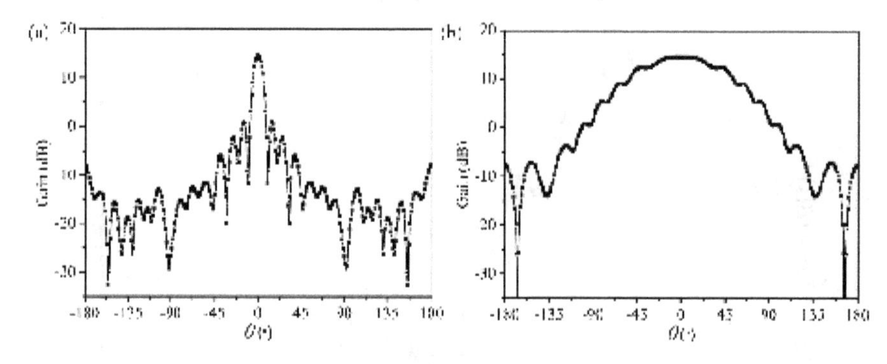

Figure 12.
Far-field radiation patterns of the 1 × 8 array with d = 0.7λ_0 in the planes of (a) φ = 0° and (b) φ = 90° at the frequency of 193.5 THz.

for a 1-D array along the x-axis with uniform element spacing d, the relationship between the beam steering angle θ and the phase difference of light fed into the two adjacent antennas ψ_x follows the equation of [25]

$$kd \sin\theta = \psi_x, \tag{1}$$

where the parameter k represents wave vector in the free space.

We further study the beam steering characteristics of the 1 × 8 array in**Figure 9** by using the methods of electromagnetic simulation and theoretical calculation. At 193.5 THz, the far-field radiation pattern of the 1 × 8 array in the plane of $\varphi = 0°$ with various phase difference of the input light fed in two adjacent antennas ψ_x is simulated, as shown in **Figure 13**. The simulation results show that when the phase difference ψ_x increases from −180° to +180° with an interval of 45°, the beam steers from −44° to + 44° in the $\varphi = 0°$ plane. When the light fed into each antenna has the same phase, i.e., $\psi_x = 0°$, the plasmonic nanoantenna array radiates light in the direction of $\theta = 0°$ with the highest gain and the narrowest beam width. As beam steering angle increases, the main lobe width of the radiation beam increases, and the gain decreases. When the phase difference is ±180.0°, i.e., $\psi_x = 180°$, the peak

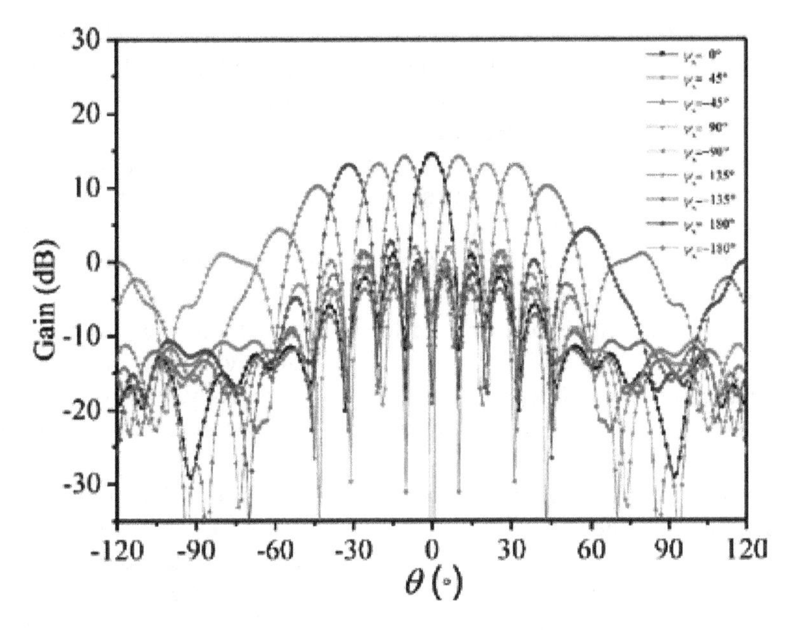

Figure 13.
Far-field beam steering of the 1 × 8 array with d = 0.7λ_0 in the plane of φ = 0° with various phase differences at 193.5 THz.

value of the grating lobe of the radiation light is the same as that of the main lobe, and the beam steering angle is ±44.0°, which is almost consistent with the theoretical value of ±45.6° calculated by Eq. (1). **Figure 14** gives the far-field radiation pattern of the 1 × 8 array at 193.5 THz, when ψ_x = 45°. As can be seen from **Figure 14**, the deflection angle of the main lobe is 10.48°. Theoretically, the beam steering angle obtained from Eq. (1) is 10.28°. The small deviation between the simulation results and theoretical results result from the mutual coupling between the nanoan-tennas in the array.

In a uniformly arranged array, element spacing is the critical factor for determining the radiation characteristics of the array. Numerical simulations of the 1 × 8 array with different element spacing *d* are performed, and the influences of element spacing on the radiation characteristics are analyzed in depth.

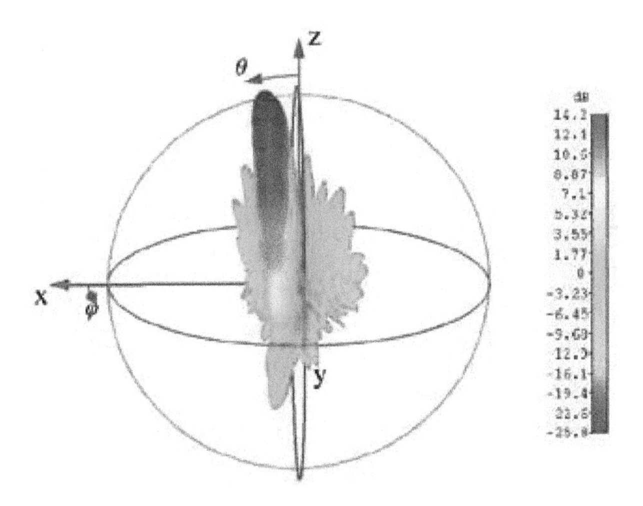

Figure 14.
Far-field radiation pattern of the 1 × 8 array with d = 0.7λ_0 at 193.5 THz when ψ_x = 45°.

The calculated S-Parameters of the 1 × 8 array with different element spacing are displayed in **Figures 15** and **16. Figure 15(a)** and **(b)** show the return loss of Port1 (S_{11}) and Port4 (S_{44}) when d varies from $0.7\lambda_0$ to $1.1\lambda_0$, respectively. In the frequency range of 187–200 THz, the return loss of Port1 is less than – 12 dB, which indicates that Port1 has good matching characteristics with different element spacing. When the element spacing is $0.7\lambda_0$, the return loss decreases sharply near 193.5 THz. The mutual coupling between the Port1 and the Port2 (S_{21}) is shown in **Figure 16.** At 193.5 THz, the mutual coupling decreases with the increase of the element spacing.

The calculated far-field radiation patterns of the 1 × 8 array with the phase difference $\psi_x = 0°$ in the $\varphi = 0°$ plane and $\varphi = 90°$ plane when d increases from $0.7\lambda_0$ to $1.1\lambda_0$ are shown in **Figure 17(a)** and **(b)**, respectively. As can be seen from **Figure 17(a)**, when the element spacing is $1.1\lambda_0$, the grating lobes appear at −62.0° and 62.5°.

The far-field radiation patterns of the 1 × 8 array with $d = 0.7\lambda_0$ at three frequencies of lower side frequency 187 THz, center frequency 193.5 THz, and upper side frequency 200 THz when $\psi_x = 0°$ are shown in **Figure .18 Figure 18(a)** and **(b)** displays the far-field radiation patterns in the planes of $\varphi = 0°$ and $\varphi = 90°$, respectively. The calculated results show that 1 × 8 array also has almost the same gains at center and side frequencies.

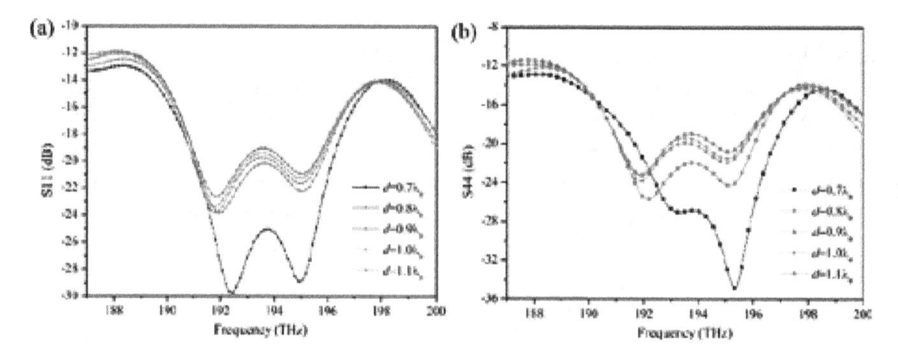

Figure 15.
Return loss of the (a) Port1 and (b) Port4.

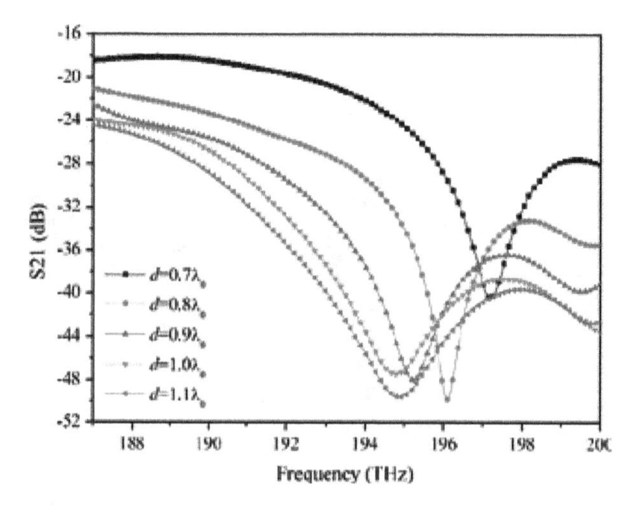

Figure 16.
Mutual coupling between the Port1 and the Port2.

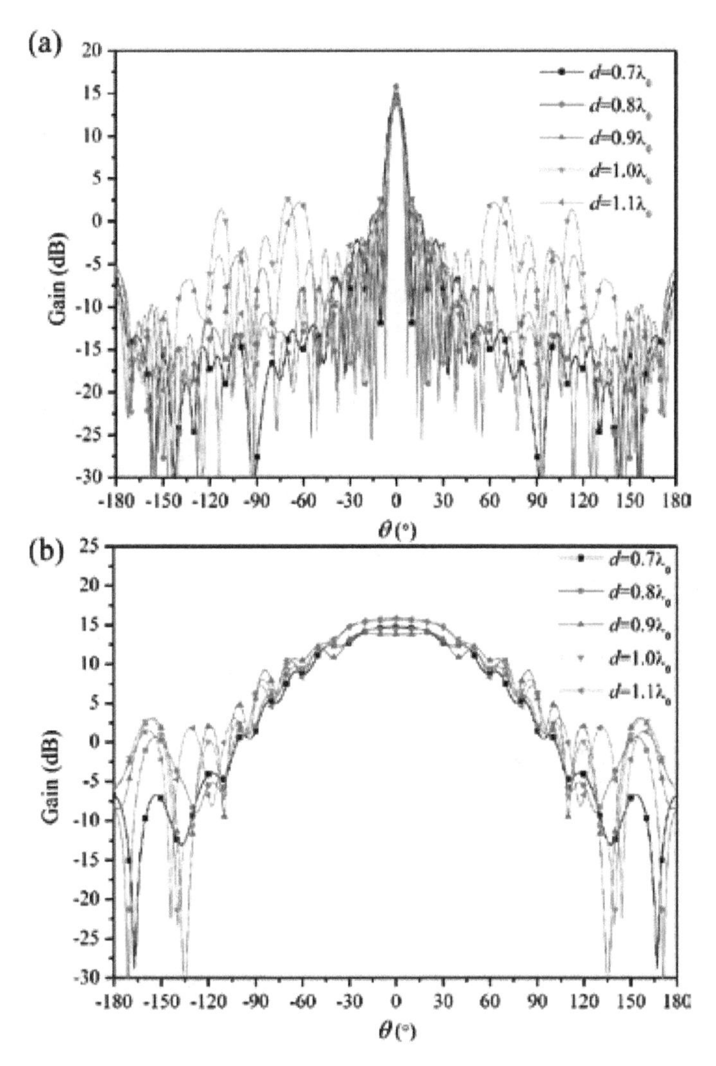

Figure 17.
Far-field radiation patterns of the 1 × 8 array with $\psi_x = 0°$ in the planes of (a) $\varphi = 0°$ and (b) $\varphi = 90°$ when d increases from $0.7\lambda_0$ to $1.1\lambda_0$ at 193.5 THz.

3.2 8 × 8 plasmonic nanoantenna array

In this section, we take the 1 × 8 array mentioned above as a sub-array and extend it along the y-axis to obtain an 8 × 8 array, as shown in **Figure 19**. In the 8 × 8 array, the spacing between two adjacent antennas is set to $0.7\lambda_0$ in both directions of x- and y-axis, that means the element spacing is $d_x = d_y = 0.7\lambda_0$. The far-field radiation pattern of a 2-D array can be calculated by using the pattern multiplication, which is given by [1]

$$E(\theta,\varphi) = S(\theta,\varphi) \times F_a(\theta,\varphi), \tag{2}$$

where $E(\theta,\varphi)$ and $S(\theta,\varphi)$ represent the far field radiation patterns of the 2-D array and the 1-D sub-array, respectively. The function $F_a(\theta,\varphi)$ indicates the array factor of the 2-D array. The array factor is determined by the 1-D sub-array spacing and the number of the 1-D sub-arrays in the 2-D array along the extend direction, and the amplitude and phase of the light fed in each antenna. According to Eq. (2), the far-field radiation pattern of the 8 × 8 array with constant amplitude and same

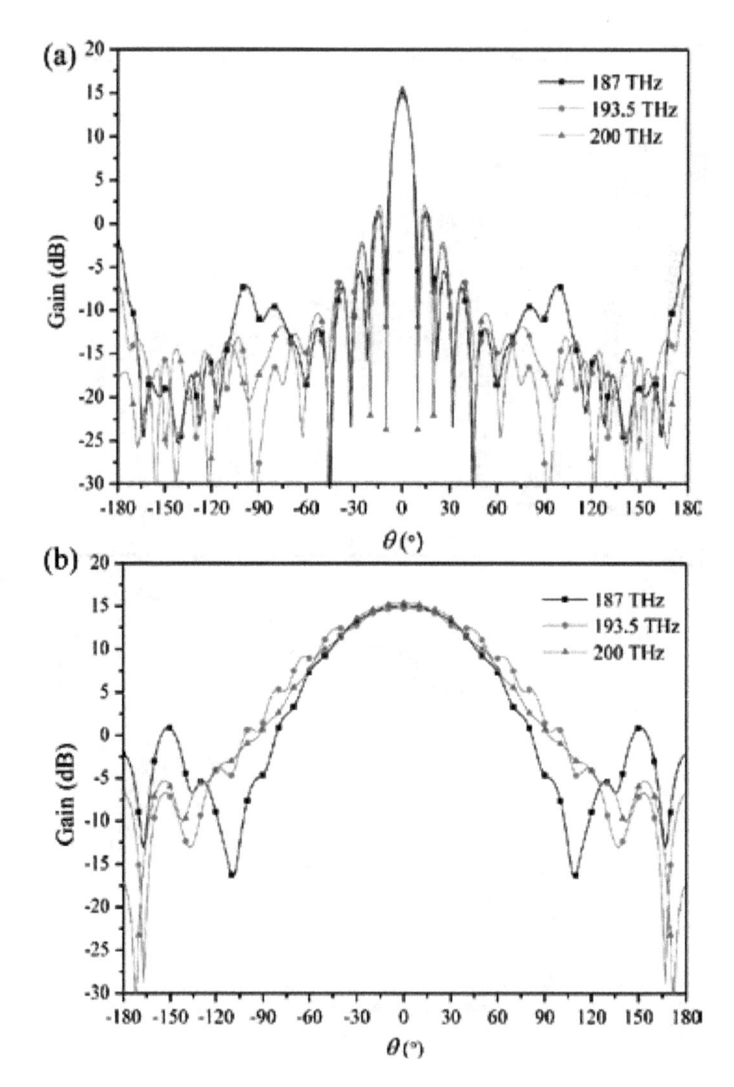

Figure 18.
Far-field radiation patterns of the 1 × 8 array with $d = 0.7\lambda_0$ *in the planes of (a)* $\varphi = 0°$ *and (b)* $\varphi = 90°$ *at three frequencies when* $\psi_x = 0°$.

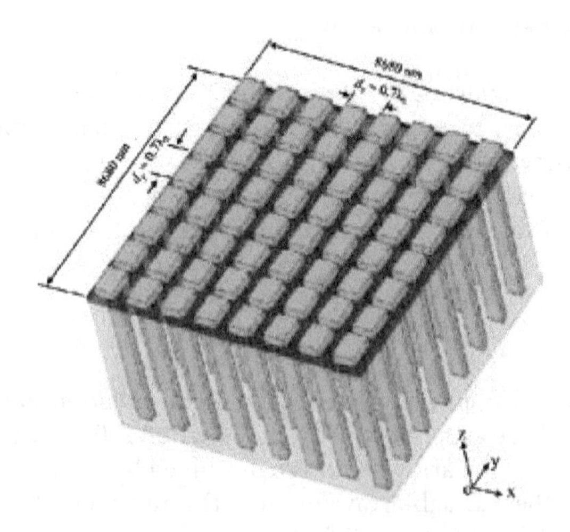

Figure 19.
Structure of the 8 × 8 array with the element spacing of $0.7\lambda_0$ *in the directions of x- and y-axis.*

phase of light in each antenna at 193.5 THz is obtained. The calculated far-field radiation patterns in the planes of φ = 0° and φ = 90° are displayed in **Figure 20(a)** and **(b)**, respectively. It can be seen that the 8 × 8 array radiates light upward with a gain of 24.2 dB and an HPBW of 9.0° × 9.0° at 193.5 THz.

Similar to the 1-D beam steering, 2-D beam steering of the 8 × 8 array is realized by changing the phases of light in the nanoantennas along the two orthogonal directions. The parameters ψ_x and ψ_y are used to represent the phase differences between two lights fed in the adjacent nanoantennas along the x- and y-directions, respectively. When the phase difference ψ_x is changed, the beam steers in the φ = 0° plane. **Figure 21** shows the far-field beam steering radiation patterns of the 8 × 8 array in the φ = 0° plane under various ψ_x, when ψ_y is fixed at 0°. It can be found that the radiation direction of the light changes in a certain range of −44.0° to +44.0°, as the phase difference ψ_x changes from −180.0° to +180.0°.

Figure 22 shows the far-field beam steering of the 8 × 8 array in the φ = 90° plane under various ψ_y, when ψ_x is fixed at 0°. When the phase difference ψ_y is

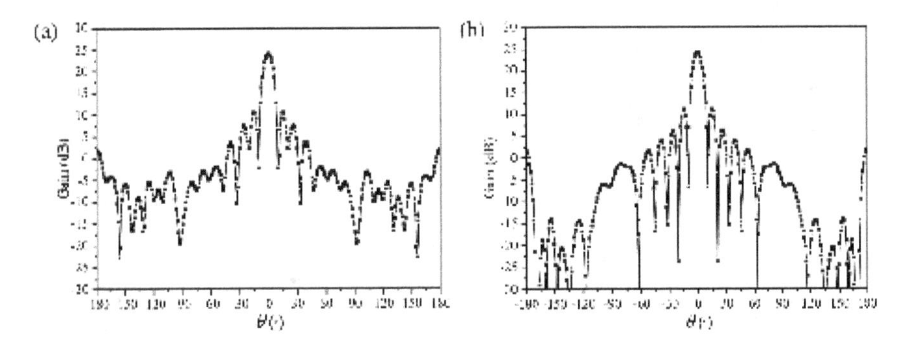

Figure 20.
Far-field radiation patterns of the 8 × 8 array with $d_x = d_y = 0.7\lambda_0$ in the planes of (a) φ = 0° and (b) φ = 90° at 193.5 THz.

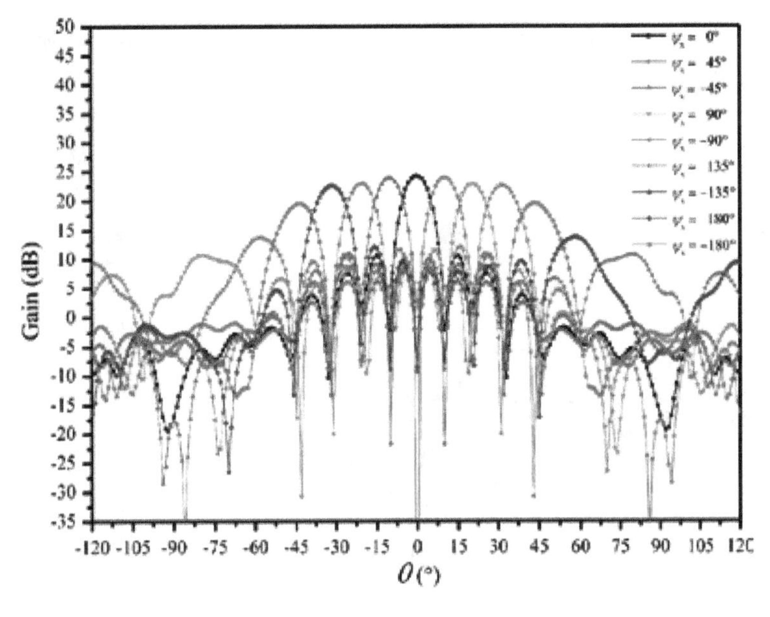

Figure 21.
Far-field beam steering in the φ = 0° plane of the designed 8 × 8 array with $d_x = d_y = 0.7\lambda_0$ at 193.5 THz when ψ_x changes from −180° to 180° and ψ_y is fixed at 0°.

changed, the beam steers in the $\varphi = 90°$ plane. As the phase difference ψ_y changes from −180.0° to +180.0°, the radiation direction of the light changes in a certain range of −45.0° to +45.0° at 193.5 THz.

Therefore, the designed 8 × 8 array can be used to realize the beam steering with a steering range of ±44.0° × ±45.0° by controlling the differences of ψ_x and ψ_y. Calculated by Eq. (1), an 8 × 8 array with the element spacing of $0.7\lambda_0$ along the x- and y-axis has a beam steering range of ±45.6° × ±45.6°, which is basically consistent with the simulation results. At 193.5 THz, a far-field radiation pattern of the designed 8 × 8 array is simulated when $\psi_x = \psi_y = 45°$, as shown in **Figure 23.** It is clearly seen that an optical beam is radiated with an angle of $\theta = 14.0°$ and $\varphi = 45°$.

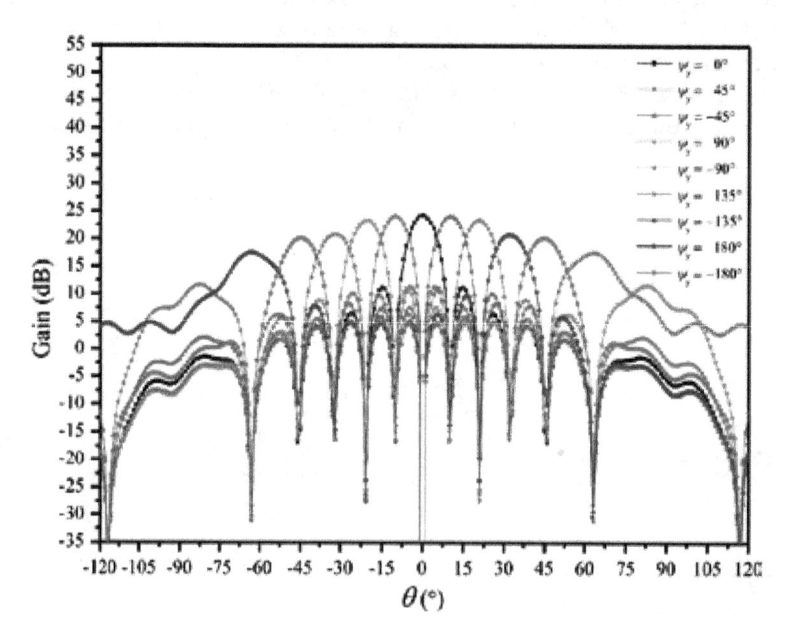

Figure 22.
Far-field beam steering in the $\varphi = 90°$ plane of the 8 × 8 array with $d_x = d_y = 0.7\lambda_0$ at 193.5 THz when ψ_y changes from −180° to 180° and ψ_x is fixed at 0°.

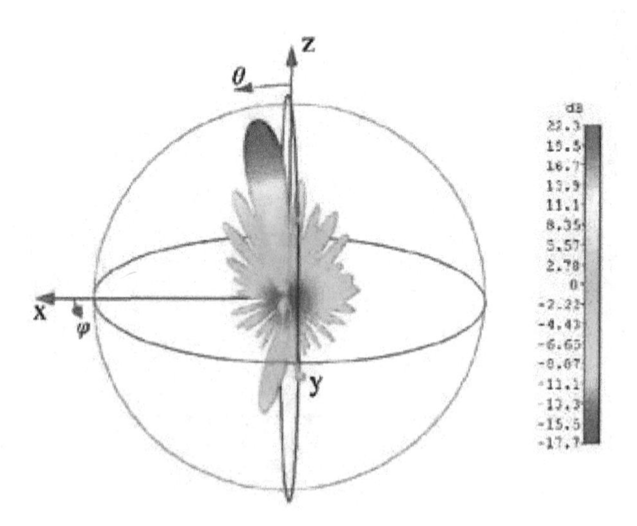

Figure 23.
Far-field radiation pattern of the designed 8 × 8 array with $d_x = d_y = 0.7\lambda_0$ at 193.5 THz when $\psi_x = \psi_y = 45°$.

4. Conclusion

In this chapter, we review the silicon-based optical nanoantennas and their applications in OPA for beam steering. In order to obtain an OPA with high gain and wide beam steering range, we propose a sub-wavelength plasmonic nanoantenna with an operating wavelength of 1550 nm. The proposed plasmonic nanoantenna consists of a silver block and a silicon block with a standard silicon waveguide for feeding light into the nanoantenna. On the basis of LSPR, the plasmonic nanoantenna radiates light vertically upward with a high gain of 8.45 dB at 1550 nm. There is a good impedance match between the plasmonic nanoantenna and the silicon waveguide in a frequency range from 176.7 to 248.5 THz. Furthermore, two nano-antenna arrays (1 × 8 and 8 × 8) with the element spacing of $0.7\lambda_0$ composed of the proposed plasmonic nanoantennas are designed, and their beam steering radiation patterns are studied in detail. The simulation results show that the 1 × 8 array can be used to realize 1-D beam steering from −44.0° to +44.0° with a gain of 14.5 dB at 1550 nm, and the 8 × 8 array can achieve a 2-D beam steering from −44.0° to +44.0° in one dimension and from −45.0° to +45.0° in the other dimension with a gain of 24.2 dB at 1550 nm.

The plasmonic nanoantenna we proposed is a good candidate for the exten-sion of the nanoantenna array used in a large-scale OPA. Utilizing the proposed plasmonic nanoantenna, a 3-D array extend mode can be adopted to form an OPA with thousands of optical nanoantennas. We first make a 1-D OPA as a sub-layer, in which the optical power division network, phase shifters, and a 1-D plasmonic nanoantenna array are integrated in a plane. After that, such 1-D OPA layers are extended longitudinally. Therefore, a highly integrated OPA containing thousands of optical nanoantennas with sub-wavelength element spacing can be obtained theoretically to steer beam in a wide angle without grating lobes. However, the processing of the OPA with multilayer structure is limited by our micro/nanofabri-cation technology. We believe that with the development of micro/nanoprocessing technology, the large-scale OPA will be applied in various fields of optical com-munication, LiDAR, security monitoring, and display advertising, which will bring great benefits to human life.

Acknowledgements

The authors would like to thank Dr. Zhihui Liu for her support in review-ing. This work is supported by Innovation Funds of China Aerospace Science and Technology (No. Y-Y-Y-GJGXKZ-18, No. Z-Y-Y-KJJGTX-17) and the 2017 Open Research Fund of Key Laboratory of Cognitive Radio and Information Processing, Ministry of Education, Guilin University of Electronic Technology (No. CRKL170202).

Author details

Tao Dong[1,2*], Yue Xu[1,2] and Jingwen He[1,2]

1 State Key Laboratory of Space-Ground Integrated Information Technology, Beijing, China

2 Beijing Institute of Satellite Information Engineering, Beijing, China

*Address all correspondence to: dongtaoandy@163.com

References

[1] Sun J, Timurdogan E, Yaacobi A, Hosseini ES, Watts MR. Large-scale nanophotonic phased array. Nature. 2013;**493**:195-199. DOI: 10.1038/nature11727

[2] Abediasl H, Hashemi H. Monolithic optical phased-array transceiver in a standard SOI CMOS process. Optics Express. 2015;**23**:6509-6519. DOI: 10.1364/OE.23.006509

[3] Sabouri S, Jamshidi K. Design considerations of silicon nitride optical phased array for visible light communications. IEEE Journal of Selected Topics in Quantum Electronics. 2018; **24**:1-7. DOI: 10.1109/JSTQE.2018.2836991

[4] Poulton CV, Byrd MJ, Russo P, Timurdogan E, Khandaker M, Vermeulen D, et al. Long-range LiDAR and free-space data communication with high-performance optical phased arrays. IEEE Journal of Selected Topics in Quantum Electronics. 2019;**25**:1-8. DOI: 10.1109/JSTQE.2019.2908555

[5] Raval M, Yaacobi A, Watts MR. Integrated visible light phased array system for autostereoscopic image projection. Optics Letters. 2018;**43**: 3678-3681. DOI: 10.1364/OL.43.003678

[6] McManamon PF, Dorschner TA, Corkum DL, Friedman LJ, Hobbs DS, Holz M, et al. Optical phased array technology. Proceedings of the IEEE. 1996;**84**:268-298. DOI: 10.1109/5.482231

[7] Van Acoleyen K, Rogier H, Baets R. Two-dimensional optical phased array antenna on silicon-on-insulator. Optics Express. 2010;**18**:13655-13660. DOI: 10.1364/OE.18.013655

[8] Kwong D, Hosseini A, Covey J, Zhang Y, Xu X, Subbaraman H, et al. On-chip silicon optical phased array for two-dimensional beam steering. Optics Letters. 2014;**39**:941-944. DOI: 10.1364/OL.39.000941

[9] Christopher VP, Ami Y, Zhan S, Matthew JB, Michael RW. Optical phased array with small spot size, high steering range and grouped cascaded phase shifters. In: Advanced Photonics Congress 2016 (IPR, NOMA, Sensors, Networks, SPPCom, SOF); 18-20 July 2016; Vancouver. Washington: OSA; 2016. IW1B.2

[10] Doylend JK, Heck MJR, Bovington JT, Peters JD, Coldren LA, Bowers JE. Two-dimensional free-space beam steering with an optical phased array on silicon-on-insulator. Optics Express. 2011;**19**:21595-21604. DOI: 10.1364/OE.19.021595

[11] Hutchison DN, Sun J, Doylend JK, Kumar R, Heck J, Kim W, et al. High-resolution aliasing-free optical beam steering. Optica. 2016;**3**:887-890. DOI: 10.1364/OPTICA.3.000887

[12] Poulton CV, Byrd MJ, Raval M, Su Z, Li N, Timurdogan E, et al. Large-scale silicon nitride nanophotonic phased arrays at infrared and visible wavelengths. Optics Letters. 2017;**42**:21-24. DOI: 10.1364/OL.42.000021

[13] Zhang Y, Ling Y, Zhang K, Gentry C, Sadighi D, Whaley G, et al. Sub-wavelength-pitch silicon-photonic optical phased array for large field-of-regard coherent optical beam steering. Optics Express. 2019;**27**:1929-1940. DOI: 10.1364/OE.27.001929

[14] Guclu C, Boyraz O, Capolino F. Theory of optical leaky-wave antenna integrated in a ring resonator for radiation control. Journal of Lightwave Technology. 2017;**35**:10-18. DOI: 10.1109/JLT.2016.2626982

[15] Zhao C, Zhang H, Zheng Z, Peng C, Hu W. Silicon optical-phased-array prototypes using electro-optical phase shifters. In: 2017 Conference on Lasers and Electro-Optics (CLEO); 14-19 May 2017; San Jose. Washington: OSA; 2017. SM1O.4

[16] Yoo B, Megens M, Sun T, Yang W, Chang-Hasnain CJ, Horsley DA, et al. A 32 × 32 optical phased array using polysilicon sub-wavelength high-contrast-grating mirrors. Optics Express. 2014;**22**:19029-19039. DOI: 10.1364/OE.22.019029

[17] Fatemi R, Abiri B, Khachaturian A, Hajimiri A. High sensitivity active flat optics optical phased array receiver with a two-dimensional aperture. Optics Express. 2018;**26**:29983-29999. DOI:10.1364/OE.26.029983

[18] Fatemi R, Khachaturian A, Hajimiri A. A nonuniform sparse 2-D large-FOV optical phased array with a low-power PWM drive. IEEE Journal of Solid-State Circuits. 2019;**54**:1200-1215. DOI:10.1109/JSSC.2019.2896767

[19] Iluz Z, Boag A. Wide-angle scanning optical linear phased array. In: 2015 IEEE International Conference on Microwaves, Communications, Antennas and Electronic Systems (COMCAS); 2-4 November 2015; Tel Aviv. New York: IEEE; 2015. pp. 1-2

[20] Yousefi L, Foster AC. Waveguide-fed optical hybrid plasmonic patch nano-antenna. Optics Express. 2012; **20**:18326- 18335. DOI: 10.1364/OE.20.018326

[21] Kim M, Li Z, Yu N. Controlling light propagation in optical waveguides using one dimensional phased antenna arrays. In`: 2014 Conference on Lasers and Electro-Optics (CLEO); 8-13 June 2014; San Jose. New York: IEEE; 2015. FW3C.4

[22] Nia BA, Yousefi L, Shahabadi M. Integrated optical-phased array nanoantenna system using a plasmonic rotman lens. Journal of Lightwave Technology. 2016;**34**:2118-2126. DOI: 10.1109/JLT.2016.2520881

[23] Xu Y, Dong T, He J, Wan Q. Large scalable and compact hybrid plasmonic nanoantenna array. Optical Engineering. 2018;**57**:087101. DOI: 10.1117/1.OE.57.8.087101

[24] Johnson PB, Christy RW. Optical constants of the noble metals. Physical Review B. 1972;**6**:4370-4379. DOI: 10.1103/PhysRevB.6.4370

[25] Eibert TF, Volakis JL. Antenna Engineering Handbook. 4th ed. New York: McGraw-Hill; 2007. pp. 10-968

9

Thermal Collective Excitations in Novel Two-Dimensional Dirac-Cone Materials

Andrii Iurov, Godfrey Gumbs and Danhong Huang

Abstract

The purpose of this chapter is to review some important, recent theoretical discoveries regarding the effect of temperature on the property of plasmons. These include their dispersion relations and Landau damping rates, and the explicit dependence of plasmon frequency on chemical potential at finite temperatures for a diverse group of recently discovered Dirac-cone materials. These novel materials cover gapped graphene, buckled howycomb lattices (such as silicene and germanene), molybdenum disulfide and other transition-metal dichalcogenides, especially the newest dice and α-$\mathcal{T}_3$ materials. The most crucial part of this review is a set of implicit analytical expressions about the exact chemical potential for each of considered m aterials, which greatly affects the plasmon dispersions and a lot of many-body quantum-statistical properties. We have also obtained the nonlocal plasmon modes of graphene which are further Coulomb-coupled to the surface of a thick conducting substrate, while the whole system is kept at a finite temperature. An especially rich physics feature is found for α-$\mathcal{T}_3$ materials, where each of the above-mentioned properties depends on both the hopping parameter α and temperature as well.

Keywords: finite temperature effects, plasmon dispersion, 2D materials

1. Introduction

Graphene, a two-dimensional (2D) carbon layer with a hexagonal atomic structure [1–3], has recently attracted outstanding attention from both academic scientists doing fundamental researches and engineers working on its technical applications [4]. Now, the scientific community is actively investigating the innovative semiconductors beyond graphene, with intrinsic spin-orbit interaction and tunable bandgap [5].

A remarkable feature of graphene is the absence of the bandgap in its energy dispersions. In spite of the obvious advantage of such bandstructure for novel electronic devices, electrons in graphene could not be confined due to the well-known Klein paradox [6]. To resolve this issue, graphene may be replaced with a material with a buckled structure and substantial spin-orbit interaction, such as silicene and germanene.

A new quasi-two-dimensional structure which has recently gained popularity among device scientists, is molybdenum disulfide monolayer, a honeycomb lattice which consists of two different molybdenum and sulfur atoms. It reveals a large direct band gap, absence of inversion symmetry and a substantial spin-orbit coupling. A summary of all recently fabricated materials beyond graphene is given in **Figure 1**. The last relevant example is black phosphorous (phosphorene) with a strong anisotropy of its composition and electron energy dispersion. Even though we do not study plasmons in phosphorene in the present chapter, there have been some crucial publications on that subject [7, 8].

Plasmons, or self-sustained collective excitations of interacting electrons in such low-dimensional materials, are especially important, since they serve as the basics for a number of novel devices and their applications [9, 10] in almost all fields of modern science, emerging nanofabrication and nanotechnology. Propagation and detection of plasmonic excitation in hybrid nanoscale devices can convert to or modify existing electromagnetic field or radiation [11–14]. Localized surface plasmons are particularly of special interest considering their interactions with other plasmon modes in closely-located optoelectronic device as well as with imposed electromagnetic radiation [15].

Finite-temperature plasmons are of special interest for possible device applications. Among them is the possibility to increase the frequency (or energy) of a plasmon by an order of magnitude or even more, specifically, as a consequence of the raised temperature. As it was shown in Ref. [16], the dispersion of a thermal plasmon is given as $\sim \sqrt{q k_B T}$, where q is the wave vector. This dispersion relation reveals the fact that the plasmon energy is monotonically increased with temperature and could be moved to the terahertz range and even above, which is crucial for imaging and spectroscopy.

At the same time, the damping rate, or broadening of the frequency, of such thermal plasmons varies as $\simeq 1/\sqrt{T}$, which means that we are dealing with a long-lived plasmon with low or even nearly zero Landau damping. Both plasmon

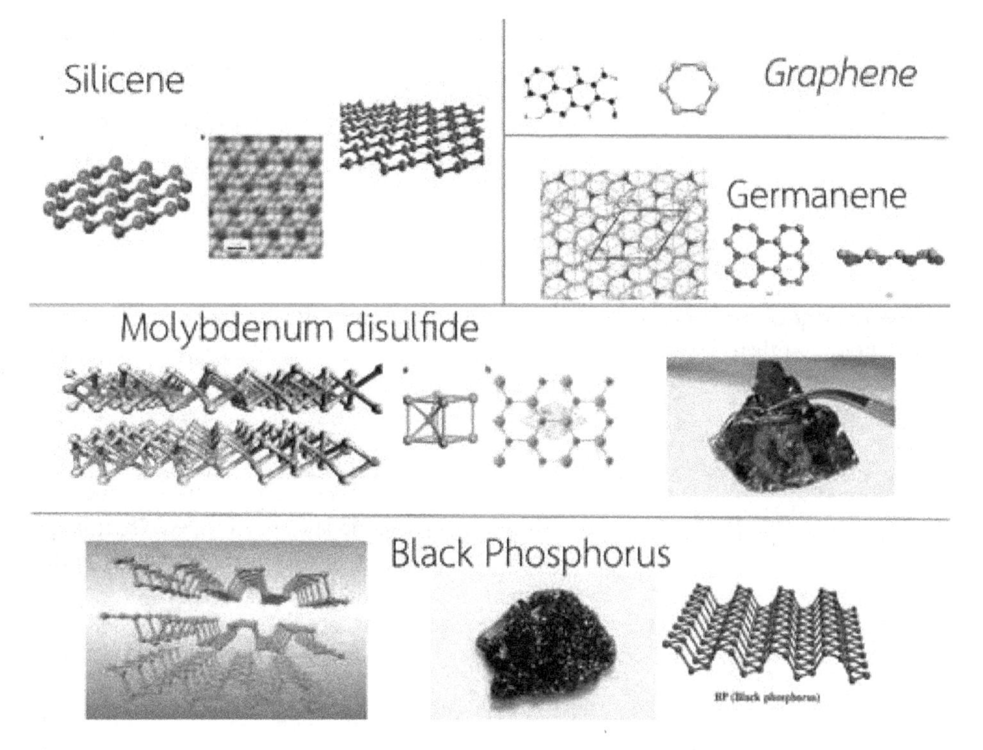

Figure 1.
Recently discovered two-dimensional materials: graphene and beyond graphene.

frequency and the corresponding damping rate at finite temperature could be adjusted by electron doping, and could also be determined for an intrinsic material, where the chemical potential at $T = 0K$ is located exactly at the Dirac point, while for zero temperature, intrinsic plasmons in graphene do not exist.

In this chapter, we will consider thermal behavior of plasmons, their dispersions and damping rates. By equipping with this information, it is possible to predict in advance the thermal properties of an electronic device designed for a particular temperature range. In spite of a number of reported theoretical studies on this subject [16–18], there is still a gap on demonstrating experimentally these unique thermal collective features of 2D materials. Therefore, our review can serve as an incentive to address this issue.

2. Novel two-dimensional materials beyond graphene

All the novel 2D materials considered here could be effectively assigned to an individual category based on their existing (or broken) symmetries and degeneracy in their low-energy band structure. We started with graphene having a bandgap Δ_0 and single-particle energy bands $\varepsilon_{\pm}(k) = \pm\sqrt{(\hbar v_F k)^2 + \Delta_0^2}$, which are symmetric with respect to the Dirac point. Moreover, there is a total spin-valley degeneracy $g = g_s g_v = 4$ for electrons and holes in each band.

Silicene and germanene, which represent buckled honeycomb lattices, possess subbands depending on valley and spin indices, and therefore are only doubly-degenerate. The electron-hole symmetry is broken for molybdenum disulfide and other transition-metal dichalcogenides (TMDC's). For these situations, even though there exists a single electron-hole index $\gamma = \pm 1$, the energy of corresponding states does not have opposite values for each wave number, even at the valley point. In contrast to the electron states, the hole subbands reveal a splitting, as shown in **Figure 2**. All these partially broken symmetries strongly affect the chemical potential of 2D materials as well as its finite-temperature many-body properties. Black phosphorous, apart from all previously discussed broken symmetries, further acquires a preferred spatial direction in its atomic structure which leads to a strong anisotropy of its electronic states and band structures.

2.1 Buckled honeycomb lattices

The energy dispersions of buckled honeycomb lattices, obtained from a Kane-Mele type Hamiltonian, appear as two inequivalent doubly-degenerate pairs of subbands with the same Fermi velocity $v_F = 0.5 \times 10^6$ m/s and are given by

$$\varepsilon_{\xi,\sigma}^{\gamma}(k) = \gamma\sqrt{(\xi\sigma\Delta_z - \Delta_{SO})^2 + (\hbar v_F k)^2}, \tag{1}$$

where $\gamma = \pm 1$ labels symmetric electron and hole states. Here, two bandgaps [19, 20] $\Delta_< = |\Delta_{SO} - \Delta_z|$ and $\Delta_> = \Delta_{SO} + \Delta_z$ are attributed to an intrinsic spin-orbit gap $\Delta_{SO} = 0.5 - 3.5$ meV [21–24] and a tunable asymmetry bandgap Δ_z proportional to applied electric field $\mathcal{E}_z$. The band structure, however, depends only on one composite index $\nu = \sigma\xi$, a product of spin σ and valley ξ index. At $\mathcal{E}_z = 0$, two gaps become the same. As $\mathcal{E}_z \neq 0$, $\Delta_<$ and $\Delta_>$ change in opposite ways, and electrons stay in a topological insulator (TI) state. Additionally, $\Delta_<$ decreases with $\mathcal{E}_z$ until reaching zero, corresponding to a new valley-spin polarized metal. On the other hand, if $\mathcal{E}_z$ further increases, both $\Delta_<$ and $\Delta_>$ will be enhanced, leading to a

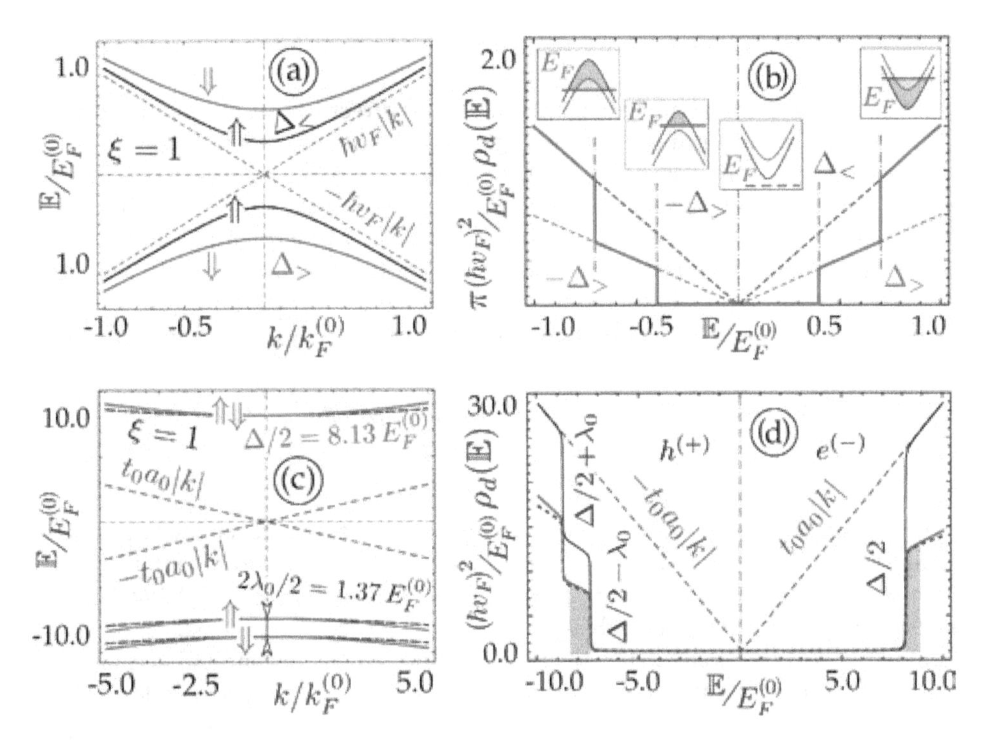

Figure 2.
Energy dispersions and density of states (DOS) $\rho_d(\mathbb{E})$ of silicene [(a) and (b)] and molybdenum disulfide MoS_2 [(c) and (d)], where $E_F^{(0)}$ and $k_F^{(0)}$ are Fermi energy and wave number, respectively. For MoS_2, its dispersions and DOS, corresponding to parabolic band approximation in Eq. (23), are also shown for comparison.

standard band insulator (BI) state for electrons. As we will see below, all of the single electronic and collective properties of buckled honeycomb lattices depend on both bandgaps $\Delta_{<,>}$, and therefore could be tuned by varying a perpendicular electric field to create various types of functional electronic devices.

The wave function of silicene, corresponding to eigenvalue equation in Eq. (1), takes the form [25]

$$\Psi^{\gamma}(k) = \begin{bmatrix} \Psi^{\gamma}_{\xi=1,\sigma=+1}(k) \\ \Psi^{\gamma}_{\xi=1,\sigma=-1}(k) \end{bmatrix},$$
$$\Psi^{\gamma}_{\xi,\sigma}(k) = \sqrt{\frac{\gamma}{2\varepsilon^{\gamma}_{\xi,\sigma}(k)}} \begin{bmatrix} \sqrt{|\varepsilon^{\gamma}_{\xi,\sigma}(k) + \Delta_0^{\xi,\sigma}|} \\ \\ \gamma\sqrt{|\varepsilon^{\gamma}_{\xi,\sigma}(k) - \Delta_0^{\xi,\sigma}|}\,e^{i\theta_k} \end{bmatrix}, \tag{2}$$

where $\theta_k = \tan^{-1}(k_y/k_x)$ and $\Delta_0^{\xi,\sigma} = |\xi\sigma\Delta_z - \Delta_{SO}|$.

Germanene, another representative of buckled honeycomb lattices [26–30], demonstrates substantially higher Fermi velocities and an enhanced intrinsic bandgap ~ 23 meV. For a free-standing germanene, first-principles studies have revealed a buckling distances $\sim 0.640.74$ Å [31, 32].

2.2 Molybdenum disulfide and transition-metal dichalcogenides

MoS_2 is a typical representative of transition-metal dichalcogenide (TMDC) monolayers. TMDC's are semiconductors with the composition of TC_2 type, where

T refers to a transition-metal atom, such as Mo or W, while C corresponds to a chalcogen atom (S, Se or Te).

MoS_2 displays broken inversion symmetry and direct bandgaps. Its most crucial distinction from the discussed buckled honeycomb lattices is its broken symmetry between the electrons and holes so that the corresponding energy bands are no longer symmetric with respect to the Dirac point, but could still be classified by a single index $\gamma = \pm 1$. The absence of this particle-hole symmetry is expected to have a considerable effect on the plasmon branches at both zero and finite temperatures through the thermal convolution of the corresponding quantum states.

Specifically, the energy bands of MoS_2 can be described by a *two-band* model, i.e.,

$$\varepsilon_{\gamma}^{\xi,\sigma}(k) \simeq \frac{1}{2}\xi\sigma\lambda_0 + \frac{\alpha\hbar^2}{4m_e}k^2 + \frac{\gamma}{2}\sqrt{\left[(2t_0a_0)^2 + (\Delta - \xi\sigma\lambda_0)\beta\hbar^2/m_e\right]k^2 + (\Delta - \xi\sigma\lambda_0)^2}, \tag{3}$$

where $\Delta = 1.9$ eV is a gap parameter leading to an extremely large direct bandgap $\simeq 1.7$ eV. There is also substantial internal spin-orbit coupling $\lambda_0 = 0.042\Delta$, $t_0a_0 = 4.95 \times 10^{-29}$ J m is a Dirac-cone term, where $t_0 = 0\ 884\ \Delta$ is the electron hopping parameter while $a_0 = 1.843$ Å is the lattice constant. The t_0a_0 term is ≈ 0.47 compared to $\hbar v_F$ in graphene. Similarly to silicene and germanene, the energy dispersions of TMDC's depend on one composite valley-spin index $\nu = \sigma\xi$. There are also other less important but still non-negligible $\sim k^2$ mass terms with $\alpha = 2.21 = 5.140\beta$, and m_e represents the mass of a free electron.

In practical, we will neglect the $\simeq k^4$ terms, $\simeq t_1a_0^2$ trigonal warping term and anisotropy, which indeed have tiny or no effect on the density of states of the considered material, but would make our model much more complicated. The above dispersions could be presented in a form similarly to those for gapped graphene, i.e., $\varepsilon_{\gamma}^{\nu}(k) = \mathbb{E}_0^{\nu}(k) + \gamma\sqrt{\left[\Delta_0^{\nu}(k)\right]^2 + (t_0a_0k)^2}$, where $\Delta_0^{\nu}(k) = \hbar^2k^2\beta/4m_e + \Delta/2 - \nu\lambda_0/2$ is a k-dependent "gap term" and the band shift is $\mathbb{E}_0^{\nu}(k) = \hbar^2k^2\alpha/4m_e + \nu\lambda_0/2$. A set of somewhat cumbersome analytical expressions for the components of the wave functions corresponding to dispersions in Eq. (3) can be found from Ref. [25].

Using Eq. (3), we can verify that the degeneracy of two hole subbands ($\gamma = -1$), corresponding to $\nu = \pm 1$, will be lifted and two subband will be separated by $\lambda_0 \simeq 79.8$ meV. However, this is not the case for two corresponding electron states ($\gamma = 1$). Consequently, the electron-hole asymmetry exists even at $k = 0$ and becomes even more pronounced at finite k values. One can clearly see this difference by comparing **Figure 2(b)** with **Figure 2(d)**.

3. Thermal plasmons in graphene and other materials

One of the most important features in connection with plasmons at zero and finite temperatures is its dispersion relations, i.e., dependence of the plasmon frequency ω on wave number q. Physically, these complex relations can be determined from the zeros of a dielectric function $\epsilon_T(q,\omega)$, [19, 33] given by

$$\epsilon_T(q,\omega) = 1 - v(q)\Pi_T(q,\omega\,|\,\mu(T)) = 0, \tag{4}$$

where $v(q) = 2\pi\alpha_r/q \equiv e^2/2\epsilon_0\epsilon_r q$ is the 2D Fourier-transformed Coulomb potential, $\alpha_r = e^2/4\pi\epsilon_0\epsilon_r$, and ϵ_r represents the dielectric constant of the host material.

The dielectric function introduced in Eq. (4) is determined directly by the finite-temperature *polarization function*, or *polarizability*, $\Pi_T(q,\omega\,|\,\mu(T))$, which is, in turn, related to its *zero-temperature counterpart*, $\Pi_0(q,\omega\,|E_F)$, by an integral convolution with respect to different Fermi energies [34], given by

$$\Pi_T(q,\omega\,|\,\mu(T)) = \frac{1}{2k_BT}\int_0^\infty d\eta\,\frac{\Pi_0(q,\omega\,|\,\eta)}{1+\cosh\left[(\mu-\eta)/k_BT\right]}, \tag{5}$$

where the integration variable η stands for the electron Fermi energy at $T=0$. This equation is derived for electron doping with $\eta = E_F > 0$. We note that, in order to evaluate this integral, one needs to know in advance how the chemical potential $\mu(T)$ varies with temperature T. Such a unique T dependence reflects a specific selection of a convolution path for a particular material band structure, which we will discuss in Section 4.

The zero-temperature polarizability, which is employed in Eq. (5), is quite similar for all 2D materials considered here. The only difference originates from the degeneracy level of the low-energy band structure, such as $g = g_v g_s = 4$ for graphene with either finite or zero bandgap. We begin with the expression of the partial polarization function with two inequivalent doubly-degenerate pairs of subbands labeled by a composite index ν

$$\begin{aligned}\Pi_0^{(\nu)}(q,\omega\,|\,E_F) = \frac{1}{4\pi^2}\int d^2\boldsymbol{k}\sum_{\gamma,\gamma'=\pm1}\left\{1+\gamma\gamma'\frac{\boldsymbol{k}\cdot(\boldsymbol{k}+\boldsymbol{q})+\Delta_\nu^2}{\varepsilon_\gamma^\nu(k)\,\varepsilon_{\gamma'}^\nu(|\boldsymbol{k}+\boldsymbol{q}|)}\right\}\\ \times\frac{\Theta_0\left[E_F-\varepsilon_\gamma^\nu(k)\right]-\Theta_0\left[E_F-\varepsilon_{\gamma'}^\nu(|\boldsymbol{k}+\boldsymbol{q}|)\right]}{\hbar(\omega+i0^+)+\varepsilon_\gamma^\nu(k)-\varepsilon_{\gamma'}^\nu(|\boldsymbol{k}+\boldsymbol{q}|)},\end{aligned} \tag{6}$$

where $\Theta_0(x)$ stands for a unit-step function, and $\gamma = \pm1$ stands for the electron or hole state with energy dispersions above or below the Dirac point. Moreover, the index ν, which equals to $\sigma\xi = \pm1$ for buckled honeycomb lattices or molybdenum disulfide, specifies two different pairs of degenerate subbands from Eq. (1) or Eq. (3).

Finally, the full polarization function at zero temperature is obtained as

$$\Pi_0(q,\omega|E_F) = \sum_{\nu=\pm1}\Pi_0^{(\nu)}(q,\omega|\,E_F). \tag{7}$$

If the dispersions of low-energy subbands do not depend on the valley or spin indices ξ and σ, summation in Eq. (7) simply gives rise to a factor of two, as we have obtained for graphene.

Integral transformation in Eq. (5), which is used to obtain the finite-temperature polarization function from its zero-temperature counterparts with different Fermi energies, was first introduced in Ref. [34]. It could be derived in a straightforward way by noting that the only quantity which substantially depends on temperature in Eq. (6) is the Fermi-Dirac distribution function $n_F\left[\varepsilon_\gamma^\nu(k)-\mu(T)\right]$. It changes to the unit-step functions $\Theta_0\left[E_F-\varepsilon_\gamma^\nu(k)\right]$ at $T=0$, as used in Eq. (6). As the temperature T increases from zero, the distribution function in Eq. (6) evolves into [35, 36]

$$\begin{aligned}n_F\left[\varepsilon_\gamma^\nu(k)-\mu(T)\right] &= \frac{1}{2}\left[1-\tanh\left(\frac{\varepsilon_\gamma^\nu(k)-\mu(T)}{2k_BT}\right)\right]\\ &= \int_0^\infty d\eta\,\frac{\Theta_0\left[\mu(T)-\varepsilon_\gamma^\nu(k)\right]}{4k_BT\cosh^2\{[\mu(T)-\eta]/2k_BT\}}.\end{aligned} \tag{8}$$

For accessible temperatures, the energy dispersions $\varepsilon_\gamma^\nu(k)$, corresponding wave functions and their overlap factors are all temperature independent. As a result, the polarization function is expected to be modified by the same integral transformation, or a convolution, as each of the Fermi-Dirac distribution function in the numerator of Eq. (6).

We first look at intrinsic plasmons with $E_F = 0$ at $T = 0$. In this case, $\mu(T)$ also remains at the Dirac point for any temperature T. As T increases to $k_BT \gg E_F > 0$, on the other hand, $\Pi_0(q, \omega | E_F)$ for gapless graphene gives rise to a plasmon dispersion $\omega_p \simeq qT$ in the long-wave limit and the damping rate is $\sim q^{3/2}/\sqrt{T}$. As a result, the plasmon mode becomes well defined [6] for $q < 16\epsilon_0\epsilon_r k_BT/\pi e^2$.

Additionally, finite -T polarization function $\Pi_T(q, \omega|\mu_T, \Delta_\beta)$ of a 2D material is directly related to its optical conductivity $\sigma_O^{(T)}(\omega|\mu_T, \Delta_\beta)$ through [19]

$$\sigma_O^{(T)}(\omega|\mu_T, \Delta_\beta) = i\omega e^2 \lim_{q\to 0} \frac{\Pi_T(q, \omega|\mu_T, \Delta_\beta)}{q^2}, \tag{9}$$

where we introduce the notation $\mu_T \equiv \mu(T), \Pi_T(q, \omega | \mu_T, \Delta_\beta) \sim q^2$ as $q \to 0$ for each of our considered 2D materials, regardless of their band structure, as given by Eq. (47) for $T = 0$. This conclusion holds true even for finite T and makes the optical conductivity independent of q, and therefore the $q \to 0$ limit in Eq. (9) becomes finite.

From Eq. (9), we find explicitly that

$$\begin{aligned}
\mathrm{Im}\left[\sigma_O^{(0)}(\omega | E_F, \Delta_\beta)\right] &= \frac{e^2}{4\pi\hbar} \sum_{\beta=\pm 1} \left\{ \frac{4E_F}{\hbar\omega}\left[1 - \left(\frac{\Delta_\beta}{E_F}\right)^2\right] + \left[1 + \left(\frac{2\Delta_\beta}{\hbar\omega}\right)^2\right] \ln\left|\frac{2E_F - \hbar\omega}{2E_F + \hbar\omega}\right| \right\}, \\
\mathrm{Re}\left[\sigma_O^{(0)}(\omega | E_F, \Delta_\beta)\right] &= \frac{e^2}{4\hbar}\Theta(\hbar\omega - 2E_F) \sum_{\beta=\pm 1}\left[1 + \left(\frac{2\Delta_\beta}{\hbar\omega}\right)^2\right].
\end{aligned} \tag{10}$$

Here, the *state-blocking effect* due to Pauli exclusion principle directly results in the diminishing of the real part of the optical conductivity at zero temperature for $\hbar\omega < 2E_F$. However, if $T > 0$, such state-blocking effect will not exist [37–40] due to

$$\Theta\left(E_F - \varepsilon_\beta^\gamma(k)\right) \Rightarrow \frac{1}{2}\left\{1 - \tanh\left[\frac{\varepsilon_\beta^\gamma(k) - \mu_T}{2k_BT}\right]\right\}. \tag{11}$$

Furthermore, for gapless ($\Delta_\beta = 0$) but doped ($E_F > 0$) graphene in the high-T limit we obtain from Eq. (10)

$$\begin{aligned}
\sigma_O^{(T)}(\omega | \mu_T, \Delta_\beta = 0) \simeq \frac{e^2}{\hbar}\Bigg\{ &\frac{\hbar\omega}{16k_BT}\left[1 - \frac{1}{3}\left(\frac{\hbar\omega}{4k_BT}\right)^2\right] \\
&+ i\,\frac{2\ln 2k_BT}{\pi\hbar\omega}\left[1 + 2\ln 2\left(\frac{E_F}{4\ln 2k_BT}\right)^4\right]\Bigg\},
\end{aligned} \tag{12}$$

where we have used the high - T limit [17] for the chemical potential $\mu_T \approx (E_F^2/4\ln 2k_BT)$. In either case above, we have to present analytical expression for $\mu(T)$ as a function of T so as to gain the explicit T dependence of optical conductivity. From Eq. (12) we conclude that Im $\left[\sigma_O^{(T)}(\omega | \mu_T, \Delta_\beta = 0)\right]$ depends weakly on E_F.

On the other hand, for gapped ($\Delta_\beta = \Delta_0$) but undoped ($E_F = 0$) graphene at high T ($k_B T \gg \Delta_0$ and $\hbar\omega$), we get its optical conductivity [41]

$$\mathrm{Re}\left[\sigma_O^{(T)}(\omega|\mu_T, \Delta_0)\right]_{E_F=0} = \frac{e^2}{16\hbar}\left(\frac{\hbar\omega}{k_B T}\right)\left(1 - \frac{\Delta_0}{\hbar\omega}\right),$$

$$\mathrm{Im}\left[\sigma_O^{(T)}(\omega|\mu_T, \Delta_0)\right]_{E_F=0} = \frac{4e^2}{\pi\hbar}\left(\frac{k_B T}{\hbar\omega}\right)\left\{2\ln 2 - \left(\frac{\Delta_0}{k_B T}\right)^2\left[\mathbb{C}_0 - \ln\left(\frac{\Delta_0}{2k_B T}\right)\right]\right\}, \tag{13}$$

where the constant $\mathbb{C}_0 \simeq 0.79$ appears due to a finite bandgap. [36].

4. Chemical potential at finite temperatures

As we have seen from Section 3, we need know $\mu(T)$ as a function of T explicitly so as to gain T dependence of polarization function, plasmon, transport and optical conductivities, or any other quantities related to collective behaviors of 2D materials at finite temperatures [42].

The density of states (DOS), which plays an important tool in calculating electron (or hole) Fermi energy E_F and chemical potential $\mu(T)$, is defined as

$$\rho_d(\mathbb{E}) = \sum_{\gamma=\pm1}\sum_{\xi,\sigma=\pm1}\int\frac{d^2\boldsymbol{k}}{(2\pi)^2}\,\delta\left(\mathbb{E} - \varepsilon^\gamma_{\xi,\sigma}(k)\right), \tag{14}$$

where $\delta(x)$ is Dirac delta function. Using Eq. (14), we immediately obtain a piece-wise linear function for silicene [43]

$$\rho_d(\mathbb{E}) = \frac{1}{\pi}\sum_{\gamma=\pm1}\frac{\gamma\mathbb{E}}{\hbar^2 v_F^2}\sum_{i=<,>}\Theta_0\left(\frac{\mathbb{E}}{\gamma} - \Delta_i\right). \tag{15}$$

This result is equivalent to the DOS of graphene except that there are no states within the bandgap region, as demonstrated by two unit-step functions $\Theta_0(|\mathbb{E}| - \Delta_<)$ and $\Theta_0(|\mathbb{E}| - \Delta_>)$.

Finally, the chemical potential $\mu(T)$ can be calculated using the *conservation of the difference of electron and hole densities*, [17] $n_e(T)$ and $n_h(T)$, for all temperatures including $T = 0$, i.e.,

$$n = n_e(T) - n_h(T) = \int_0^\infty d\mathbb{E}\rho_d(\mathbb{E}) f_{\gamma=1}(\mathbb{E}, T) - \int_{-\infty}^0 d\mathbb{E}\rho_d(\mathbb{E})\left[1 - f_{\gamma=1}(\mathbb{E}, \mathbb{T})\right], \tag{16}$$

where $f_{\gamma=1}(\mathbb{E}, T) = \{1 + \exp\,[(\mathbb{E} - \mu(T))/k_B T]\}^{-1}$ is the Fermi function for electrons in thermal equilibrium. The hole distribution function is just $f_{\gamma=-1}(\mathbb{E} < 0, T) = 1 - f_{\gamma=1}(\mathbb{E}, T)$.

At $T = 0$, it is straightforward to get the Fermi energy E_F from Eq. (16) for silicene

$$E_F^2 - \frac{1}{2}\left(\Delta_>^2 + \Delta_<^2\right) = (\hbar v_F)^2 \pi n. \tag{17}$$

where we have assumed that both subbands are occupied for simplicity. The discussions of other cases can be found from Ref. [18]. Consequently, minimum electron density required to occupy the upper subband of silicene is $n_c = 2\Delta_{SO}\Delta_z/\pi\hbar^2 v_F^2$.

On the other hand, by applying Eq. (16), in combination with DOS in Eq. (15), for silicene, a transcendental (non-algebraic) equation [43, 44] could be obtained for $\mu(T)$, that is

$$\left(\frac{\hbar v_F}{k_B T}\right)^2 n = \frac{1}{\pi}\sum_{\gamma=\pm 1}\gamma \sum_{i=<,>}\left\{-\mathrm{Li}_2\left(-\exp\left[\frac{\gamma\mu(T)-\Delta_i}{k_B T}\right]\right) + \frac{\Delta_i}{k_B T}\ln\left(1+\exp\left[\frac{\gamma\mu(T)-\Delta_i}{k_B T}\right]\right)\right\}, \tag{18}$$

where $\mathrm{Li}_2(x)$ is a polylogarithm or dilogarithm function, defined mathematically by

$$\mathrm{Li}_2(z) = -\int_0^z dt\left[\frac{\ln(1-t)}{t}\right]. \tag{19}$$

Interestingly, the right-hand side of Eq. (18) contains terms corresponding to both pristine and gapless graphene, using which we find from Ref. [17].

$$\frac{1}{2(k_B T)^2}E_F^2 = -\sum_{\gamma=\pm 1}\gamma\,\mathrm{Li}_2\left\{-\exp\left[\frac{\gamma\mu(T)}{k_B T}\right]\right\}, \tag{20}$$

as well as a well-known analytical expression for $\mu_0(T)$ of 2D electron gas with Schrödinger-based electron dynamics

$$\mu_0(T) = k_B T\log\left(1+\exp\left[\frac{\pi\hbar^2 n_0}{m^* k_B T}\right]\right). \tag{21}$$

An advantage of Eq. (18) is that it could be solved even without taking an actual integration. In fact, one can either readily solve it numerically using some standard computational algorithms, or introduce an analytical approximation to the sought solution near specific temperature assigned.

Numerical results for $\mu(T)$ of silicene are presented in **Figure 3**. In all cases, $\mu(T)$ decreases with increasing T from zero. However, it is very important to notice that

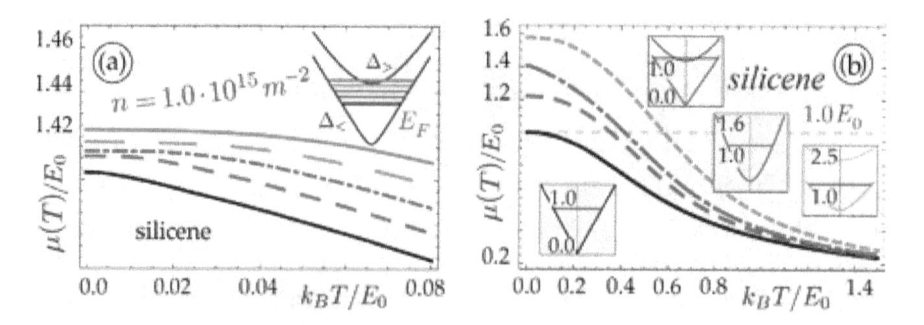

Figure 3.
Temperature dependence of the chemical potential μ(T) for silicene with two inequivalent energy subbands with various bandgaps and a fixed doping density n = 1 × 10^{11} cm^{-2}. Panel (a) highlights the situation close to T = 0, while panel (b) shows the whole temperature range. Here, E_0 is the Fermi energy of graphene.

$\mu(T)$ never reaches zero or changes its sign in the systems with an electron–hole symmetry due to increasing contribution from holes in Eq. (16). All of displayed results in **Figure 3** depend on individual bandgaps $\Delta_<$ and $\Delta_>$. The special case with $\Delta_< = \Delta_>$ corresponds to gapped graphene, for which plasmon modes at $T = 0$ were studied in Ref. [33]. The graphene Fermi wave number is $k_F = \sqrt{\pi n}$, irrelevant to its bandgap. The general relation between k_F and n in 2D materials is given by $(2\pi)^2 n = g\pi k_F^2$. The experimentally allowable electron (or hole) doping is within the range of $n = 10^{10} - 10^{12}$ cm^{-2}, leading to $k_F = 10^6 - 10^7$ cm^{-1}. For two pair of inequivalent subbands, such as in silicene or MoS_2, there are two different Fermi wave numbers for these subbands. Moreover, the numerically calculated $\mu(T)$ as functions of T for electron and hole doping are presented in **Figure 4**. In this case, however, there exists no electron-hole symmetry, and therefore the resulting $\mu(T)$ can be zero and change its sign as T increases, in contrast to the results in **Figure 3**.

Eq. (18) could also be applied to a wide range of 2D materials if its DOS has a linear dependence on energy $\mathbb{E}$. Particularly, it is valid for calculating the finite-T chemical potential of TMDC's with an energy dispersion presented in Eq. (3). However, we are aware that some terms in Eq. (3) for TMDC's, which might be insignificant for dispersions of other 2D materials, become essential in DOS because of very large bandgap and mass terms around $k = 0$. As an estimation, for $k/k_0 \approx 5.0$, the correction term $\simeq \beta\Delta k^4$ must be taken into account. Meanwhile, the highest accessible doping $n = 10^{13}$ cm^{-2} only gives rise to a Fermi energy $E_F \sim \lambda_0$, comparable to spin-orbit coupling.

Now, we turn to calculate $\mu(T)$ as a function of T for MoS_2 with a much more complicated band structure. After taking into account the $\sim k^2$ mass terms, we are able to write down [18]

$$\rho_d(\mathbb{E}) = \frac{1}{2\pi\hbar^2}\sum_{\gamma,\nu=\pm1}\left|\frac{\alpha+\gamma\beta}{4m_e} + \frac{\gamma(t_0 a_0)^2}{\hbar^2(\Delta-\nu\lambda_0)}\right|^{-1}\Theta_0\left(\gamma\left[\mathbb{E}-\frac{\nu\lambda_0}{2}\right]-\frac{1}{2}(\Delta-\nu\lambda_0)\right), \tag{22}$$

where the calculation is based on a parabolic-subband approximation, i.e.,

$$\varepsilon_\gamma^\nu(k) = \frac{1}{2}[\nu\lambda_0(1-\gamma)+\gamma\Delta] + \left[\frac{\hbar^2}{4m_e}(\alpha+\gamma\beta) + \frac{\gamma(t_0 a_0)^2}{\Delta-\nu\lambda_0}\right]k^2. \tag{23}$$

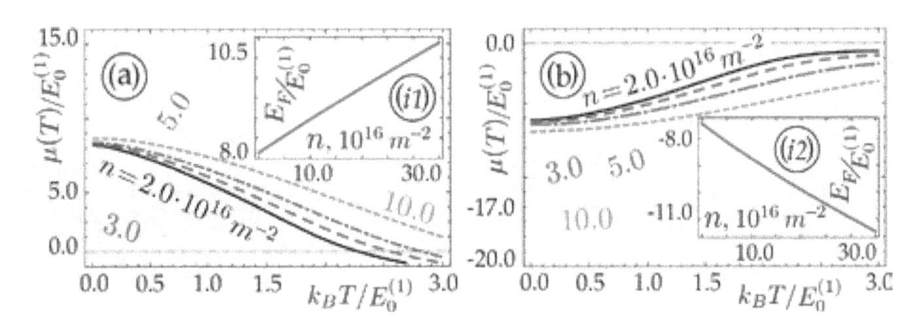

Figure 4.
Temperature dependence of the chemical potential μ(T) for molybdenum disulfide for cases of electron (a) and hole (b) doping with various doping densities. μ(T) might change its sign in contrast to the previously considered graphene and silicene. The two insets demonstrate how the Fermi energy depends on the electron and hole doping densities, correspondingly.

From Eq (22), we further seek an explicit expression for DOS in the form a piecewise-linear function of energy $\mathbb{E}$: $\rho_d(\mathbb{E}) = A_i + B_i\mathbb{E}$. A complete set of expressions for DOS of MoS_2 has been reported in Ref. [18]. Here, we merely provide and discuss these DOS expression around the lower hole subband with $\mathbb{E} \approx -\Delta/2 - \lambda_0$, yielding

$$\rho_d(\mathbb{E}) = c_0^{(1)} + \left[\mathbb{E} - \left(\frac{\Delta}{2} + \lambda_0\right)\right] c_1^{(1)},$$
$$c_0^{(1)} = \frac{1}{2\pi} \sum_{\nu=\pm 1} \frac{\Delta - \nu\lambda_0}{(a_0 t_0)^2 + (\beta - \alpha)(\Delta - \nu\lambda_0)},$$
$$c_1^{(1)} = \frac{1}{\pi} \sum_{\nu=\pm 1} \frac{\left[(a_0 t_0)^2 + \hbar^2\beta/(4m_e)(\Delta - \nu\lambda_0)\right]^2 \delta\varepsilon}{\left\{(a_0 t_0)^2 + \hbar^2/(4m_e)[(\beta - \alpha)(\Delta - \nu\lambda_0)]\right\}^3} < 0, \tag{24}$$

or numerically,

$$c_0^{(1)} = 0.233 \frac{1}{t_0 a_0^2} = 15.17 \frac{E_0}{(\hbar v_F)^2},$$
$$c_1^{(1)} = -0.458 \frac{1}{(t_0 a_0)^2} = -2.077 \frac{1}{(\hbar v_F)^2}. \tag{25}$$

The calculated numerical results for DOS in all regions are listed in **Table 1**. All introduced coefficients $c_0^{(i)}, c_1^{(i)}$ for $i = 1, 2, 3$ can be deduced from the calculated parameters A_i and B_i using a similar correspondence as in Eq. (24).

The critical doping density which is required to populate the lower hole subband in MoS_2 is found to be

$$n_c = \frac{2}{\pi} \frac{\lambda_0 \Delta}{(t_0 a_0)^2} = 1.0 \times 10^{14}\ \text{cm}^{-2}. \tag{26}$$

Therefore, for most experimentally accessible densities $n \leq 10^{13}\ \text{cm}^{-2}$, the lower hole subband still could not be populated at $T = 0$.

Next, we would evaluate both sides of Eq. (16) for MoS_2. As an example, we consider electron doping with density $n_e > 0$. The electron Fermi energy E_F^e is determined by the following relation

$$n_e = \frac{c_1^{(3)}}{2}\left[(E_F^e)^2 - \frac{\Delta^2}{4}\right] + c_0^{(3)}\left(E_F^e - \frac{\Delta}{2}\right). \tag{27}$$

From Eq. (27), we can easily find the electron Fermi energy $E_F > 0$ as

Range index	Energy range	γ	ν	$A_i[1/(t_0 a_0^2)]$	$B_i[1/(t_0 a_0)^2]$
$i = 1$	$\mathbb{E} < -\Delta/2 - \lambda_0$	−1	+1	0.0174	−0.169
$i = 2$	$\lvert\mathbb{E} + \Delta/2\rvert < \lambda_0$	−1	−1	0.043	−0.308
$i = 3$	$\mathbb{E} > \Delta/2$	+1		0.078	+0.179

Table 1.
Linearized density of states (DOS) $\rho_d(\mathbb{E}) = A_i + B_i\mathbb{E}$ of MoS_2 for all three energy regions. Here, the DOS within the gap region, $-\Delta/2 + \lambda_0 < \mathbb{E} < \Delta/2$, is zero.

$$E_F^e = \frac{1}{c_1^{(3)}} \left[-c_0^{(3)} + \sqrt{\left(c_0^{(3)} + c_1^{(3)} \frac{\Delta}{2} \right)^2 + 2n_e c_1^{(3)}} \right]. \tag{28}$$

In a similar way, for hole doping with density n_h and the Fermi energy E_F^h located between two hole subbands (region 2), we find

$$|E_F^h| = \frac{1}{c_1^{(2)}} \left\{ \sqrt{\left[c_0^{(2)} - c_1^{(2)} \left(\frac{\Delta}{2} - \lambda_0 \right) \right]^2 - 2n_h c_1^{(2)}} - c_0^{(2)} \right\}, \tag{29}$$

where $c_1^{(2)} < 0$ and $\mathbb{E} < -\Delta/2 + \lambda_0$. From Eq. (29) we easily find the doping density

$$n_h = \left(\frac{\Delta}{2} + E_F^h - \lambda_0 \right) \left[c_0^{(2)} + \frac{c_1^{(2)}}{2} \left(\frac{\Delta}{2} - \left(E_F^h + \lambda_0 \right) \right) \right]. \tag{30}$$

The right-hand side of Eq. (16) for TMDC's could be expressed as a combination of electron and hole contributions $\mathcal{I}_e(\Delta|T) - \mathcal{I}_h(\Delta, \lambda_0|T)$. Here, we will introduce two self-defined functions

$$\begin{aligned} \mathcal{A}_0(\mathbb{E}, T) &= \left[1 + \exp\left(\frac{\mathbb{E} - \mu(T)}{k_B T} \right) \right]^{-1}, \\ \mathcal{A}_1(\mathbb{E}, T) &= \mathbb{E} \mathcal{A}_0(\mathbb{E}, T) = \mathbb{E} \left[1 + \exp\left(\frac{\mathbb{E} - \mu(T)}{k_B T} \right) \right]^{-1}, \end{aligned} \tag{31}$$

so that

$$\mathcal{I}_e(\Delta|T) = \sum_{j=0}^{1} c_j^{(3)} \int_{\Delta/2}^{\infty} d\mathbb{E} \, \mathcal{A}_j(\mathbb{E}, T). \tag{32}$$

For convenience, we introduce another function $\mathcal{R}_p(T, X)$

$$\mathcal{R}_p(T, X) = \int_0^{\infty} d\xi \frac{\xi^p}{1 + \exp\left(\xi - X/k_B T \right)}, \tag{33}$$

where $\xi = (\mathbb{E} - \Delta_<)/k_B T$. Consequently, we are able to rewrite Eq. (32) as

$$\mathcal{I}_e(\Delta\,|\,T) = k_B T \left(c_0^{(3)} + \frac{\Delta}{2} \right) \mathcal{R}_0\left(T, \mu(T) - \frac{\Delta}{2} \right) + c_1^{(3)} (k_B T)^2 \mathcal{R}_1\left(T, \mu(T) - \frac{\Delta}{2} \right), \tag{34}$$

where two terms with $p = 0, 1$ are physically related to a 2D electron gas. Explicitly, Eq. (33) leads to

$$\begin{aligned} \mathcal{R}_0(T, X) &= \ln\left(1 + \exp\left(\frac{X}{k_B T} \right) \right), \\ \mathcal{R}_1(T, X) &= -\mathrm{Li}_2\left(-\exp\left(\frac{X}{k_B T} \right) \right), \end{aligned} \tag{35}$$

Using these self-defined functions and their notations, we finally arrive at the "*hole term*" $\mathcal{I}_h(\Delta, \lambda_0|T)$ in Eq. (16)

$$\int_{-\infty}^{0} d\mathbb{E}\rho_d(|\mathbb{E}|)[1-f_1(\mathbb{E},\mathrm{T})] = \int_{-\infty}^{-\Delta/2+\lambda_0} d\mathbb{E}\left(c_0^{(2)} - c_1^{(2)}\mathbb{E}\right)\left[1+\exp\left(\frac{\mu(T)-\mathbb{E}}{k_BT}\right)\right]^{-1}$$
$$+\int_{-\infty}^{-\Delta/2-\lambda_0} d\mathbb{E}\left(\delta c_0^{(1)} - \delta c_1^{(1)}\mathbb{E}\right)\left[1+\exp\left(\frac{\mu(T)-\mathbb{E}}{k_BT}\right)\right]^{-1}$$
$$\equiv \sum_{j=1}^{4} \mathcal{I}_h^{(j)}(\Delta, \lambda_0\,|\,T), \tag{36}$$

where $\delta c_i^{(1)} = c_i^{(1)} - c_i^{(2)}$ for $i = 0, 1$, and

$$\mathcal{I}_h^{(1)}(\Delta, \lambda_0\,|\,T) = k_BT\left(\frac{\Delta}{2} - \lambda_0 + c_2^{(0)}\right)\mathcal{R}_0(T, -[\mu(T) + \Delta/2 - \lambda_0]),$$
$$\mathcal{I}_h^{(2)}(\Delta, \lambda_0\,|\,T) = c_2^{(1)}(k_BT)^2\mathcal{R}_1(T, -[\mu(T) + \Delta/2 - \lambda_0]),$$
$$\mathcal{I}_h^{(3)}(\Delta, \lambda_0\,|\,T) = k_BT\left(\frac{\Delta}{2} + \lambda_0 + \delta c_1^{(0)}\right)\mathcal{R}_0(T, -[\mu(T) + \Delta/2 + \lambda_0]),$$
$$\mathcal{I}_h^{(4)}(\Delta, \lambda_0\,|\,T) = \delta c_1^{(1)}(k_BT)^2\mathcal{R}_1(T, -[\mu(T) + \Delta/2 + \lambda_0]). \tag{37}$$

Here, both $\mathcal{I}_e(\Delta\,|\,T)$ in Eq. (34) and $\mathcal{I}_h(\Delta, \lambda_0\,|\,T)$ in Eq. (36) comprise a finite-temperature part for the right-hand side of Eq. (16). Its left-hand side has already been given by Eq. (27). From these results, it is clear that there exists no symmetry between the electron and hole states at either zero or finite T. Finally, $\mu(T)$ of TMDC's could be computed from a transcendental equation in Eq. (18), similarly to finding $\mu(T)$ for silicene.

By using the calculated $\mu(T)$, the plasmon dispersions and their Landau damping, determined from Eqs. (4) and (5), are displayed in **Figure 5** for silicene at different T. Comparison of panels (a) and (b) indicates that the T dependence of plasmon damping is not uniform even on a fixed convolution path $\mu(T)$. The doping

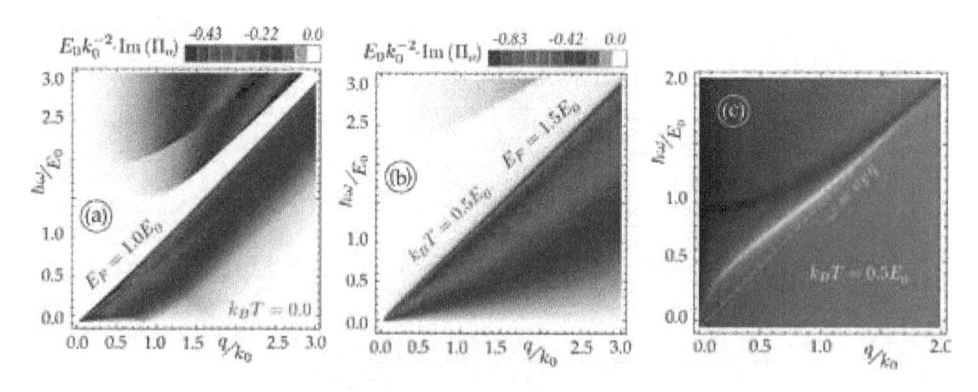

Figure 5.
Particle-hole modes and plasmon branch for extrinsic (doped) silicene layer at a finite temperature. Panels (a) and (b) show two comparative graphs for $\mathrm{Im}[\Pi_0(q, \omega|\mu(T))]$ *at zero and finite* T*, respectively, while plot (c) presents the finite-*T *plasmon branch with* $\mu(T)$ *calculated from Eq. (16).*

density, on the other hand, widens the plasmon damping-free regions. Therefore, both the thermal and doping effects are found to compete with each other in dominating the plasmon dampings through selecting different convolution paths $\mu(T)$ with various doping densities or Fermi energies. Furthermore, the plasmon energy in (c) is pushed up slightly by increasing doping density at finite T.

5. Dice lattice and α-$\mathcal{T}_3$ materials

In addition to graphene and silicene, another type of Dirac-cone materials is the one with fermionic states in which multiple Dirac points evolve into a middle flat band. One of the first fabricated materials with such a flat band is a dice or a $\mathcal{T}_3$ lattice, for which its atomic composition consists of hexagons similarly to graphene, but with an additional atom at the center of each hexagon. In a dice lattice, the bond coupling between a central site and three nearest neighbors is the same as that between atoms on corners, while for an α-$\mathcal{T}_3$ model the ratio α between hub-rim and rim-rim hopping coefficients can vary [45, 46] within the range of $0 < \alpha < 1$.

The low-energy electronic states of α-$\mathcal{T}_3$ materials are specified by a 3×3 pseudospin - 1 Dirac Hamiltonian, which results in three solutions for the energy dispersions and includes one completely flat and dispersionless band with $\varepsilon_0(k) \equiv 0$. The other two bands are equivalent to Dirac cone $\varepsilon_\pm(k) = \pm\hbar v_F k$ in graphene with the same Fermi velocity $v_F = 10^8$ cm/s. All of three bands touch at the corners of the first Brillouin zone, and therefore the band structure becomes metallic. In addition, the flat band has been shown to be stable against external perturbations, magnetic fields and structure disorders [47].

The α-$\mathcal{T}_3$ model was initially considered only as a theoretical contraption, an interpolation between graphene and a dice lattice. As parameter $\alpha \to 0$, this structure approaches graphene and a completely decoupled system of the hub atoms at the centers of each hexagon. A bit later, first evidence of really existing or fabricated materials with α-$\mathcal{T}_3$ electronic structure began mounting up. This includes Josephson arrays, optical arrangement based on the laser beams, Kagome and Lieb lattices with optical waveguides, $Hg_{1-x}Cd_x Te$ for a specific electron doping density, dielectric photonic crystals having zero-refractive index and a few others [48, 49]. So far, α-$\mathcal{T}_3$ model is believed to be the most promising innovative low-dimensional systems, and is one of the mostly investigated material in modern condensed matter physics. The most important technological application of α-$\mathcal{T}_3$ rests on the availability of materials with various α values, i.e., with small and large rim-hub hopping coefficients, ranging from $\alpha = 0$ for graphene up to $\alpha = 1$ for a dice lattice.

The low-energy Dirac-Weyl Hamiltonian for the α-$\mathcal{T}_3$ model is [45]

$$\hat{\mathbb{H}}_\xi^\phi(\boldsymbol{k}) = \hbar v_F \begin{bmatrix} 0 & k_-^\xi \cos\phi & 0 \\ k_+^\xi \cos\phi & 0 & k_-^\xi \sin\phi \\ 0 & k_+^\xi \sin\phi & 0 \end{bmatrix}, \tag{38}$$

where $\boldsymbol{k} = (k_x\ k_y)$ is the electron wave vector $k_\pm^\xi = \xi k_x \pm i k_y$ $\xi = \pm 1$ corresponds to two different valleys, and v_F is the Fermi velocity. Here, the parameter α is related to the geometry phase $\phi = \tan^{-1}\alpha$ which directly enters into the Hamiltonian in Eq. (38). The phase ϕ possesses a fixed, one-to-one correspondence to the Berry phase of electrons in α-$\mathcal{T}_3$ model. In particular, for $\alpha = 1$ or $\phi = \pi/4$ we get a dice lattice with its Hamiltonian given by [50].

$$\hat{\mathbb{H}}_{\xi}^{d}(\boldsymbol{k}) = \frac{\hbar v_F}{\sqrt{2}} \begin{bmatrix} 0 & k_{-}^{\xi} & 0 \\ k_{+}^{\xi} & 0 & k_{-}^{\xi} \\ 0 & k_{+}^{\xi} & 0 \end{bmatrix}. \tag{39}$$

Three energy bands from Hamiltonian in Eq. (38) or Eq. (39) are $\varepsilon_0^{\gamma}(k) = \gamma \hbar v_F k$ for valence ($\gamma = -1$), conduction ($\gamma = +1$) and flat ($\gamma = 0$) bands. These energy bands are degenerate with respect to ξ and phase ϕ. The corresponding wave functions for the valence and conduction bands take the form

$$\Psi_0^{\gamma=\pm 1}(\boldsymbol{k}|\,\xi,\phi) = \frac{1}{\sqrt{2}} \begin{bmatrix} \xi \cos\phi \, \mathrm{e}^{-i\xi\theta_{\mathbf{k}}} \\ \gamma \\ \xi \sin\phi \, \mathrm{e}^{i\xi\theta_{\mathbf{k}}} \end{bmatrix}, \tag{40}$$

where $\tan\theta_{\mathbf{k}} = k_y/k_x$. Meanwhile, for the flat band, we find

$$\Psi_0^{\gamma=0}(\boldsymbol{k}\,|\xi,\phi) = \begin{bmatrix} \xi \sin\phi \, \mathrm{e}^{-i\xi\theta_{\mathbf{k}}} \\ 0 \\ -\xi \cos\phi \, \mathrm{e}^{i\xi\theta_{\mathbf{k}}} \end{bmatrix}. \tag{41}$$

Here, the components of wave functions in Eqs. (40) and (41) depend on valley index ξ and phase ϕ, which leads to the same dependence on all collective properties of an α-$\mathcal{T}_3$ materials, including plasmon dispersion.

Now we turn to deriving plasmon branches and their damping rates at finite T in α-$\mathcal{T}_3$ model. The computation procedure is quite similar to that in the case of two non-equivalent doubly degenerate subband pairs, including silicene, germanene and MoS_2 discussed in Section 4.

For α-$\mathcal{T}_3$ model, the finite-T polarization function $\Pi_T(q,\omega\,|\,\mu(T))$ can be obtained by an integral transformation of its zero-temperature counterpart $\Pi_0(q,\omega|E_F)$, as presented in Eq. (5). In this case, the zero-T counterpart $\Pi_0(q,\omega|E_F)$ is calculated as

$$\Pi_0(q,\omega|E_F) = \frac{1}{\pi^2} \sum_{\gamma,\gamma'=0,\pm 1} \int d^2\boldsymbol{k}\,\mathbb{O}_{\gamma,\gamma'}(\boldsymbol{k},\boldsymbol{k}+\boldsymbol{q}\,|\phi) \times \frac{\Theta_0(E_F - \varepsilon_\gamma(k)) - \Theta_0(E_F - \varepsilon_{\gamma'}(|\boldsymbol{k}+\boldsymbol{q}|))}{\hbar(\omega + i0^+) + \varepsilon_\gamma(k) - \varepsilon_{\gamma'}(|\,\boldsymbol{k}+\boldsymbol{q}\,|)}. \tag{42}$$

Structurally, Eq. (42) looks quite similarly to Eq. (6) for buckled honeycomb lattices and TMDC's. The most significant difference comes as the existence of an additional flat band with $\gamma = 0$ so that the summation index runs over ± 1 and 0 instead of two. On the other hand, the overall expression for $\Pi_0(q,\omega|E_F)$ in Eq. (42) is simplified because the 4-fold degeneracy of each energy band independent of valley and spin index.

Here, we would limit our consideration to the case of electron doping with $n > 0$ and apply the random-phase approximation theory only for that case. For electron doping with $n > 0$, we can neglect the transitions within the valence band and also the transitions between the flat and valences bands due to full occupations of these electronic states. On the other hand, the overlap of initial and final electron transition states is defined by [51] $\mathbb{O}_{\gamma,\gamma'}^{\xi}\left(\boldsymbol{k},\boldsymbol{k}'|\phi,\lambda_0\right)$ with respect to the initial $\Psi_{\gamma}^{\xi}(\boldsymbol{k},\lambda_0)$ and the final $\Psi_{\gamma'}^{\xi}(\boldsymbol{k}',\lambda_0)$ states with a momentum transfer $\boldsymbol{q} = \boldsymbol{k}' - \boldsymbol{k}$, i.e.,

$$\mathbb{O}^{\xi}_{\gamma,\gamma'}(\boldsymbol{k},\boldsymbol{k}+\boldsymbol{q}\,|\phi,\lambda_0) = |\mathbb{S}^{\xi}_{\gamma,\gamma'}(\boldsymbol{k},\boldsymbol{k}+\boldsymbol{q}\,|\,\phi,\lambda_0|)^2,$$
$$\mathbb{S}^{\xi}_{\gamma,\gamma'}(\boldsymbol{k},\boldsymbol{k}+\boldsymbol{q}|\phi,\lambda_0) = \left\langle \Psi^{\xi}_{\gamma}(\boldsymbol{k},\lambda_0)\,|\Psi^{\xi}_{\gamma'}(\boldsymbol{k}+\boldsymbol{q},\lambda_0)\right\rangle, \tag{43}$$

where $\beta_{\mathbf{k},\mathbf{k}'} = \theta_{\mathbf{k}'} - \theta_{\mathbf{k}}$ is the scattering angle between two electronic states and $k' = \sqrt{k^2 + q^2 + 2kq\,\cos\beta_{\mathbf{k},\mathbf{k}'}}$. Moreover, we find from Eq. (43) [52]

$$\mathbb{O}^{\xi}_{\gamma,\gamma'}(\boldsymbol{k},\boldsymbol{k}+\boldsymbol{q}|\phi,\lambda_0) = \frac{1}{4}\left[(1+\cos\beta_{\mathbf{k},\mathbf{k}'})^2 + \cos^2(2\phi)\sin^2\beta_{\mathbf{k},\mathbf{k}'}\right] \tag{44}$$

for an arbitrary value of ϕ or α. It is easy to verify the known results $(1+\cos\beta_{\mathbf{k},\mathbf{k}'})/2$ for graphene and $(1+\cos\beta_{\mathbf{k},\mathbf{k}'})^2/4$ for a dice lattice as two limiting cases of our general result in Eq. (44) as $\alpha \to 0$ or $\alpha \to 1$, respectively. Furthermore, we find from Eq. (44) that the overlap does not depend on valley index ξ, even though individual wave function does, and then this index can be dropped. However, the valley-dependence in $\mathbb{O}^{\xi}_{\gamma,\gamma'}(\boldsymbol{k},\boldsymbol{k}+\boldsymbol{q}\,|\phi,\lambda_0)$ persists if α-$\mathcal{T}_3$ material is irradiated by circularly- or elliptically-polarized light. This incident radiation permits creating an valleytronic filter or any other types of valleytronic electron device.

Density plots for Landau damping with $\mathrm{Im}[\Pi_O(q,\omega|\mu(T))] \neq 0$ is presented in **Figure 6**, where we find plasmon branch will be completely free from damping within the region determined by $\hbar\omega/E_0 \leq 1$ and $q/k_0 \leq 1$, independent of geometry phase ϕ. On the other hand, another region with $\hbar\omega \leq \hbar v_F q$ (below the diagonal) becomes always Landau damped. Increasing T is able to increase greatly the damping in the region below the diagonal, as seen in **Figure 6(c)**.

In a correspondence to the damping of plasmons presented in **Figure 6**, we show in **Figure 7** the density plots for plasmon dispersions at $T = 0$ in (a), (b) and $k_B T = E_F$ in (c), (d). Comparing **Figure 7(a)** with **Figure 7(c)** we have clearly seen the thermal suppression of Landau damping for plasmon mode entering into a high-frequency region beyond $\hbar\omega = E_F$. To visualize a full plasmon dispersion clearly, we also include damped counterpart in **Figure 7(b)** and **(d)** at $T = 0$ and $k_B T = E_F$, respectively, where a significant enhancement of plasmon energy appears for large q values, moving upwards from the diagonal.

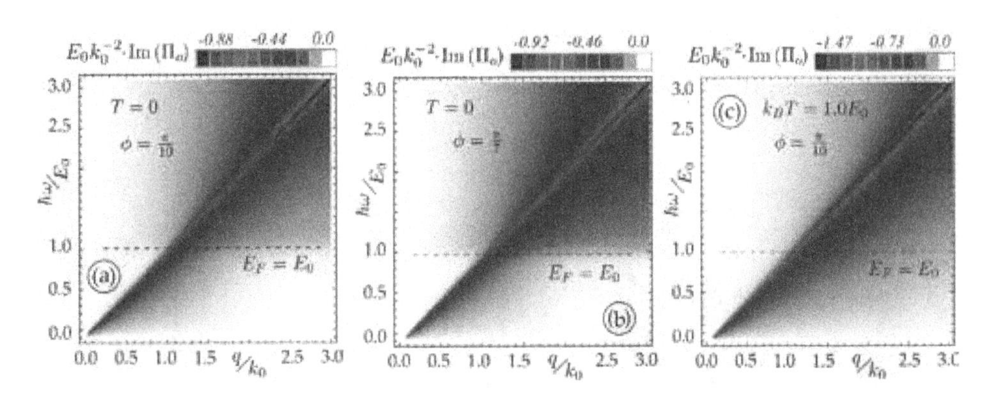

Figure 6.
Particle-hole modes, determined by non-zero $\mathrm{Im}[\Pi_O(q,\omega|\mu(T))]$ *within the* q-ω *plane, for an* α-$\mathcal{T}_3$ *layer with* $\phi = \pi/10$ *(in (a), (c)) and* $\phi = \pi/7$ *in (b). Panels (a) and (b) correspond to* $T = 0$, *while plot (c) is for* $k_B T = 1.0\,E_F$.

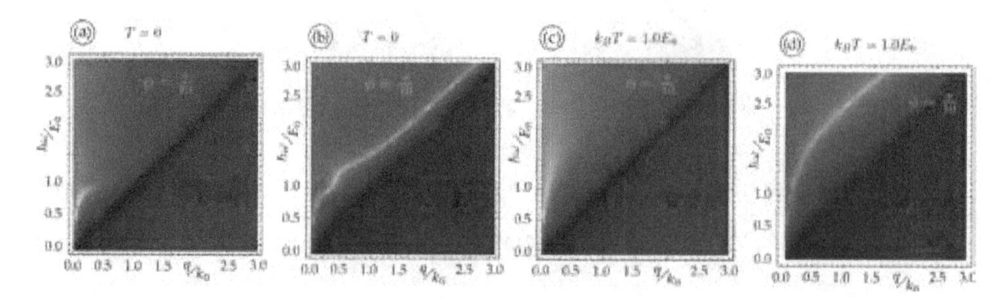

Figure 7.
Plasmon branches for an isolated α-$\mathcal{T}_3$ layer with $\phi = \pi/10$. Panels (a) and (c) only show undamped plasmons, while (b) and (d) display full plasmon branches including damped ones. Left panels (a) and (b) corresponds to $T = 0$, *while* $k_B T = 1.0\,E_F$ *for right panels (c) and (d).*

6. Plasmons in α-$\mathcal{T}_3$ layer coupled to conducting substrate

In the last part of THIS CHAPTER, WE WOULD LIKE TO FOCUS ON finite-T plasmons in a so-called nanoscale-hybrid structure consisting of a 2D layer, such as, graphene, silicene or a dice lattice, which is Coulomb-coupled to a large, conducting material. Physically, the Coulomb coupling between the 2D layer and the conductor results in a strong hybridization of graphene plasmon and localized surface-plasmon modes. This structure, which is referred to as an *open system*, could be realized experimentally or even by a device fabrication.

Our schematics for an open system is shown in **Figure 8**. The dynamical screening to the Coulomb interaction between electrons in a 2D layer and in metallic substrate is taken into account by anonlocal and dynamical inverse dielectric function $\mathcal{K}(\boldsymbol{r}, \boldsymbol{r}'; \omega)$, as demonstrated in Refs. [53, 54]. This nonlocal inverse dielectric function is connected to a dielectric function $\epsilon(\boldsymbol{r}, \boldsymbol{r}'; \omega)$ in Eq. (4) by

$$\int d^3\boldsymbol{r}'\, \mathcal{K}(\boldsymbol{r}, \boldsymbol{r}'; \omega)\, \epsilon(\boldsymbol{r}', \boldsymbol{r}''; \omega) = \delta(\boldsymbol{r} - \boldsymbol{r}''), \qquad (45)$$

and the resonances in $\mathcal{K}(\boldsymbol{r}, \boldsymbol{r}'; \omega)$ reveal the nonlocal hybridized plasmon modes supported by both 2D layer and the conducting surface as a single quantum system.

By using the Drude model for metallic substrate, the dielectric function can be written as $\epsilon_B(\omega) = 1 - \Omega_p^2/\omega^2$, where $\Omega_p = \sqrt{n_0 e^2/\epsilon_0 \epsilon_b m^*}$ is the bulk-plasma frequency for the conductor, n_0 electron concentration and m^* is the effective mass of electrons. Drude model describes electron screening in the long-wavelength limit. Based on the previously developed mean-field theory [53, 55, 56], we are able to

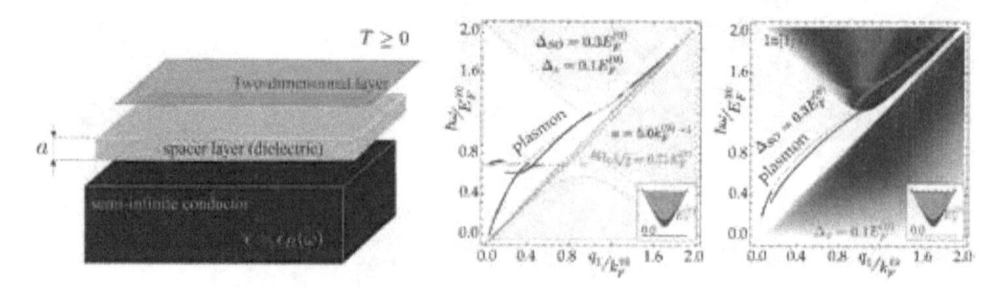

Figure 8.
Schematics for a silicene-based open system and numerical results for the two plasmon branches and their damping in this system with $\Delta_{SO} = 0.3 E_0$ *and* $0.1 E_0$, *where* $E_0 = \hbar v_F \sqrt{\pi n_0} = 54.6$ *meV.*

calculate plasmon dispersions in this 2D open system. For this, the plasmon dispersions are obtained from the zeros of the so-called *dispersion factor* $\mathbb{S}_C(q,\omega)$, instead of the dielectric function in Eq. (4). $\mathbb{S}_C(q,\omega)$ for this open system is given by [25, 54, 57]

$$\mathbb{S}_C(q,\omega|E_F) = 1 - \frac{2\pi\alpha_r}{q}\,\Pi_0(q,\omega|E_F)\left[1 + \frac{\Omega_p^2}{2\omega^2 - \Omega_p^2}\,\exp\left(-2qa\right)\right], \tag{46}$$

where a is the separation between the 2D layer and the conducting surface. Most important, we should emphasize that the second term in Eq. (46) does not have a full analogy with polarization function of an isolated layer, and the resulting plasmon dispersions in open system represents a hybridized plasmon mode with the environment. Therefore, these plasmon dispersions are expected to be sensitive to Coulomb coupling to el ectrons in the conducting substrate through a factor $\sim$ exp $(-2qa)$ in Eq. (46), similarly to what we have found for coupled double graphene layers [16]. The strong Coulomb coupling leads to a *linear dispersion* of plasmon in this open system [54, 58], which is in contrast with well-known $\sim\sqrt{q}$ dependence in all 2D materials.

As a special example, let us consider a silicene 2D layer with two bandgaps $\Delta_{<,>}$ and an electron doping density n. We start with seeking for a non-interacting polarization function in the long-wave limit $q \ll k_F^{<,>}$ for doping density n and assume a high-enough n to keep the Fermi level $E_F = \sqrt{\left(\hbar v_F k_F^\beta\right)^2 + \Delta_\beta^2} > \Delta_>$ above the large bandgap. Under this assumption, we get

$$\Pi_0(q,\omega|E_F) = \frac{q^2}{\pi\hbar^2\omega^2}\sum_{\beta=>,<} k_F^\beta \left|\frac{\partial\mathbb{E}_\beta(k)}{\partial k}\right|_{k=k_F^\beta} = \frac{E_F}{\pi}\left(2 - \frac{\Delta_<^2}{E_F^2} - \frac{\Delta_>^2}{E_F^2}\right)\frac{q^2}{\hbar^2\omega^2}, \tag{47}$$

where $k_F^\beta = \sqrt{2\pi n_\beta}$ are two different Fermi wave numbers associated with a single Fermi energy E_F, and n_β is the electron density for each subband satisfying $n = n_< + n_>$.

In the limit of $a \to \infty$, the plasmon branch of an isolated silicene layer can be recovered from Eq. (46), yielding

$$\omega_p^2(q) = \frac{4\alpha_r}{\hbar^2 E_F}\left(E_F^2 - \frac{\Delta_>^2 + \Delta_<^2}{2}\right)q \equiv \Xi q, \tag{48}$$

where for convenience we introduced a coefficient

$$\Xi(E_F,\Delta_\beta) = \frac{4\alpha_r}{\hbar^2 E_F}\left(E_F^2 - \frac{\Delta_>^2 + \Delta_<^2}{2}\right). \tag{49}$$

We notice from Eq. (48) that $\omega_p(q) \sim \sqrt{q}$, disregarding of the energy bandgaps $\Delta_{<,>}$ or doping density n. On the other hand, the Fermi energy E_F for silicene is given by Eq. (17).

Furthermore, using the notation defined by Eq. (49), we get from Eqs. (46) and (47) that

$$1 - \Xi(E_F,\Delta_\beta)\,\frac{q}{\omega^2}\left[1 + \frac{\Omega_p^2}{2\omega^2 - \Omega_p^2}\,\exp\left(-2qa\right)\right] = 0, \tag{50}$$

which leads to a bi-quadratic equation

$$2\left(\frac{\omega^2}{\Omega_p^2}\right)^2 - \left[1+\Xi(E_F,\Delta_\beta)\frac{2q}{\Omega_p^2}\right]\left(\frac{\omega^2}{\Omega_p^2}\right) + \Xi(E_F,\Delta_\beta)\frac{q}{\Omega_p^2}[1-\exp(-2qa)] = 0. \tag{51}$$

Eq. (51) gives rise to two solutions

$$\frac{4}{\Omega_p^2}\omega_{p,\pm}^2 = \left(1+\Xi(E_F,\Delta_\beta)\frac{2q}{\Omega_p^2}\right) \pm \sqrt{\left(1+\Xi(E_F,\Delta_\beta)\frac{2q}{\Omega_p^2}\right)^2 - \frac{8q\Xi(E_F,\Delta_\beta)}{\Omega_p^2}[1-\exp(-2qa)]}, \tag{52}$$

where $\pm$ terms correspond to in-phase and out-of-phase plasmon modes, respectively. Two hybrid plasmon modes in Eq. (52) become

$$\begin{aligned} \omega_{p,+}(q) &\simeq \frac{\Omega_p}{\sqrt{2}} + \frac{\Xi}{\sqrt{2}\Omega_p}q - \frac{\Xi\left(\Xi+4a\Omega_p^2\right)}{2\sqrt{2}\Omega_p^3}q^2 + \mathcal{O}(q^3), \\ \omega_{p,-}(q) &\simeq q\sqrt{2a\Xi} - \frac{\Xi\sqrt{2a\Xi}}{\Omega_p^2}q^2 + \mathcal{O}(q^3). \end{aligned} \tag{53}$$

In Eq. (53), both plasmon branches contain a linear $\sim q$ term, and $\omega_{p,+}(q)$ approaches a constant as $q \to 0$, i.e., an optical mode for plasmons. Two independent bandgaps, Δ_{SO} and Δ_z, together with doping density n, play a crucial role on shaping the plasmon dispersions, as well as the particle-hole mode damping regions. The outer boundaries of a particle-hole mode region specify an area within the q-ω plane in which the plasmon modes become damping free and are solely determined by $\Delta_<$, while the group velocity of plasmon mode depends on both $\Delta_<$ and $\Delta_>$. Since each bandgap could be experimentally tuned by applying a perpendicular electric field, we acquire a full control of both plasmon dispersions and their damping-free regions at the same time.

Numerical results for thermal plasmons in open system are presented in **Figure 9**. Similarly to what we have found for graphene and silicene, there are two plasmon branches, both of which depend linearly on q with a finite slope as $q \ll k_0$.
The acoustic-plasmon branch starts from the origin, while the optical-plasmon branch from $\Omega_p/\sqrt{2}$. The dispersion of each branch also varies with parameter α (or $\phi = \tan^{-1}\alpha$), which is observed for the upper branch, as shown in (a) and (c) of **Figure 9**. In addition, we see a much smaller slope for the lower plasmon branch in **Figure 9(b)** and **(d)** due to enhanced Coulomb coupling with a reduced separation a. The finite-temperature upper plasmon branches in **Figure 9 (e)** and **(f)** are shifted up greatly, as it is expected to be true for all finite-temperature plasmons, which is further accompanied by enhanced damping below the diagonal as seen in **Figure 6(c)**. Meanwhile, the lower plasmon branch seems much less affected by finite temperatures, as demonstrated by both upper and lower rows of **Figure 9** for different separations, except for enhanced damping in **Figure 9(f)** below the surface-plasmon energy $\hbar\Omega_p/\sqrt{2}$.

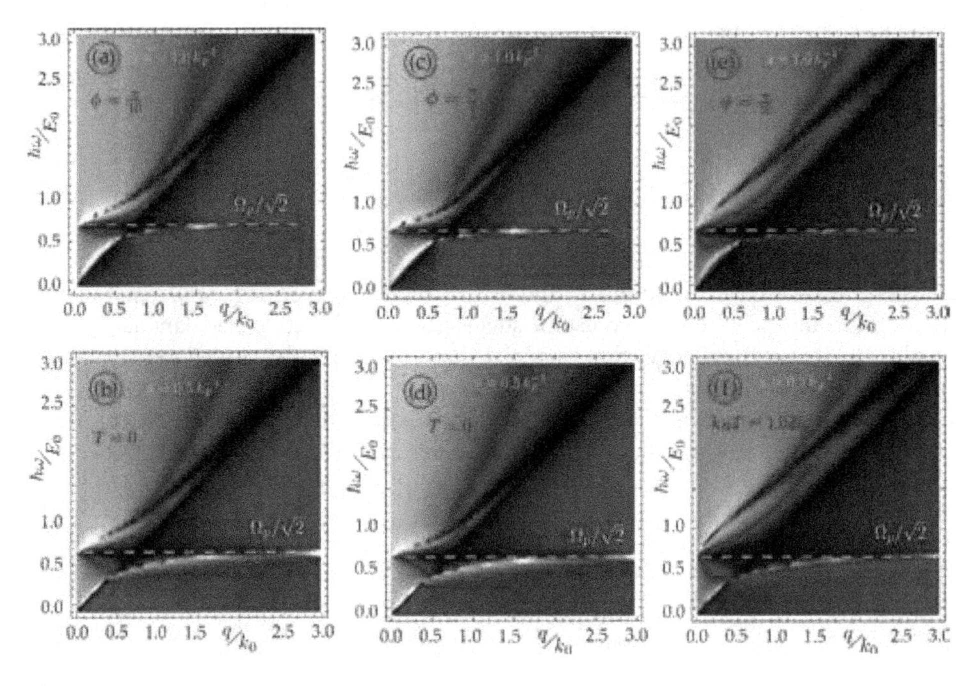

Figure 9.
Nonlocal hybridized plasmon dispersions for α-$\mathcal{T}_3$ layer coupled to a closely-located surface of a semi-infinite conductor. Panels (a)–(d) are for $T = 0$, *while plots (e) and (f) for* $k_B T = E_0$. *All the upper-row plots correspond to the separation* $a = 1.0\,k_F^{-1}$, *and the lower-row ones to* $a = 0.5k_F^{-1}$. *Additionally, middle-column plots, (b) and (d), correspond to* $\phi = \pi/7$, *and all other columns to* $\phi = \pi/10$.

7. Summary and remarks

In conclusion, we have developed a general theory for finite-temperature polarization function, plasmon dispersions and their damping for all known innovation 2D Dirac-cone materials with various types of symmetries and bandgaps. We have also derived a set of explicit transcendental equations determining the chemical potential as a function of temperature, which serves as a key part in calculating finite-temperature polarization function through the so-called thermal convolution path. The selection of a particular path with a specific $\mu(T)$ could be employed for studying the temperature dependence of plasmon modes in each of the considered 2D materials. The fact that a chemical potential keeps its sign is true only for materials with symmetric energy bands of electrons and holes, but can cross the zero line for TMDC's with asymmetric electron and hole bands.

Using the calculated finite-temperature polarization function, we have further found the dispersions of hybrid plasmon-modes in various types of open systems including a 2D material coupled to a conducting substrate. The obtained plasmon dispersions in these 2D-layer systems are crucial for measuring spin-orbit interaction strength and dynamical screening to Coulomb interaction between electrons in 2D materials, as well as for designing novel surface-plasmon based multi-functional near-field opto-electronic devices.

We have generalized our developed theory for 2D materials further to most recently proposed α-$\mathcal{T}_3$ lattices, in which the characteristic parameter α is the ratio of hub-rim to hub-hub hopping coefficients and can vary from 0 to 1 continuously corresponding to different material properties. For α-$\mathcal{T}_3$ materials, we have obtained the hybrid plasmon modes for different α values at both zero and finite temperatures and demonstrated that the resulting hybridized plasmon dispersions could be tuned sensitively by geometry phase, temperature, and separation between α-$\mathcal{T}_3$ layer and conducting surface. Such tunability has a profound influence on performance of α-$\mathcal{T}_3$ material based quantum electronic devices.

Acknowledgements

A.I. thanks Liubov Zhemchuzhna for helpful and fruitful discussions, and Drs. Armando Howard, Leon Johnson and Ms. Beverly Tarver for proofreading the manuscript and providing very useful suggestions on the style and language. G.G. would like to acknowledge the financial support from the Air Force Research Laboratory (AFRL) through grant FA9453-18-1-0100 and award FA2386-18-1-0120. D.H. thanks the supports from the Laboratory University Collaboration Initiative (LUCI) program and from the Air Force Office of Scientific Research (AFOSR).

Author details

Andrii Iurov[1*], Godfrey Gumbs[2,3] and Danhong Huang[4]

1 Department of Physics and Computer Science, Medgar Evers College of the City University of New York, Brooklyn, NY, USA

2 Department of Physics and Astronomy, Hunter College of the City University of New York, NY, USA

3 Donostia International Physics Center (DIPC), San Sebastian, Basque Country, Spain

4 Air Force Research Laboratory, Space Vehicles Directorate, Kirtland Air Force Base, NM, USA

*Address all correspondence to: aiurov@mec.cuny.edu; theorist.physics@gmail.com

References

[1] Novoselov K, Geim AK, Morozov S, Jiang D, Katsnelson M, Grigorieva I, et al. Two-dimensional gas of massless Dirac fermions in graphene. Nature. 2005;**438**:197

[2] Geim AK, Novoselov KS. The rise of graphene. Nature Materials. 2007;**6**:183

[3] Neto AC, Guinea F, Peres N, Novoselov KS, Geim AK. The electronic properties of grapheme. Reviews of Modern Physics. 2009;**81**:109

[4] Grigorenko A, Polini M, Novoselov K. Graphene plasmonics. Nature Photonics. 2012;**6**:749

[5] Agarwal A, Vitiello MS, Viti L, Cupolillo A, Politano A. Plasmonics with two-dimensional semiconductors: From basic research to technological applications. Nanoscale. 2018;**10**:8938

[6] Katsnelson M, Novoselov K, Geim A. Chiral tunneling and the Klein paradox in graphene Nature Physics. 2006;**2**:620

[7] Low T, Roldán R, Wang H, Xia F, Avouris P, Moreno LM, et al. Plasmons and screening in monolayer and multilayer black phosphorus. Physical Review Letters. 2014;**113**:106802

[8] Soleimanikahnoj S, Knezevic I. Tunable electronic properties of multilayer phosphorene and its nanoribbons. Journal of Computational Electronics. 2017;**16**:568

[9] Koppens FH, Chang DE, Garcia de Abajo FJ. Graphene plasmonics: A platform for strong light-matter interactions. Nano Letters. 2011;**11**:3370

[10] Garcia de Abajo FJ. Graphene plasmonics: Challenges and opportunities. ACS Photonics. 2014; **1**:135

[11] Politano A, Chiarello G. Plasmon modes in graphene: Status and prospect. Nanoscale. 2014;**6**:10927

[12] Politano A, Radović I, Borka D, Mišković Z, Yu H, Farias D, et al. Dispersion and damping of the interband π plasmon in graphene grown on Cu (111) foils. Carbon. 2017;**114**:70

[13] Politano A, Chiarello G. Quenching of plasmons modes in air-exposed graphene-Ru contacts for plasmonic devices. Applied Physics Letters. 2013; **102**:201608

[14] Politano A, Marino AR, Chiarello G. Effects of a humid environment on the sheet plasmon resonance in epitaxial grapheme. Physical Review B. 2012;**86**: 085420

[15] Iurov A, Huang D, Gumbs G, Pan W, Maradudin A. Effects of optical polarization on hybridization of radiative and evanescent field modes. Physical Review B. 2017;**96**:081408

[16] Das Sarma S, Li Q. Intrinsic plasmons in two-dimensional Dirac materials. Physical Review B. 2013;**87**: 235418

[17] Hwang EH, Das Sarma S. Screening-induced temperature-dependent transport in two-dimensional graphene. Physical Review B. 2009;**79**:165404

[18] Iurov A, Gumbs G, Huang D, Balakrishnan G. Thermal plasmons controlled by different thermal-convolution paths in tunable extrinsic Dirac structures. Physical Review B. 2017;**96**: 245403

[19] Tabert CJ, Nicol EJ. Dynamical polarization function, plasmons, and screening in silicene and other buckled honeycomb lattices. Physical Review B. 2014;**89**:195410

[20] Tabert CJ, Nicol EJ. AC/DC spin and valley Hall effects in silicene and germanene. Physical Review B. 2013;**87**: 235426

[21] Kane CL, Mele EJ. Quantum spin Hall effect in grapheme. Physical Review Letters. 2005;**95**:226801

[22] Ezawa M. A topological insulator and helical zero mode in silicene under an inhomogeneous electric field. New Journal of Physics. 2012;**14**:033003

[23] Ezawa M. Valley-polarized metals and quantum anomalous Hall effect in silicone. Physical Review Letters. 2012; **109**:055502

[24] Liu C-C, Feng W, Yao Y. Quantum spin Hall effect in silicene and two-dimensional germanium. Physical Review Letters. 2011;**107**:076802

[25] Iurov A, Gumbs G, Huang D, Zhemchuzhna L. Controlling plasmon modes and damping in buckled two-dimensional material open systems. Journal of Applied Physics. 2017;**121**: 084306

[26] Chang H-R, Zhou J, Zhang H, Yao Y. Probing the topological phase transition via density oscillations in silicene and germanene. Physical Review B. 2014;**89**: 201411

[27] d'Acapito F, Torrengo S, Xenogiannopoulou E, Tsipas P, Velasco JM, Tsoutsou D, et al. Evidence for Germanene growth on epitaxial hexagonal (h)-AlN on Ag(111). Journal of Physics: Condensed Matter. 2016;**28**: 045002

[28] Walhout C, Acun A, Zhang L, Ezawa M, Zandvliet H. Scanning tunneling spectroscopy study of the Dirac spectrum of germanene. Journal of Physics: Condensed Matter. 2016;**28**: 284006

[29] Derivaz M, Dentel D, Stephan R, Hanf M-C, Mehdaoui A, Sonnet P, et al. Continuous germanene layer on Al (111). Nano Letters. 2015;**15**:2510

[30] Bampoulis P, Zhang L, Safaei A, Van Gastel R, Poelsema B, Zandvliet HJW. Germanene termination of Ge_2Pt crystals on Ge(110). Journal of Physics: Condensed Matter. 2014;**26**:442001

[31] Acun A, Zhang L, Bampoulis P, Farmanbar M, van Houselt A, Rudenko A, et al. Germanene: The germanium analogue of grapheme. Journal of Physics: Condensed Matter. 2015;**27**: 443002

[32] Zhang L, Bampoulis P, van Houselt A, Zandvliet H. Two-dimensional Dirac signature of germanene. Applied Physics Letters. 2015;**107**:111605

[33] Pyatkovskiy P. Dynamical polarization, screening, and plasmons in gapped graphene. Journal of Physics: Condensed Matter. 2008;**21**:025506

[34] Maldague PF. Many-body corrections to the polarizability of the two-dimensional electron gas. Surface Science. 1978;**73**:296

[35] IurovA, GumbsG, HuangD, SilkinV. Temperature-dependent plasmons and their damping rates for graphene with a finite energy bandgap. arXiv preprint arXiv:1504.04552. 2015

[36] Iurov A, Gumbs G, Huang D, Silkin V. Plasmon dissipation in gapped graphene open systems at finite temperature. Physical Review B. 2016; **93**:035404

[37] Falkovsky L, Varlamov A. Space-time dispersion of graphene conductivity. The European Physical Journal B. 2007;**56**:281

[38] Gusynin V, Sharapov S. Transport of Dirac quasiparticles in graphene: Hall

and optical conductivities. Physical Review B. 2006;**73**:245411

[39] Singh A, Bolotin KI, Ghosh S, Agarwal A. Nonlinear optical conductivity of a generic two-band system with application to doped and gapped grapheme. Physical Review B. 2017;**95**:155421

[40] Falkovsky L, Pershoguba S. Optical far-infrared properties of a graphene monolayer and multilayer. Physical Review B. 2007;**76**:153410

[41] Iurov A, Gumbs G, Huang D. Temperature-and frequency-dependent optical and transport conductivities in doped buckled honeycomb lattices. Physical Review B. 2018;**98**:075414

[42] Iurov A, Gumbs G, Huang D. Temperature-dependent collective effects for silicene and germanene. Journal of Physics: Condensed Matter. 2017;**29**:135602

[43] Tsaran VY, Kavokin A, Sharapov S, Varlamov A, Gusynin V. Entropy spikes as a signature of Lifshitz transitions in the Dirac materials. Scientific Reports. 2017;**7**:10271

[44] Gorbar EV, Gusynin VP, Miransky VA, Shovkovy IA. Magnetic field driven metal-insulator phase transition in planar systems. Physical Review B. 2002;**66**:045108

[45] Raoux A, Morigi M, Fuchs J-N, Piéchon F, Montambaux G. From dia- to paramagnetic orbital susceptibility of massless fermions. Physical Review Letters. 2014;**112**:026402

[46] Iurov A, Gumbs G, Huang D. Peculiar electronic states, symmetries, and Berry phases in irradiated materials. Physical Review B. 2019;**99**:205135

[47] Oriekhov D, Gorbar E, Gusynin V. Electronic states of pseudospin-1 fermions in dice lattice ribbon. Low Temperature Physics. 2018;**44**:1313

[48] Dey B, Ghosh TK. Photoinduced valley and electron-hole symmetry breaking in α-T_3 lattice: The role of a variable Berry phase. Physical Review B. 2018;**98**:075422

[49] Gorbar E, Gusynin V, Oriekhov D. Electron states for gapped pseudospin-1 fermions in the field of a charged impurity. Physical Review B. 2019;**99**: 155124

[50] Malcolm J, Nicol E. Frequency-dependent polarizability, plasmons, and screening in the two-dimensional pseudospin-1 dice lattice. Physical Review B. 2016;**93**:165433

[51] Iurov A, Zhemchuzhna L, Dahal D, Gumbs G, Huang D. Quantum-statistical theory for laser-tuned transport and optical conductivities of dressed electrons in α-T3 materials. Physical Review B. arXiv preprint arXiv: 1905.01703. 2019 [accepted for publication]

[52] Huang D, Iurov A, Xu H-Y, Lai Y-C, Gumbs G. Interplay of Lorentz-Berry forces in position-momentum spaces for valley-dependent impurity scattering in lattices. Physical Review B. 2019;**99**: 245412

[53] Horing NJM. Coupling of graphene and surface plasmons. Physical Review B. 2009;**80**:193401

[54] Gumbs G, Iurov A, Horing N. Nonlocal plasma spectrum of graphene interacting with a thick conductor. Physical Review B. 2015;**91**:235416

[55] Horing NJM. Aspects of the theory of grapheme. Philosophical Transactions of the Royal Society of London A: Mathematical, Physical and Engineering Sciences. 2010;**368**:5525

[56] Horing NJM, Kamen E, Cui H-L. Inverse dielectric function of a bounded solid-state plasma. Physical Review B. 1985;**32**:2184

[57] Horing N, Iurov A, Gumbs G, Politano A, Chiarello G. Recent progress on nonlocal graphene/surface plasmons. In: Low-dimensional and Nanostructured Materials and Devices. 2016. Available from: https://link. springer.com/chapter/ 10. 1007/978-3- 319-25340-4_9

[58] Kramberger C, Hambach R, Giorgetti C, Rümmeli M, Knupfer M, Fink J, et al. Linear plasmon dispersion in single-wall carbon nanotubes and the collective excitation spectrum of grapheme. Physical Review Letters. 2008;**100**:196803

10

Simple Preparations for Plasmon-Enhanced Photodetectors

Yu Liu, Junxiong Guo, Jianfeng Jiang, Wenjie Chen, Linyuan Zhao, Weijun Chen, Renrong Liang and Jun Xu

Abstract

Localized surface plasmon resonance (LSPR), known as the collective oscillation of electrons and incident light in metallic nanostructures, has been applied in high performance photodetectors over the past few years. But the preparation process is complex and expensive due to the introduction of electron beam lithography (EBL) for preparing nanostructures. In the past few months, we have demonstrated two simple methods to prepare plasmon-enhanced photodetectors: (i) Au nanoparticles (Au NPs) solution were directly spun coated onto the WS_2-based photodetectors. The performance has been enhanced by the LSPR of Au NPs, and reached an excellent high responsivity of 1050 A/W at the wavelength of 590 nm. (ii) Au NPs were deposited on MoS_2 by magnetron sputtering. The spectral response of pure MoS_2 was located in visible light and which was extended to near-infrared region (700–1600 nm) by Au NPs. Further, the responsivity reaches up to 64 mA/W when the incident light is 980 nm. In this book chapter, more details for developing those two simple methods and the discussion of the enhanced mechanism are performed, which can be very useful for the next generation photodetection.

Keywords: surface plasmon resonance, Au nanoparticles, enhanced photodetectors, simple preparations, high responsivity

1. Introduction

Photodetectors are one of the most important devices in photonic chips, which show a great potential in optical communication, flame sensing, environmental monitoring, and astronomical studies [1]. A lot of semiconductors have been exploited for photodetection such as silicon, GaN, PbS, InGaAs and HgCdTe, operating from ultraviolet to far-infrared region [2]. It is significant to decrease the dimension of photodetectors down to the sub-nano scale for the next generation highly integrated photonic chips.

Two-dimensional (2D) materials such as graphene and layered transition metal dichalcogenides (TMDs) have attractive electronic and optical properties such as flexibility, transparency and metal-oxide-semiconductor (CMOS) compatibility [3–7]. Tungsten disulfide (WS_2), a typical member of the TMDs group, has a higher carrier mobility than other TMD materials due to smaller electron effective mass [8]. Moreover, WS_2 enjoys excellent thermal stability for extensive applications [9].

Hence, WS_2-based high stable photodetectors can be used in many attractive applications such as extreme environment detection. Furthermore, Two-dimensional stacked molybdenum disulfide (MoS_2) has attracted many research interests for applications in optoelectronic devices, due to its outstanding merits of electronic and optical properties, especially photodetectors. But the few layers 2D material-based photodetectors suffer from low photoresponsivity mainly due to the poor optical absorption in the atomic layered materials.

A number of 2D material-based photodetectors have been enhanced using resonance microcavity, PbS quantum dots, tunneling effect, heterojunctions or perovskite [10–14]. The narrow band absorption or complex preparation process limits the application of these attractive methods. Recently, localized surface plasmon resonance (LSPR)-enhanced photodetector has been demonstrated, and we have also provided effective ways to enhance the efficiency of light-harvesting [15–17]. LSPR can be excited by Ag or Au nanoparticles (NPs) such as deep metallic grating, nanodisk array, bowtie array, hybrid antenna and fractal metasurface [18–20]. Furthermore, patterning 2D materials into periodic structure can also excite plasmon resonance [21]. The methods for preparing the nanostructure involve electron beam lithography, hydrothermal synthesis, and template-based electrochemical method, which are generally complicated, expensive and may deleterious to the device.

For the current existing problems, we have demonstrated two methods for easy-preparation and high-performance 2D material-based photodetectors [22, 23]. (i) Au NPs solution was directly spun coated onto WS_2-based photodetectors. The performance has been enhanced by the LSPR of Au NPs, and reach quite high responsivity of 1050 A/W at the wavelength of 590 nm. The diameters and distance of Au NPs will affect the resonant wavelength and absorption of the device. (ii) We have demonstrated a MoS_2 plasmonic photodetector by depositing Au NPs on MoS_2 sheet using magnetron sputtering without need of template, which shows a significant improvement of photo-response in near-infrared region. The spectral response of pure MoS_2 was in visible light and which was extended to near-infrared region (700–1600 nm). Furthermore, the responsivity reaches up to 64 mA/W when the incident light is 980 nm. Detailed preparation and discussion of the mechanism are performed in the chapter.

2. The preparation and theoretical mechanism of spun-coated WS_2-based photodetectors

2.1 The preparation process

The WS_2 film was grown on sapphire substrate by chemical vapor deposi-tion (CVD) method and transferred to Si/SiO_2 substrate by wet transfer method. The molecular configurations of WS_2 are shown in **Figure 1a**. Raman spectra (**Figure 1b**), PL spectrum (**Figure 1c**) and atomic force microscopy (AFM) files (**Figure 1d**) reveal monolayer feature of WS_2 film.

The 3D and cross-section view of the photodetector is shown in **Figure 2a, b**, respectively. The light incident normally from the top of the device. The fabrication was carried out by the following steps. First of all, a 200-nm-thick molybdenum (Mo) layer was deposited and patterned on WS_2 layer. Afterwards, the WS_2 was patterned (size: 30 × 100 μm) by photolithography and plasma etching to form active area. Finally, Au NPs solution was spun coating onto the channel and dried in air. Nearly spherical Au NPs can be seen in the low-resolution transmis-sion electron microscopy (TEM) image (**Figure 2c**). The mean diameter of Au

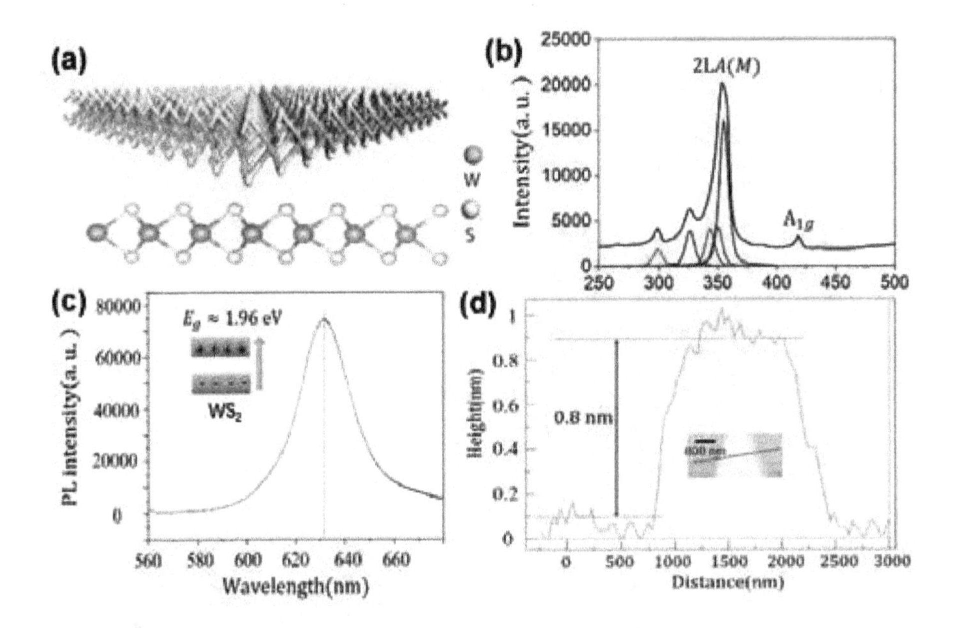

Figure 1.
The microscopic molecular structures and characterization of monolayer WS_2 film [22]. (a) Schematic molecular structure of 1 L-WS_2. The blue and yellow balls present sulfur and tungsten, respectively. (b) The Raman spectrum consisted of several characteristic peaks of the WS_2 film on Si/SiO_2 substrate acquired with laser excitation of λ = 532 nm. (c) PL spectra of 1 L WS_2 layer. The band gap is about 1.96 eV as shown in the inset. (d) The AFM height profiles and corresponding AFM image of 1 L-WS_2. The thickness is about 0.8 nm.

NPs is ~20 nm (**Figure 2d**) as shown in the statistical analysis of the TEM images. Typical low-resolution (**Figure 2e**) and high-resolution (**Figure 2f**) scanning electron microscope (SEM) images of the Au NPs are also presented. The Au NPs are well distributed on the top of the WS_2 film.

2.2 The performance of the WS_2-based photodetector

The drain-source current (I_{DS}) under illumination at room temperature without Au NPs are shown in **Figure 3a**. It is concluded that I_{DS}increase as V_{DS} increase from 0 to 2 V. The irradiance power is 20.5 mW/cm^2 in all of these three wavelength (590, 740 and 850 nm). I_{DS} decrease with the increase of wavelength. The ratio of the on/off current (I_{DS}/I_{dark}) reached nearly 10^3 under 590 nm light illumination. The responsivity can be calculated by

$$R = I_{Ph}/P \quad (1)$$

where I_{Ph} is the photocurrent, P is the irradiance power. The responsivity at V_{DS} = 2 V is illustrated in**Figure 3b**. The responsivity decrease with the increase of the power which are typical for photodetectors [24, 25].R reached 35 A/W at the wavelength of 590 nm, and reached 1.8 A/W at the wavelength of 850 nm when irradiance power is both 0.2 mW/cm^2. The performance of the photodetector decoated with Au NPs is shown in**Figure 3c**. The drain-source current are measured at the irradiance power of 20.5 mW/cm^2. The enhancedI_{DS} also reached the highest value at 590 nm and decreased with the increase of the wavelength (λ).

The current gain is defined as

$$G = I_{pe}/I_{ph} \quad (2)$$

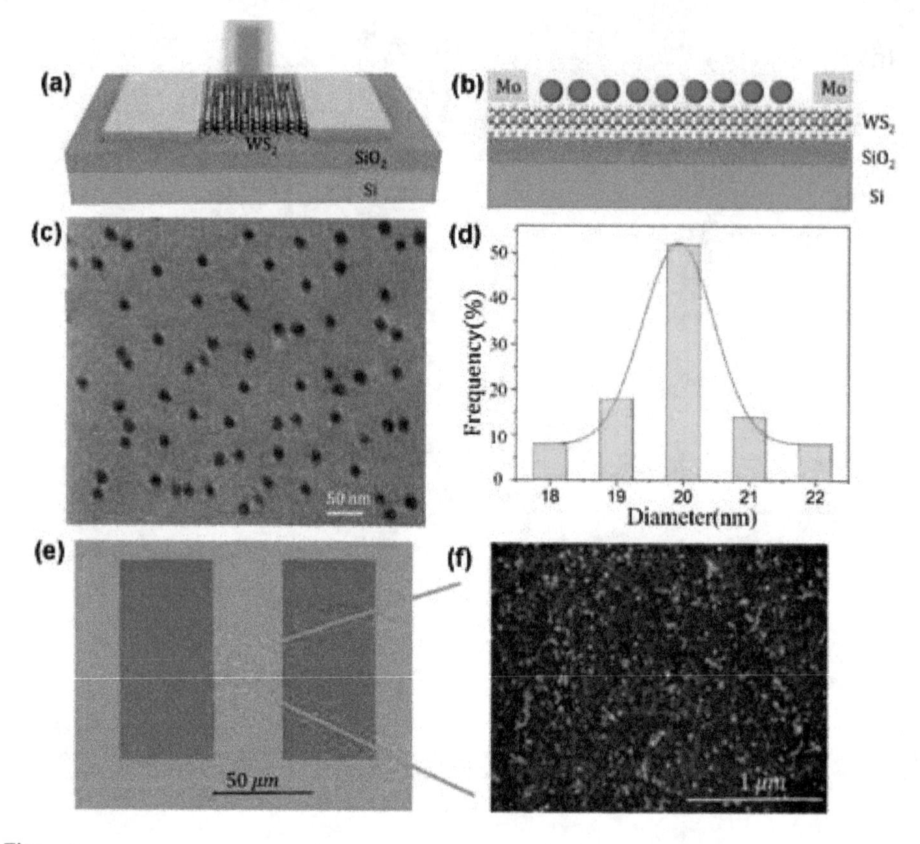

Figure 2.
Characterization of the fabricated photodetector [22]. (a) The schematic 3D view of 1 L-WS_2-based photodetector was presented. The drain/source electrodes are fabricated by Mo. Au NPs (red balls) were spun coated on the channel. (b) The cross-section view of the photodetector. (c) TEM image of Au NPs. The Au NPs are well distributed in the solution. (d) The statistics size distribution of Au NPs based on the TEM image. We can conclude that the mean size of Au NPs is ~20 nm. (e) Low-resolution and (f) high-resolution SEM images of the photodetector. Clear electrodes and An NPs can be seen from these images.

where I_{pe} is the enhanced photocurrent of the photodetector. The current gain reveals the improvement of the device by Au NPs as shown in **Figure 3d** when P =0.2 mW/cm^2 and V_{DS} = 2 V. The photoresponsivity was enhanced ~30 times and reached 1050 A/W when λ = 590 nm. The photoresponsivity was enhanced ~11 times and reached 55 A/W when λ = 740 nm. And the photoresponsivity was enhanced ~5 times at near infrared light (λ = 850 nm) and reached 8 A/W. In general, the switching behaviour, which reflect the response speed and high-frequency characteristic, is very important for photodetectors. The detectors also need to quickly refresh in some applications such as instant display. **Figure 3e** presents the switching behavior at near infrared light (λ = 850 nm) of the WS_2 photodetector. The photodetector shows a good repeatability during on-off cycles. Moreover, the on-off characteristic in a period is shown in **Figure 3f**. The rise and decay time are about 100 and 200 ms, respectively.

2.3 The theoretical mechanism

To explain the mechanism of the enhancement by Au NPs, we take finite-difference time-domain (FDTD) method to investigate the distribution of electric field of Au NPs. According to the Förster's expression for energy W transferred from donor to acceptor [24, 25].

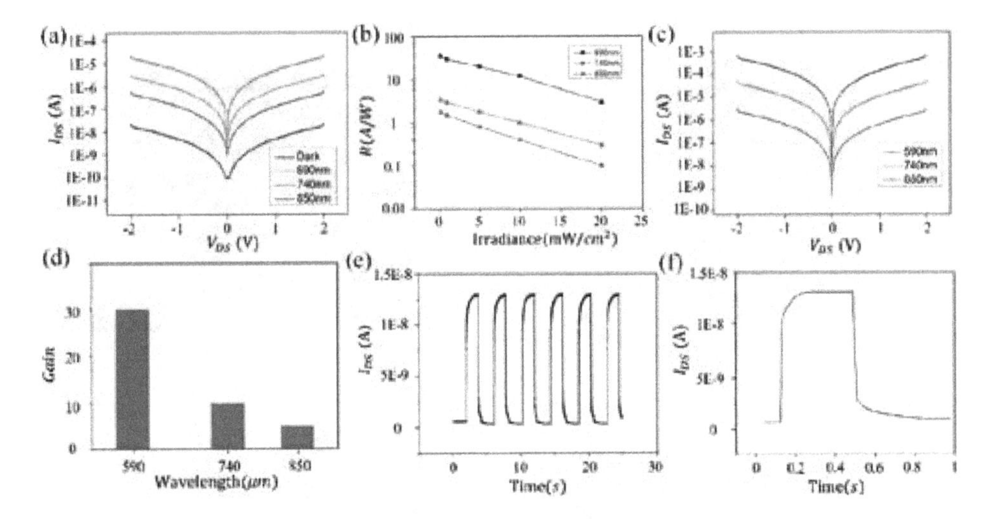

Figure 3.
Visible to NIR light response of the fabricated photodetector [22]. (a) The drain-source current (I_{DS}) changed by the drain-source bias from −2 to 2 V. The optical power is 20.5 mW/cm² for these three wavelength. (b) The photo-responsivity changed as a function of the illumination power when V_{DS} = 2 V. (c) The drain-source current of the photodetector with Au NPs at the same irradiance of 20.5 mW/cm². (d) The current gain of the photodetectors with and without Au NPs when the irradiance is 0.2 mW/cm² and the voltage is 2 V. (e) The on/off behavior of the photodetector when λ = 850 nm. (f) The on-off characteristic of the Au NPs coated photodetector in a period time. The rise time is about 100 ms and the decay time is about 200 ms.

$$\frac{W}{W_d} = \frac{9}{8\pi}\int \frac{d\omega}{k^4} f_d(\omega)\,\sigma_a(\omega)\,|D|^2 \tag{3}$$

where W_d is the donor's energy, $f_d(\omega)$ and k are the spectral function and wave vector of the source, $\sigma_a(\omega)$ is the absorption of acceptor, $D = q/r^3$ is the coupling coefficient, and r is the distance. The performance could be changed by key parameters of Au NPs such as diameter d and distance between two particles s. In order to give an intuitive description, simplified models were built. The distance between the edges of two adjacent nanospheres, s, was fixed on 10 nm. The diameter d was fixed as 20 nm.

The resonant wavelength can be acquired by SPPs dispersion equation.

$$\beta = k_0\sqrt{\frac{\varepsilon_d + \varepsilon_m}{\varepsilon_d \varepsilon_m}} \tag{4}$$

where k_0 is the vacuum wavevector, ε_d is the relative permittivity of dielectric, ε_m is the dielectric function of gold, which can be represented by Drude model. The mismatch of SPPs and incident wavevector can be compensated by the gold nanoparticles, which can be approximately presented by

$$\beta = \sqrt{\varepsilon_d}\,k_0 \sin\theta + \frac{2\pi n}{\lambda_g}\left(\hat{x} + \hat{y}\right) \tag{5}$$

where θ, λ_g, n are the horizontal angle of incident wave vector, the grating period, and an integer, respectively. From Eq. (5), we can see that the absorption wavelength depends only on the dimension of Au NPs.

The results for the illumination at λ =590, 740, and 850 nm are shown in **Figures 4a–c**, respectively. It is clear that the intense electromagnetic fields were introduced by LSPR of Au NPs. The electric field near the Au NPs was significantly enhanced and stronger than the rest region, revealing that electromagnetic energy was compactly confined by the Au NPs. The electromagnetic field was enhanced more significant when the illumination was under λ = 590 nm. The

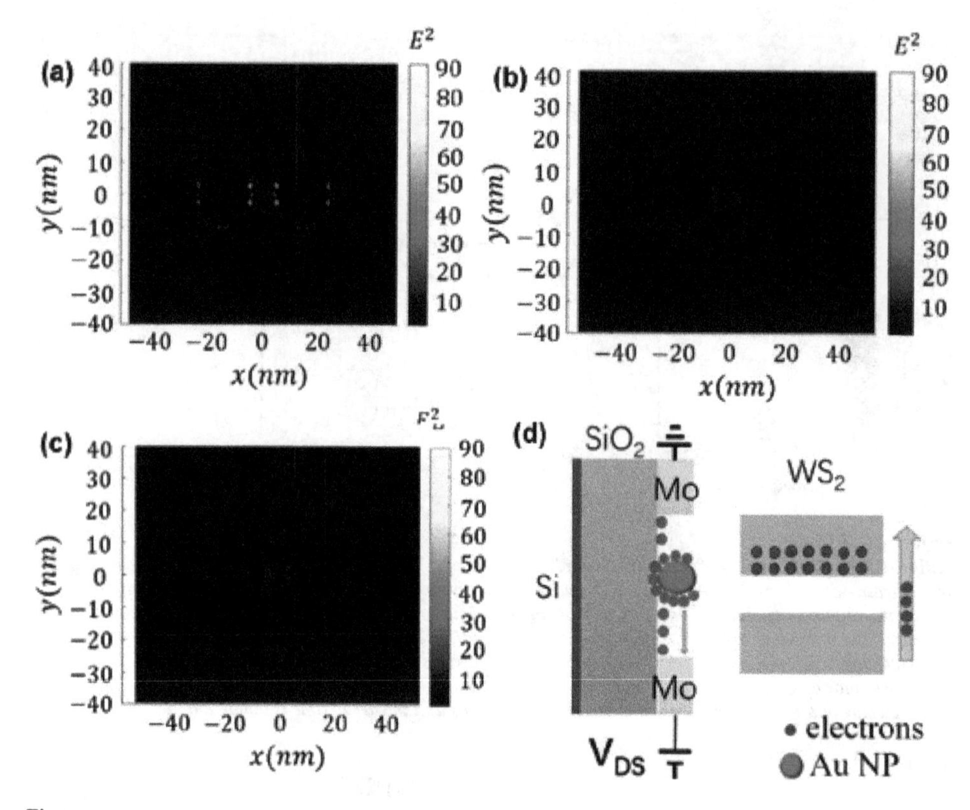

Figure 4.
LSPR and carrier transfer of the presented enhanced photodetector [22]. Cross-section distribution of the square of electric field (|E| 2) near Au NPs under the illumination at the wavelength of (a) 590 nm, (b) 740 nm, and (c) 850 nm. (d) The charge transfer between Au NPs and WS_2 film.

enhanced electric fields could excite the generation of the carriers in the WS_2 film, resulting a prominent photoresponse. The highest responsivity obtained at λ = 590 nm (**Figure 3b, d**) is consistent with the most intense LSPR at λ = 590 nm (**Figure 4a**). The generation and transportation of the electrons are shown in **Figure 4d**. The photons were absorbed by WS_2 film and the excited electrons were driven by the drain-source voltage. There are more electrons around the Au NPs as **Figure 4d** shows.

3. The preparation and theoretical mechanism of magnetron-sputtering-based MoS_2 photodetectors

3.1 The preparation process

Figure 5a, b shows the schematic and optical image of our designed MoS_2 plasmonic photodetector by introducing Au NPs, respectively. The few-layered MoS_2 sheet was obtained using mechanical exfoliation method. Then, we trans-ferred the exfoliated MoS_2 to the SiO_2/Si substrate which contacted with Au/Ni electrodes. Next, we fabricated the Au NPs on exfoliated MoS_2 sheet using magnetron sputtering technique. This facile method offers the convenience of without need of template compared with other reported MoS_2 plasmonic photodetectors.

Figure 6 shows the morphology of as-prepared materials was characterized using an AFM. As shown in Section A1 in **Figure 6a,b**, it indicates the thickness of

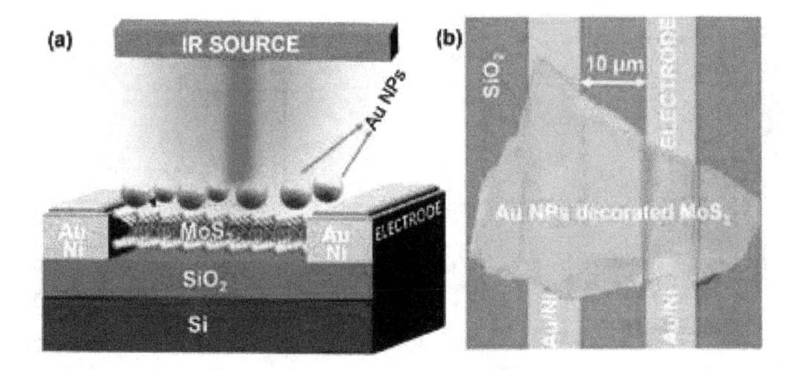

Figure 5.
(a) Schematic and (b) optical image of MoS_2 photodetector decorated with Au NPs [23].

exfoliated bare MoS_2 is just about 6 nm (about 10 folds). In comparison, the surface morphology of the MoS_2 after depositing with Au NPs by magnetron sputtering was shown in **Figure 6c–f**. It clearly exhibits the physical size and particle distribu-tion of Au NPs can be easily tuned by sputtering technique. When the sputtering current increased from 30 mA (LPP1, **Figure 6d**) to 35 mA (HPP1, **Figure 6e**) with deposition period fixed to 1 s, we obtained the controllable Au NPs with lateral size increasing from ~3 to ~5 nm and vertical height increasing from ~5 to ~8 nm, respectively. For another, if we fixed the applied current to 30 mA and prolonged the deposit period to 2 s (LPP2, **Figure 6f**), the Au NPs maintain almost same physi-cal size as LPP1 but the gap of adjacent deposited Au NPs sharply drops compared with LPP1.

3.2 The structure of the MoS_2-based photodetector

In order to investigate chemical composition of the prepared materials, the X-ray photoelectron spectroscopy (XPS) was employed. As shown in **Figure 7**, The peaks at 229.2 and 232.3 eV correspond to the doublet of Mo $3d_{5/2}$ and Mo $3d_{3/2}$, respectively. And the peaks of 226.3, 162.1 and 163.2 eV of the binding energy attach to the S 2s, S $2p_{3/2}$ and S $2p_{1/2}$, respectively [26, 27]. For Au 4f, the peak positions of 83.6 and 87.2 eV bind to the Au $4f_{7/2}$ and Au $4f_{5/f}$, indicating the Au NPs are directly introduced into the exfoliated MoS_2 sheet [28]. More importantly, the banding energies of Mo and S in Au decorated MoS_2 maintain the same values as that of bare MoS_2 sheet, indicating the introduction of Au NPs has non-influence on the crystal structures of exfoliated MoS_2 sheet.

Further, we used Raman spectroscopy of 532 nm laser to confirm the structural properties of the fabricated devices. For MoS_2, the difference of E^1_{2g} and A_{1g}, Δ, corresponding to in-plane and out of plane energy vibrations, is used to index the layer number of obtained MoS_2. **Figure 8a** shows the Δ of bulk MoS_2 and our exfoli-ated MoS_2 are 27.8 and 25.3 cm^{-1}, respectively, indicating that the thickness of bare MoS_2 is about 10 layers [29], which is highly consistent with the AFM results. After decorating Au NPs with MoS_2, it exhibits the Δ maintains nearly same as bare MoS_2 but the intensities obviously increase, shown in **Figure 8b, c**.

3.3 The performance of the MoS_2-based photodetector

The photoelectric performance of the fabricated photodetector was studied at room temperature, which applied a 980 nm laser source with controllable incident power. In order to produce the laser beam pulses, we combined an oscilloscope to

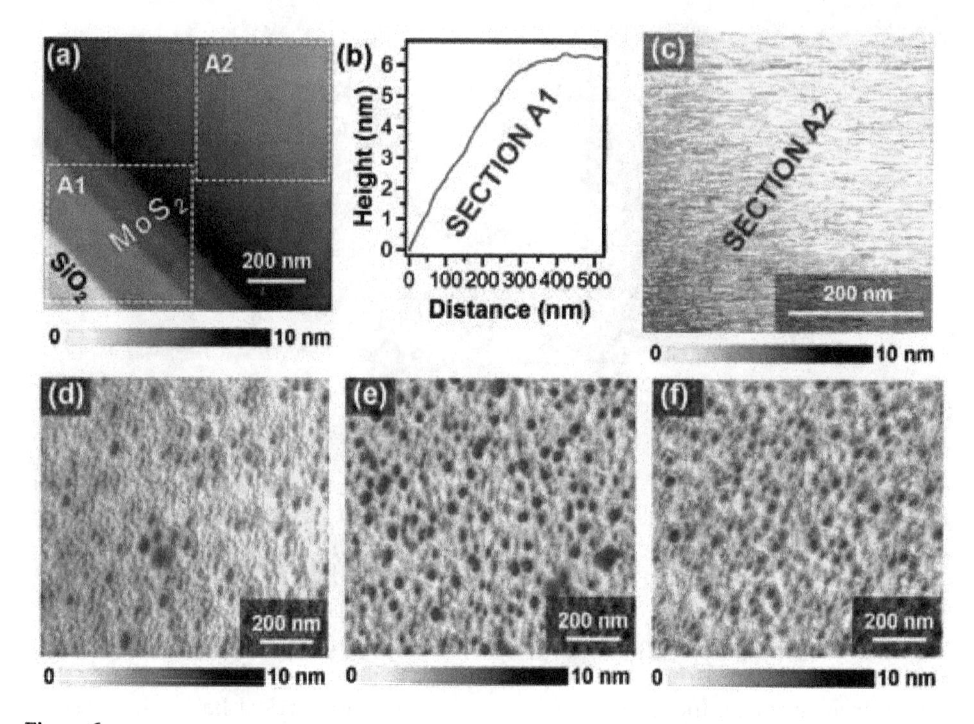

Figure 6.
AFM images of as-prepared bare MoS_2 and MoS_2 modified with Au NPs [23]. (a) Height of exfoliated bare MoS_2 sheet. (b) Height vs. distance plot of bare MoS_2, correspond to the Section A1 in (a). (c) Morphology of bare MoS_2, correspond to the Section A2 in (a). Height of Au NPs decorated MoS_2 sheet for (d) LPP1, (e) HPP1 and (f) LPP2.

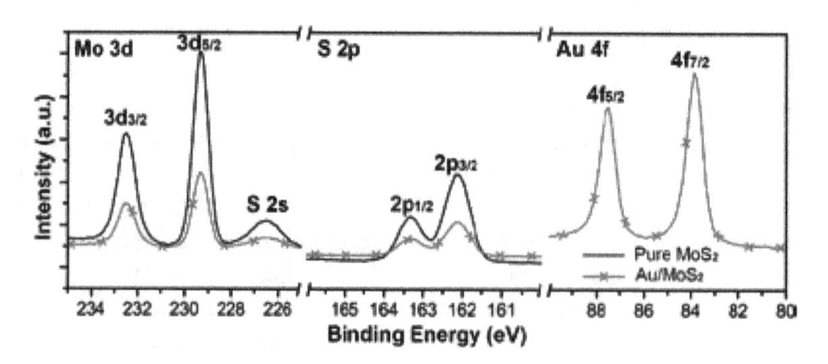

Figure 7.
XPS plots of bare MoS_2 and Au NPs decorated MoS_2 [23].

the incident laser source. **Figure 9a** shows the photocurrent plots of photodetector, where the illumination power and bias voltage are 1.60 mW and 2 V, respectively. Obviously, the Au NPs/MoS_2 heterostructure-based photodetector exhibits an ultra-high photocurrent (8.6 nA) compared with that of bare MoS_2-based photodetector (0.59 nA).

Figure 9b shows the plots of photocurrent vs. applied bias voltage ranging from 0.1 to 15 V. We obtained a photocurrent up to ~480 nA, when the applied incident laser power and bias voltage were 7.50 mW and 15 V, yielding an improved responsibility of 64 mA/W. Moreover, the *I-V* plots indicate the photocurrent owns a good linear relationship with applied bias voltage, when the illumination intensities tuned from 0.85 to 7.5 mW. For another, **Figure 9c** shows the dependence plots of photocurrent (I_{ph}, nA) on laser power irradiation (P_{in}, μW). It is found that the

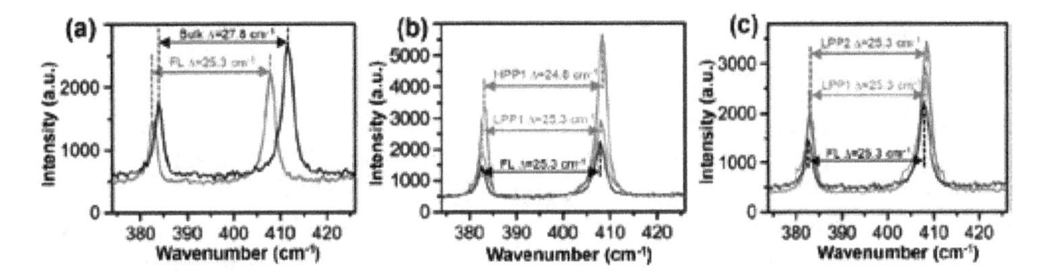

Figure 8.
Raman spectroscopy of the prepared materials. (a) Typical spectra of bulk MoS_2 and exfoliated MoS_2. Raman shift of exfoliated bare MoS_2 sheet and Au NPs decorated MoS_2 with different (b) sputtering electric current (LPP1, HPP1) (c) deposition period (LPP1, LPP2) [23].

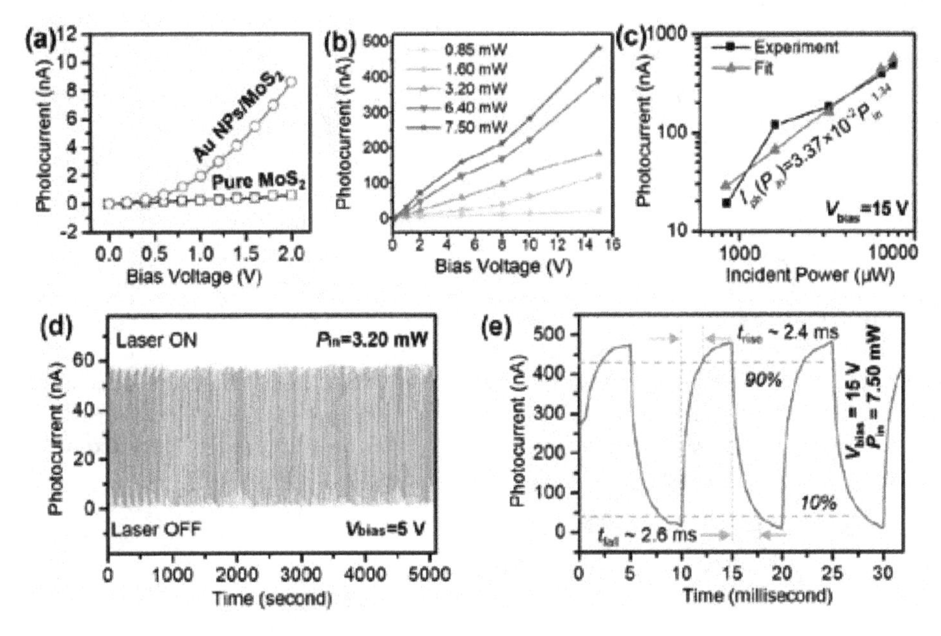

Figure 9.
Photoelectrical performances of Au NPs/MoS_2-based photodetector [23]. (a) I-V plots of photodetectors based on bare MoS_2 and Au NPs decorated MoS_2. (b) I-V scatters of Au NPs/MoS_2-based photodetector with different laser power irradiation ranging from 0.85 to 7.50 mW. (c) Photocurrent as a function of laser power under bias of 15 V. (d) Stability of the fabricated Au NPs/MoS_2-based photodetector. (e) Time response of the fabricated Au NPs/MoS_2-based photodetector.

photocurrent follows a nonlinear dependence to the incident power intensity, $aP_{in}{}^{b}$, where a and b are constant for different bias voltage. For example, when the bias voltage are 15 V, the fitting a and b are 3.37×10^{-2} and 1.34, respectively.

With respect to the stability of the photodetector, we performed the extended duration photocurrent measurements by periodically switching the incident laser under illumination of 3.20 mW at bias of 5 V, and the periods of both on and off state are 5 s. **Figure 9d** shows the photocurrents over 500 circles of continuous operation, exhibiting a well stability. Moreover, in order to characterize the response time for detecting infrared wavelengths of our designed device, we applied an oscilloscope in the process of laser excitation to produce laser pulses with a duration of 10 ms. **Figure 9e** shows the time-response of Au decorated MoS_2-based device under illumination of 7.50 mW at bias of 15 V. It indicates that the rise time (t_{rise}) and fall time (t_{fall}) are 2.4 and 2.6 ms, respectively, which are signifi-cantly superb to that of other reported MoS_2 photodetectors.

3.4 The operational mechanism

In order to study the potential mechanism of our fabricated photodetector, we simulated the electric field distribution of Au decorated MoS_2 using finite element method in the infrared region. We assumed that the Au NPs with a gap of 6 nm and a diameter of 5 nm under illumination of a linearly polarized plane wave with an electric field amplitude of 1 Vm^{-1} for the simulated model. And the thicknesses of MoS_2 sheet and incident laser wavelength were 6 and 980 nm, respectively. **Figure 10a, b** shows the cross-section of the simulated electric field distribution of Au NPs decorated MoS_2. Benefiting from the LSPR effect excited by the Au NPs, when the diameter matched to the incident wavelengths, the intensities of electric field at interfaces of air/Au/MoS_2, up to ~3.96×10^5 V/m, are obviously higher than other districts. In comparison, interfaces of air/MoS_2 show a poor intensity of electric field, only about 6.72×10^4 V/m. This tendency can be also observed in the recent researches of Au NPs guided MoS_2 sheets for photo-detection [30]. We further experimentally proved the absorption by using UV-visible-NIR spectroscopy analysis. **Figure 10c** shows the absorption spectrum of bare MoS_2 sheet Au NPs decorated MoS_2. The normalized absorptance plots indicate that the Au NPs/MoS_2 is obviously enhanced than that of bare MoS_2 ranging from 700 to 1600 nm.

The above results and discussion clearly unveil the introduction of Au NPs plays a key role in enhancing the light matter interactions of MoS_2 with infrared wavelengths. The significantly improved sensitivity of the fabricated photodetector could be attributed to the LSPR effect, shown in **Figure 10d**, induced by the periodically aligned Au NPs, resulting in obvious improvement of local electric field. When the incident infrared wavelengths highly confined by the deposited Au NPs, the local electric field at the interface of Au/MoS_2 is greatly improved by the surface plasmon waves. Firstly, the Au surface plasmons effectively excite a

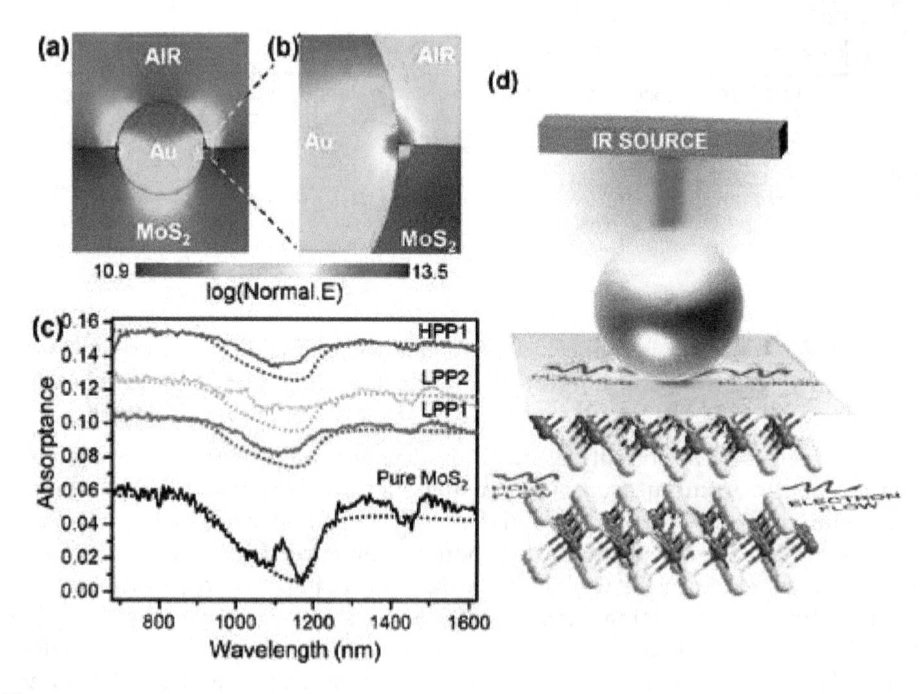

Figure 10.
Cross-section of the simulated electric field distribution for (a) Au NPs/MoS_2 sensing layer and (b) air/Au/MoS_2 interfaces. (c) Absorptance plots of bare MoS_2 sheet and Au NPs decorated MoS_2 using sputtering method. The solid and dashed lines correspond to the experimental and fitted results, respectively. (d) Schematic representation for the operational mechanism of the fabricated Au NPs/MoS_2 device [23].

coupling effect at the interface of Au/MoS_2 for absorbing photons, yielding a much more photo-induced carriers to improve the photo sensing [30, 31]. Moreover, the additional local electric field generated by the LSPR effect of Au NPs accelerates the photogenerated carriers to separate for producing photocurrent [32, 33]. This facile method, tuning Au NPs by sputtering method to excite LSPR effect for fabricating the unique device structure, is expected to be practical applications in other 2D materials such as WS_2 and $MoSe_2$ [34–36], thus offers a new route on a variety of high-performance optoelectronic devices.

4. Conclusion

High performance photodetectors are very important in a lot of applications. We have successfully developed two simple methods to prepare plasmon-enhanced photodetectors. (i) Au nanoparticles (Au NPs) solution were directly spun coated onto the WS_2-based photodetectors. The performance has been enhanced by the LSPR of Au NPs, and reached an excellent high responsivity of 1050 A/W at the wavelength of 590 nm. (ii) Au NPs were deposited on MoS_2 by magnetron sputtering. The spectral response of pure MoS_2 was located in visible light and which was extended to near-infrared region (700–1600 nm) by Au NPs. Further, the responsivity reaching up to 64 mA/W when the incident light is 980 nm. These photodetectors achieved excellent responsivity and response speed. The results not only promote the development of high-performance photodetectors, but also provide a simplified method for the fabrication of other hybrid structure devices.

Acknowledgements

The project was supported by grants from the National Basic Research Program of China (No. 2015CB351905), the National Key Research and Development Program of China (No. 2016YFA 0302300, No. 2016YFA0200400), the National Natural Science Foundation of China (No. 61306105).

Author details

Yu Liu[1†], Junxiong Guo[2*†], Jianfeng Jiang[3], Wenjie Chen[1], Linyuan Zhao[1], Weijun Chen[1], Renrong Liang[1*] and Jun Xu[1]

1 Institute of Microelectronics, Tsinghua National Laboratory for Information Science and Technology (TNList), Tsinghua University, Beijing, China

2 State Key Laboratory of Electronic Thin Films and Integrated Devices, School of Electronic Science and Engineering (National Exemplary School of Microelectronics), University of Electronic Science and Technology of China, Chengdu, China

3 School of Microelectronics, Shandong University, Jinan, China

*Address all correspondence to: guojunxiong25@163.com and liangrr@mail.tsinghua.edu.cn

† These authors contributed equally to this work.

References

[1] Sun Z, Chang H. Graphene and graphene-like two-dimensional materials in photodetection: Mechanisms and methodology. ACS Nano. 2014;**8**: 4133-4156. DOI: 10.1021/nn500508c

[2] Chen H, Liu H, Zhang Z, et al. Nanostructured photodetectors: From ultraviolet to terahertz. Advanced Materials. 2016;**28**:403-433. DOI: 10.1002/adma.201503534

[3] Castro Neto AH, Guinea F, Peres NMR, et al. The electronic properties of graphene. Reviews of Modern Physics. 2009;**81**:109-162. DOI: 10.1103/RevModPhys.81.109

[4] Geim AK, Novoselov KS. The rise of graphene. Nature Materials. 2007; **6**:183- 191. DOI: 10.1038/nmat1849

[5] Stankovich S, Dikin DA, Dommett GHB, et al. Graphene-based composite materials. Nature. 2006;**442**: 282-286. DOI: 10.1038/nature04969

[6] Stankovich S, Dikin DA, Piner RD, et al. Synthesis of graphene-based nanosheets via chemical reduction of exfoliated graphite oxide. Carbon. 2007;**45**:1558-1565. DOI: 10.1016/j.carbon.2007.02.034

[7] Zhu Y, Murali S, Cai W, et al. Graphene and graphene oxide: Synthesis, properties, and applications. Advanced Materials. 2010;**22**:3906-3924. DOI: 10.1002/adma.201001068

[8] Liu L, Kumar SB, Ouyang Y, et al. Performance limits of monolayer transition metal dichalcogenide transistors. IEEE Transactions on Electron Devices. 2011;**58**:3042-3047. DOI: 10.1109/ted.2011.2159221

[9] Huo N, Yang S, Wei Z, et al. Photoresponsive and gas sensing field-effect transistors based on multilayer WS_2 nanoflakes. Scientific Reports. 2014;**4**:5209. DOI: 10.1038/srep05209

[10] Furchi M, Urich A, Pospischil A, et al. Microcavity-integrated graphene photodetector. Nano Letters. 2012;**12**: 2773-2777. DOI:10.1021/ nl204512x

[11] Sun Z, Liu Z, Li J, et al. Infrared photodetectors based on CVD-grown graphene and pbs quantum dots with ultrahigh responsivity. Advanced Materials. 2012;**24**:5878-5883. DOI: 10.1002/adma.201202220

[12] Liu C-H, Chang Y-C, Norris TB, et al. Graphene photodetectors with ultra-broadband and high responsivity at room temperature. Nature Nanotechnology. 2014;**9**:273-278. DOI: 10.1038/nnano.2014.31

[13] Xu H, Wu J, Feng Q , et al. High responsivity and gate tunable graphene-MoS_2 hybrid phototransistor. Small. 2014;**10**:2300-2306. DOI: 10.1002/smll.201303670

[14] Lee Y, Kwon J, Hwang E, et al. High-performance perovskite-graphene hybrid photodetector. Advanced Materials. 2015;**27**:41-46. DOI: 10.1002/adma.201402271

[15] Su Y, Guo Z, Huang W, et al. Ultrasensitive graphene photodetector with plasmonic structure. Applied Physics Letters. 2016;**109**:173107. DOI: 10.1063/1.4966597

[16] Liu Y, Huang W, Gong T, et al. Ultra-sensitive near-infrared graphene photodetectors with nanopillar antennas. Nanoscale. 2017;**9**:17459- 17464. DOI: 10.1039/c7nr06009b

[17] Liu Y, Gong T, Zheng Y, et al. Ultrasensitive and plasmon-tunable graphene

photodetectors for micro-spectrometry. Nanoscale. 2018;**10**:20013-20019. DOI: 10.1039/c8nr04996c

[18] Zhao B, Zhao JM, Zhang ZM. Enhancement of near-infrared absorption in graphene with metal gratings. Applied Physics Letters. 2014;**105**: 031905. DOI: 10.1063/1.4890624

[19] Fang J, Wang D, DeVault CT, et al. Enhanced graphene photodetector with fractal metasurface. Nano Letters. 2017;**17**:57-62. DOI: 10.1021/acs. nanolett.6b03202

[20] Jiang N, Zhuo X, Wang J. Active plasmonics: Principles, structures, and applications. Chemical Reviews. 2018;**118**:3054-3099. DOI: 10.1021/acs. chemrev.7b00252

[21] Freitag M, Low T, Zhu W, et al. Photocurrent in graphene harnessed by tunable intrinsic plasmons. Nature Communications. 2013;**4**:1951. DOI: 10.1038/ncomms2951

[22] Liu Y, Huang W, Chen W, et al. Plasmon resonance enhanced WS_2 photodetector with ultra-high sensitivity and stability. Applied Surface Science. 2019;**481**:1127-1132. DOI: 10.1016/j.apsusc.2019.03.179

[23] Guo JX, Li SD, He ZB, et al. Near-infrared photodetector based on few-layer MoS_2 with sensitivity enhanced by localized surface plasmon resonance. Applied Surface Science. 2019;**483**: 1037-1043. DOI: 10.1016/j. apsusc.2019.04.044

[24] Komsa H-P, Krasheninnikov AV. Electronic structures and optical properties of realistic transition metal dichalcogenide heterostructures from first principles. Physical Review B. 2013;**88**:085318. DOI: 10.1103/PhysRevB.88.085318

[25] Fedutik Y, Temnov VV, Schoeps O, et al. Exciton - plasmon - photon conversion in plasmonic nanostructures. Physical Review Letters. 2007; **99**:136802. DOI: 10.1103/PhysRevLett.99. 136802

[26] Baker MA, Gilmore R, Lenardi C, et al. XPS investigation of preferential sputtering of S from MoS_2 and determination of MoS_x stoichiometry from Mo and S peak positions. Applied Surface Science. 1999;**150**:255-262. DOI: 10.1016/s0169-4332(99)00253-6

[27] Ambrosi A, Sofer Z, Pumera M. Lithium intercalation compound dramatically influences the electrochemical properties of exfoliated MoS_2. Small. 2015;**11**:605-612. DOI: 10.1002/smll.201400401

[28] Huang X, Li S, Huang Y, et al. Synthesis of hexagonal close-packed gold nanostructures. Nature Communications. 2011;**2**:292. DOI: 10.1038/ncomms1291

[29] Lee C, Yan H, Brus LE, et al. Anomalous lattice vibrations of single- and few-layer MoS_2. ACS Nano. 2010;**4**:2695-2700. DOI: 10.1021/nn1003937

[30] Wu ZQ, Yang JL, Manjunath NK, et al. Gap - mode surface - plasmon-enhanced photoluminescence and photoresponse of MoS_2. Advanced Materials. 2018;**30**:1706527. DOI: 10.1002/adma.201706527

[31] Miao JS, Hu WD, Jing YL, et al. Surface plasmon-enhanced photodetection in few layer MoS_2 phototransistors with Au nanostructure arrays. Small. 2015;**11**:2392-2398. DOI: 10.1002/smll.201403422

[32] Sun YH, Liu K, Hong XP, et al. Probing local strain at MX_2-metal boundaries with surface plasmon-enhanced Raman scattering. Nano Letters. 2014;**14**:5329-5334. DOI: 10.1021/nl5023767

[33] Bang S, Duong NT, Lee J, et al. Augmented quantum yield of a 2D monolayer photodetector by surface plasmon coupling. Nano Letters. 2018;**18**:2316-2323. DOI: 10.1021/acs.nanolett.7b05060

[34] Zhao W, Wang S, Liu B, et al. Exciton-plasmon coupling and electromagnetically induced transparency in monolayer semiconductors hybridized with Ag nanoparticles. Advanced Materials. 2016;**28**:2709-2715. DOI: 10.1002/adma.201504478

[35] Abid I, Chen W, Yuan J, et al. Temperature-dependent plasmon exciton interactions in hybrid Au/$MoSe_2$ nanostructures. ACS Photonics. 2017;**4**:1653-1660. DOI: 10.1021/acsphotonics.6b00957

[36] Abid I, Bohloul A, Najmaei S, et al. Resonant surface plasmon-exciton interaction in hybrid MoSe2@Au nanostructures. Nanoscale. 2016;**8**:8151- 8159. DOI: 10.1039/c6nr00829a

11

Nanowires Integrated to Optical Waveguides

Ricardo Téllez-Limón and Rafael Salas-Montiel

Abstract

Chip-scale integrated optical devices are one of the most developed research subjects in last years. These devices serve as a bridge to overcome size mismatch between diffraction-limited bulk optics and nanoscale photonic devices. They have been employed to develop many on-chip applications, such as integrated light sources, polarizers, optical filters, and even biosensing devices. Among these integrated systems can be found the so-called hybrid photonic-plasmonic devices, structures that integrate plasmonic metamaterials on top of optical waveguides, leading to outstanding physical phenomena. In this contribution, we present a comprehensive study of the design of hybrid photonic-plasmonic systems consisting of periodic arrays of metallic nanowires integrated on top of dielectric waveguides. Based on numerical simulations, we explain the physics of these struc-tures and analyze light coupling between plasmonic resonances in the nanowires and the photonic modes of the waveguides below them. With this chapter we pretend to attract the interest of research community in the development of inte-grated hybrid photonic-plasmonic devices, especially light interaction between guided photonic modes and plasmonic resonances in metallic nanowires.

Keywords: plasmonics, integrated optics, nanowires, optical waveguides, hybrid modes

1. Introduction

Plasmonics, the science of plasmons, is a research field that has been extensively studied in recent years due to its multiple applications like biosensing, optical communications, or quantum computing, to mention but a few.

Generally, the field of plasmonics is associated with two types of collective oscillations of conductive electrons at the boundaries of metallic nanostructures, known as surface plasmon polaritons (SPP) and localized surface plasmons (LSP). While SPP are referred as surface waves propagating at a dielectric-metal interface, LSP can be regarded as standing surface waves confined in metallic nanoparticles embedded in a dielectric environment [1].

As it is well known, SPP modes can only be excited when appropriate phase match conditions are fulfilled. An option to achieve this condition, is by making use of the electromagnetic near field scattered by a local defect or emitter. To this purpose, the LSP mode of a metallic nanoparticle can be excited and coupled to the SPP of a metallic substrate, giving rise to hybrid plasmon polaritons [2, 3].

In addition to these types of plasmonic oscillations, there are other resonances named plasmonic chain modes. These modes can be generated in linear arrays of closely spaced metallic nanoparticles, including nanowires, and they result from the near field coupling between adjacent nanoparticles excited at their plasmonic resonances. Due to this coupling effect, light can propagate through the periodic arrays. Thus, these periodic structures can be regarded as discrete plasmonic waveguides [4–6]. When placing a periodic array of metallic nanoparticles in a layered media, under proper excitation conditions, the plasmonic chain modes can also couple to the SPP of a metallic substrate, forming hybrid SPP-chain modes [7].

In this same sense, when placing periodic arrays of metallic nanoparticles on top of dielectric waveguides, the plasmonic chain modes can couple to the photonic modes of the waveguide [8]. These integrated structures give rise to the so-called hybrid photonic-plasmonic waveguide modes [9], and they are the main subject of interest in this chapter. We will focus our attention to integrated structures consisting of periodic arrays of metallic nanowires integrated on top of two-dimensional dielectric photonic waveguides.

To this purpose, we will bring a comprehensive explanation about the physics behind the dispersion curves of integrated hybrid photonic-plasmonic waveguiding structures. Then will be studied the propagation of electromagnetic fields through the integrated systems varying the geometric cross-section of the metallic nanowires. For a better understanding, this comprehensive study will be accompa-nied by numerical simulations, making easier to elucidate the potential applications of these outstanding structures.

2. Hybrid photonic-plasmonic waveguides

2.1 Optical waveguides

From the analysis of the chemical composition of farer stars to imaging of microscopic living cells, information transport through light is one of the main subjects of interest in optical sciences. Among the different ways to transport light can be found optical waveguides, whose principle of operation is based on the total internal reflection effect. This phenomenon consists of the complete reflection of light within a medium surrounded by media with smaller refractive index, as depicted in **Figure 1**.

The schematic in **Figure 1a** represents an asymmetric planar waveguide invariant along the out-of-plane direction, consisting of a dielectric medium of refractive index n_2 between two media of refractive index n_1 and n_3, where $n_2 > n_1 > n_3$. A s light propagates in the inner medium n_2, certain rays will present a phase difference of zero or a multiple of 2π, when they are twice reflected. This situation means that

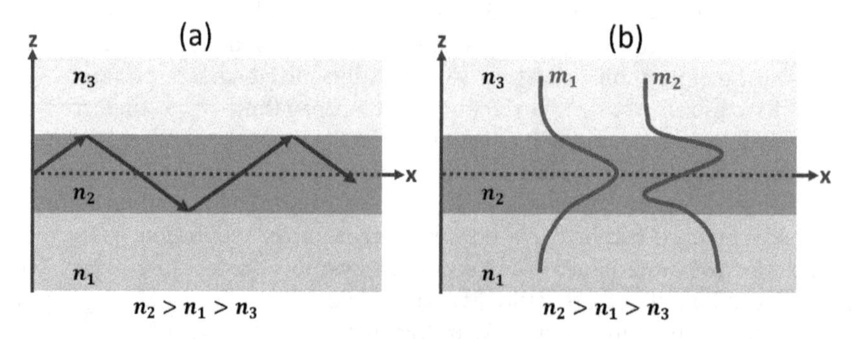

Figure 1.
Schematic representation of a planar asymmetric waveguide consisting of three dielectric media of refractive index n_1, n_2 and n_3. (a) Self-consistency condition defining the modes of the waveguide. (b) Profile of the field distributions of the first two guided modes of the waveguide.

after a round-trip the wave reproduces itself, preserving the same spatial distribution and polarization along the waveguide. Fields satisfying this self-consistency condition are known as eigenmodes or modes of the waveguide [10]. The schematic in **Figure 1b** shows the profile of the field distribution for the first two modes of the proposed waveguide.

2.2 Dispersion relation

To determine the propagation constant of the modes supported by the waveguide, let us consider a waveguide with a core of refractive index n_2 and thickness d, surrounded by two semi-infinite dielectric media of refractive index n_1 and n_3, as depicted in **Figure 2**.

For each medium, the field can be represented as a sum of propagative and counter-propagative waves along the z axis, and propagative in the x direction that can be represented as

$$\psi_m(x,z,\omega) = A_m e^{-i\alpha_m z} e^{-i\beta_m x} e^{-i\omega t} + B_m e^{i\alpha_m z} e^{-i\beta_m x} e^{-i\omega t}, \quad (1)$$

where $m = I, II, III$, A_m and B_m are the amplitudes of the propagative and counter-propagative waves, respectively, and the propagation constants α_m and β_m along the z and x axis are related through

$$\beta_m = \sqrt{\left(\frac{\omega}{c}\right)^2 \varepsilon_m(\omega) - \alpha_m^2}, \quad (2)$$

where $\varepsilon_m(\omega)$ is the dielectric constant of the m-th medium related to the refractive index by $n_m = \sqrt{\varepsilon(\omega)\mu(\omega)}$. At optical wavelengths, he magnetic permeability $\mu(\omega)$ can be considered as unit. Req. (2) is obtained from the Helmholtz and Maxwell equations [11].

At the interfaces $z = 0$ and $z = d$, the electromagnetic field should be continuum, that is to say:

$$\psi_I(x,z,\omega)|_{z=0} = \psi_{II}(x,z,\omega)|_{z=0}, \quad (3)$$

and

$$\psi_{II}(x,z,\omega)|_{z=d} = \psi_{III}(x,z,\omega)|_{z=d}. \quad (4)$$

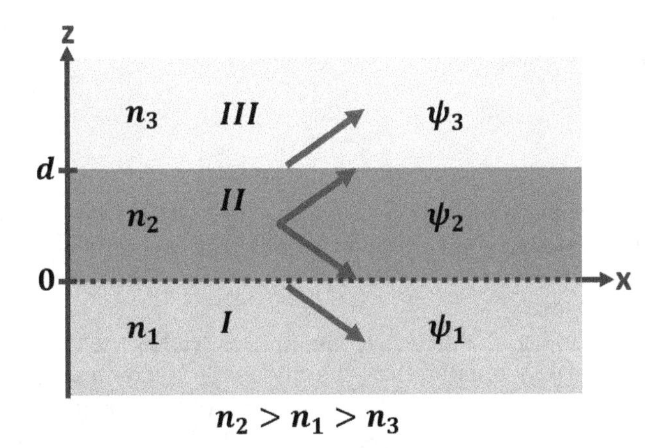

Figure 2.
Schematic representation of the field components in an asymmetric planar waveguide.

From the conservation of the tangential components of the electromagnetic field at the boundaries between two media [12] are obtained the relationships

$$\frac{1}{\nu_I}\frac{\partial\psi_I(x,z,\omega)}{\partial z}\bigg|_{z=0} = \frac{1}{\nu_{II}}\frac{\partial\psi_{II}(x,z,\omega)}{\partial z}\bigg|_{z=0}, \tag{5}$$

$$\frac{1}{\nu_{II}}\frac{\partial\psi_{II}(x,z,\omega)}{\partial z}\bigg|_{z=-d} = \frac{1}{\nu_{III}}\frac{\partial\psi_{III}(x,z,\omega)}{\partial z}\bigg|_{z=d}, \tag{6}$$

with $\nu_m = 1$ for TE polarized electromagnetic fields and $\nu_m = \varepsilon_m(\omega)$ for TM polarized fields. Substituting Eq. (1) in Eqs. (3–6), and considering that $A_I = B_{III} = 0$ because both, I and III are semi-infinite media and no back-reflections from the boundaries are present, it is obtained a two coupled equation system that can be represented in a matrix way of the form

$$\begin{bmatrix} \frac{\alpha_I}{\nu_I} - \frac{\alpha_{II}}{\nu_{II}} & \frac{\alpha_I}{\nu_I} + \frac{\alpha_{II}}{\nu_{II}} \\ e^{-i\alpha_{II}d}\left(\frac{\alpha_I}{\nu_I} + \frac{\alpha_{III}}{\nu_{III}}\right) & -e^{-i\alpha_{II}d}\left(\frac{\alpha_I}{\nu_I} - \frac{\alpha_{III}}{\nu_{III}}\right) \end{bmatrix} \begin{bmatrix} A_{II} \\ B_{II} \end{bmatrix} = \begin{bmatrix} 0 \\ 0 \end{bmatrix}. \tag{7}$$

By equating to zero the determinant of the matrix it is possible to obtain the non-trivial solutions of this eigenmode equation system, resulting in the dispersion relation of a three-layered media

$$\frac{\left(\frac{\alpha_{II}}{\nu_{II}} - \frac{\alpha_I}{\nu_I}\right)\left(\frac{\alpha_{II}}{\nu_{II}} - \frac{\alpha_{III}}{\nu_{III}}\right)}{\left(\frac{\alpha_{II}}{\nu_{II}} + \frac{\alpha_I}{\nu_I}\right)\left(\frac{\alpha_{II}}{\nu_{II}} + \frac{\alpha_{III}}{\nu_{III}}\right)} = e^{i2d\alpha_{II}}. \tag{8}$$

We must notice that Eq. (8) is a transcendental function with no analytical solution, thus, numerical methods should be employed to solve it.

When solving this dispersion relation, it is obtained the mode propagation constant, β, that depends on the optical frequency or wavelength of light and determines how the amplitude and phase of light varies along the x direction. In the same way as wavenumber can be related to the refractive index of a homogeneous medium, the propagation constant can be regarded as the wavenumber (spatial frequency) of light propagating through an effective medium composed by the inhomogeneous three-layered structure. The propagation constant is then related to the so-called effective index through the relationship

$$\beta = \frac{2\pi}{\lambda} n_{eff}, \tag{9}$$

being λ the wavelength of light in vacuum. We must notice that the effective index is only defined for a mode of the waveguide and it should not be understood as a material property. We can say then that each mode of the waveguide will "see" different effective media.

As the refractive index of a dielectric medium, as well as the dielectric constant, is a real number equal or greater than the unit ($n \geq 1$) the modes in a dielectric waveguide are diffraction limited: if the thickness of the waveguide, d, is smaller than $\lambda/(2n_{eff})$, the solutions for the dispersion Eq. (8) will lead to evanescent waves, meaning that no modes can be propagated below this limit.

2.3 Plasmonic waveguides

As previously explained, dielectric waveguides guide light modes by using the total internal reflection principle and self-consistency condition. These waveguides are diffraction limited due to the dielectric constant values. However, if the dielec-tric constant is a complex number, it would be possible to obtain solutions to the dispersion relation (Eq. 8) below the diffraction limit. This is the case of metallic materials. Hence, if at least one of the three media in the waveguide structure is metallic, it is always possible to obtain a propagative mode in the structure. The price to pay for this solution is that due to ohmic losses in metals, these modes propagate just few microns, in opposition to dielectric waveguides where light can propagate through kilometers.

These structures are known as plasmonic waveguides, and their operation principle is based on SPP mode propagation. These surface waves are the result of collective oscillations of the conductive electrons at a metal-dielectric interface induced by the electric field of an electromagnetic wave. For a system invariant in the $\hat{y}$ direction, SPP modes can only be excited if the electric field oscillates in the xz plane. Hence, only TM polarized electromagnetic fields couple to SPP modes (for TE polarized waves the electric field only oscillates along the $\hat{y}$ direction).

Different combinations of insulator (I) and metallic (M) materials can be used to define a plasmonic waveguide. In **Figure 3** are represented IIM, IMI and MIM plasmonic waveguide structures as well as the amplitude distribution of the out-of-plane electromagnetic field (H_y component) of the SPP modes. For the IIM structure, there is only one SPP mode at the interface between the metal (ε_m) and first dielectric (ε_{d_1}). For both IMI and MIM configurations, two SPP modes can be excited. They result from in-phase and out-of-phase coupling of SPP at the first and second dielectric-metal interfaces, and they are known as symmetric and antisymmetric modes, respectively.

As plasmonic waveguides allow light propagation beyond the diffraction limit, these structures have been used for the development of integrated nanophotonic devices for optical signal transportation, optical communications, biosensing and even imaging applications [13–15].

2.4 Hybrid photonic-plasmonic waveguides

From the previous waveguiding configurations, it is natural to think that modes propagating through a dielectric waveguide can be coupled to a plasmonic wave-guide. This kind of structures is named hybrid photonic-plasmonic waveguide, or simply hybrid plasmonic waveguide.

The structure depicted in **Figure 3a** can be considered as a hybrid plasmonic waveguide, but more complex multilayered systems can be designed to propagate more than one mode in these structures. For instance, in **Figure 4** are presented two

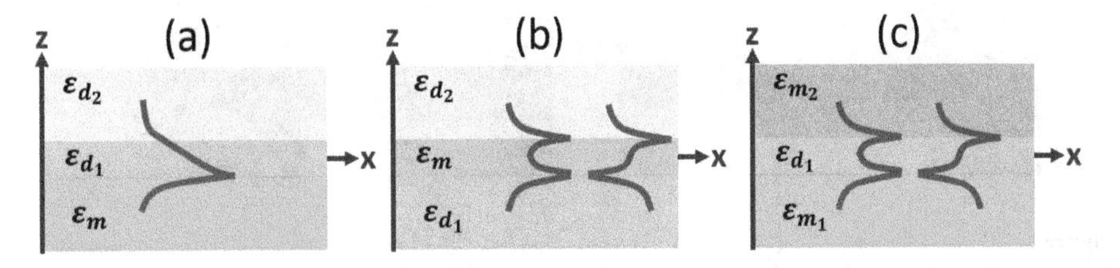

Figure 3.
Schematic representation of plasmonic waveguides for (a) IIM, (b) IMI and (c) MIM configurations and SPP modes profiles. For IMI and MIM waveguides, symmetric and antisymmetric modes are excited.

examples of hybrid plasmonic waveguides able to support symmetric and antisymmetric SPP modes coupled to photonic modes of a dielectric waveguide.

To compute the supported modes of these structures, we can make use of the dispersion relation for a N-layered medium in terms of the T-matrix that relates the amplitudes of propagative and counter-propagative waves, A_m and B_m, in the m-th medium, to those from the $m+1$ medium through the relationship [16].

$$\begin{bmatrix} A_{N+1} \\ B_{N+1} \end{bmatrix} = [\mathbb{T}] \begin{bmatrix} A_1 \\ B_1 \end{bmatrix} = \prod_{m=1}^{N} [T_m] \begin{bmatrix} A_1 \\ B_1 \end{bmatrix} = \begin{bmatrix} t_{11} & t_{12} \\ t_{21} & t_{22} \end{bmatrix} \begin{bmatrix} A_1 \\ B_1 \end{bmatrix}, \tag{10}$$

where $A_1 = B_{N+1} = 0$ (no back reflections from substrate and superstrate) and

$$[T_m] = \frac{1}{2} \begin{bmatrix} (1+\gamma)e^{-i(k_m - k_{m+1})} & (1-\gamma)e^{i(k_m + k_{m+1})} \\ (1-\gamma)e^{-i(k_m + k_{m+1})} & (1+\gamma)e^{i(k_m - k_{m+1})} \end{bmatrix}, \tag{11}$$

with $k_m = \alpha_m d_m$, $k_{m+1} = \alpha_{m+1} d_m$, $\gamma = (\alpha_m \nu_{m+1})/(\alpha_{m+1}\nu_m)$, being d_j the position of the interface between j and $j+1$ media, considering that $d_{N+1} = d_N$ and $\nu_m = 1$ for TE polarized electromagnetic fields and $\nu_m = \varepsilon_m(\omega)$ for TM polarized fields. By equating to zero the term t_{22} of the $\mathbb{T}$ matrix, we can directly obtain the propagation constant of the modes supported by the structure.

2.5 Dispersion curves and mode analysis

Before studying light propagation in complex hybrid plasmonic waveguides, it is worthily to briefly comment on the dispersion diagrams that would help to perform an analysis of the modes propagating through these waveguides. So far, we have presented the dispersion relation for a multilayered media. It is not our intention to explore the numerical methods that can be employed to solve this transcendental equation, but to analyze the information obtained from these results. The reader can look at references [17–22] to have an insight of how to solve the dispersion relation.

As an example, let us analyze the modes of a four-layered media as schematized in **Figure 4a**, consisting of a glass substrate with refractive index $n_1 = 1.5$, a silicon nitride layer (core of the photonic waveguide) of thickness $d = 300$ nm and refractive index $n_c = 2.0$, a thin gold layer of thickness $t = 40$ nm, and air superstrate ($n_2 = 1.0$). The numerical results obtained from the calculation of the dispersion relation for TM polarized fields by using the Raphson–Newton method [23] are plotted in **Figure 5**. For these calculations was considered a spectral wavelength range from 400 nm to 1 μm, and effective index range between 0.9 and 2.5. Since

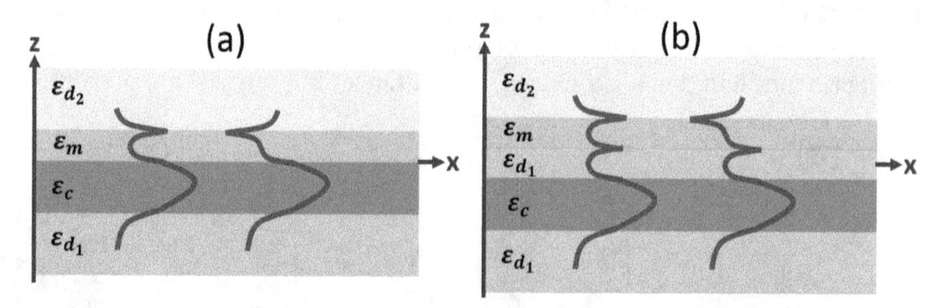

Figure 4.
Schematic representation of hybrid photonic-plasmonic waveguides and mode profiles for (a) a metallic layer placed directly on top of the dielectric waveguide, and (b) with an intermediate dielectric layer between photonic and plasmonic waveguides. In both systems, the fundamental mode of the waveguide couple to the symmetric and antisymmetric SPP modes.

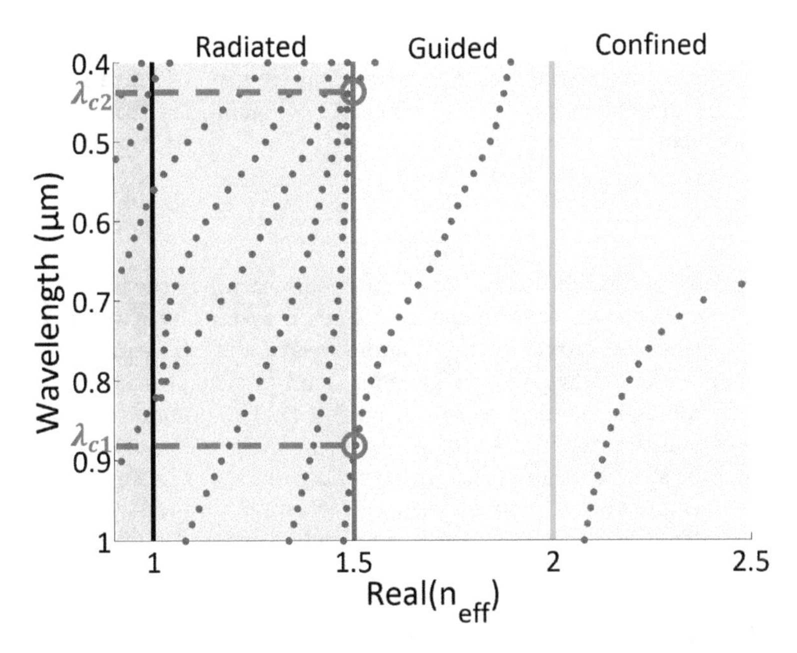

Figure 5.
Dispersion curves for a hybrid photonic-plasmonic waveguide consisting of a glass substrate ($n_1 = 1.5$), silicon nitride core of 300 nm thickness ($n_c = 2.0$), a thin gold layer of thickness 40 nm, and an air superstrate ($n_2 = 1.0$). The numerical results show a confined mode (blue region) and two guided modes (green region). At the orange region, many radiated modes were obtained, as well as many non-physical solutions (gray region).

the results are just numerical solutions, we need to understand the physical meaning for each solution.

The vertical constant lines at 1, 1.5 and 2.0 correspond to the refractive index of each dielectric medium: air superstrate, glass substrate and silicon nitride core, respectively. These vertical lines are also referred as light lines, as they are linked to the propagation constant of light traveling in that specific homogeneous medium through Eq. (9).

These light lines define four different regions. The first region, for effective index values below 1 (gray region), are numerical solutions without physical mean-ing: if the effective index is smaller than unit, the modes would travel faster than speed of light in vacuum (which obviously is not our case). The second region between the refractive index of glass and air refractive index (orange region) defines modes with effective index smaller than glass but greater than air. Hence, they are modes whose energy is propagating in the glass substrate, and they are referred as radiated modes. The third region, between the silicon nitride (core) light line and glass light line, define modes whose energy is propagating in the core of the waveguide: as the effective index is higher than glass substrate index, the energy of these modes does not propagates in glass, so the energy is confined in the core. These are guided modes. The value at which the effective index of these modes matches the refractive index of the glass substrate determines the cut-off wave-length of guided modes. For the analyzed example, these values are $\lambda_{c1} = 886$ nm and $\lambda_{c2} = 430$ nm (red circles).

The fourth region (blue colored) correspond to modes whose energy does not propagates in any of the dielectric layers: their effective index is greater than the core, substrate, and superstrate. Hence, these modes are confined to the metallic layer. These solutions correspond to propagative SPP modes and they are referred as confined modes.

In literature, different representations of the dispersion curves can be found, like propagation constant vs. frequency (usually normalized to a reference value), wavelength vs. incidence angle (used in attenuated total internal reflection measurements), among others. The representation that we use in **Figure 5** allows us to understand the dispersion curves in terms of two quantities that can be easily identified: wavelength and effective index.

2.6 Mode hybridization

To further understand the origin of the modes appearing in the hybrid photonic-plasmonic structure, let us analyze the multilayered system by parts: first we will compute the dispersion curves of the photonic waveguide (**Figure 6a**), then the modes of the plasmonic waveguide (**Figure 6b**), and finally compare them with the full hybrid photonic-plasmonic structure (**Figure 6c**). The numerical results for the dispersion relation of each one of these cases are presented in **Figure 6d**. The dimensions of the structures are the same than those used for **Figure 5**.

The blue dots and circles in guided region, correspond to the fundamental TM_0 and higher order TM_1 modes of the silicon nitride waveguide (thickness $d = 300$ nm and refractive index $n_c = 2.0$) surrounded by air superstrate ($n_{sup} = 1.0$) and glass substrate ($n_{sub} = 1.5$), as depicted in **Figure 6a**. These modes present cut-off wavelength values around $\lambda_{TM0} = 1.58$ μm and $\lambda_{TM1} = 545$ nm, respectively.

When computing the modes of a thin gold layer of thickness $t = 40$ nm on top of a glass substrate ($n_{sub} = 1.5$) and air superstrate, as depicted in **Figure 6b**, it is observed one mode in the guided region that tends to a constant value (red dots in **Figure 6d**). For this structure, we must notice that no core was present, then, the effective index of this mode is higher than the refractive index of the glass substrate, hence, it is a SPP mode confined to the metallic layer.

We can observe that both, TM_0 and SPP modes, cross each other around an effective index value of 1.798. In other words, the propagation constant for both modes are the same, so they are phase matched. This situation means that the photonic mode of the waveguide will excite the plasmonic mode of the metallic layer.

When the modes of the complete integrated structure (**Figure 6c**) are computed, two branches are observed. The first one, represented by red triangles, is a

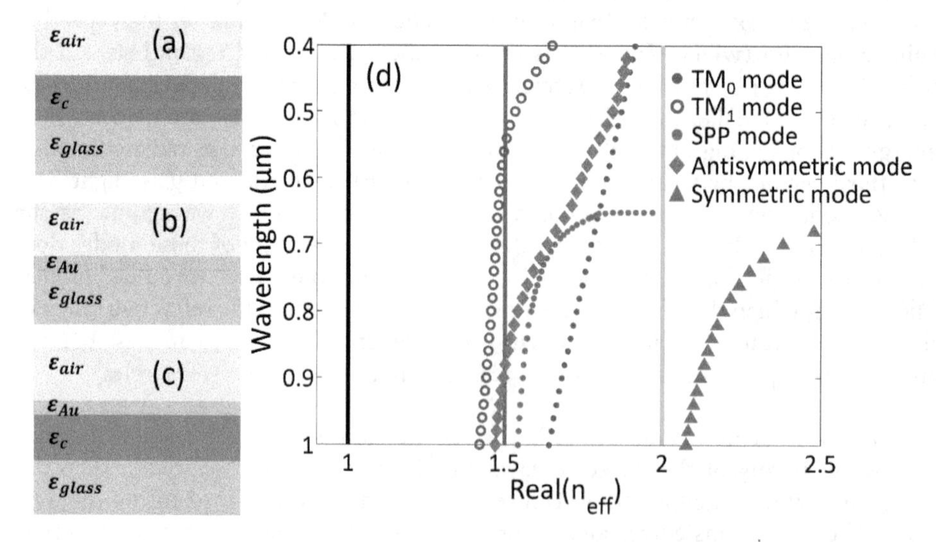

Figure 6.
Schematic representation of (a) photonic waveguide, (b) plasmonic waveguide, and (c) hybrid photonic-plasmonic waveguide. (d) Dispersion curves for the three studied structures. Due to phase matching between TM_0 and SPP modes, hybrid symmetric and antisymmetric modes arise in the full integrated structure.

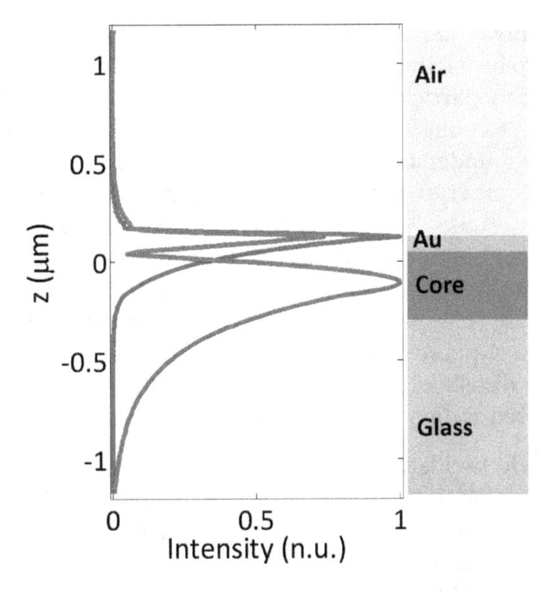

Figure 7.
Intensity profiles of the symmetric (red) and antisymmetric (green) modes of the hybrid photonic plasmonic waveguide at $\lambda = 750$ *nm. The symmetric mode is mainly confined in the metallic layer, while the antisymmetric presents amplitude in both, photonic and plasmonic waveguides.*

mode confined to the metallic layer, and the second one, relies in the guided region (green diamonds). These modes arise from the coupling of the TM_0 and SPP modes, and they are referred as hybrid modes, being the first one a symmetric mode and the second an antisymmetric mode.

It is important to say that both symmetric and antisymmetric modes are not independent, they are hybrid modes. Like in two coupled harmonic oscillators, this hybridization means that there is an energy exchange between photonic and plasmonic waveguides. For the symmetric mode, the amplitude of light in both, photonic and plasmonic waveguides, are in phase, while for the antisymmetric are out-of-phase [24, 25].

Finally, in **Figure** 7 are plotted the normalized intensity profiles of the symmetric (red curve) and antisymmetric (green curve) modes at a wavelength of $\lambda = 750$ nm. The intensity is derived from the amplitudes of Eq. 1. As expected from the dispersion curves, the intensity of the symmetric mode is mainly confined in the metallic layer (the mode solution relies in the confined region), while for the antisymmetric mode the intensity is distributed in both photonic and plasmonic waveguides, being greater the intensity in the dielectric region (the solution relies in the guided region).

3. Mode propagation in a periodic array of metallic nanowires

In general, plasmonic resonances in metallics nanostructures are divided in two kinds, namely SPP and LSP. SPP modes are propagative waves confined at the dielectric/metal interface, while LSP are standing waves or cavity modes oscillating in a nanoparticle.

As it is well known, LSP resonances depend not only on the material of the nanoparticles, but also on their shape and polarization of the incident wave: the orientation of the electric field defines the direction of the oscillation of the charges in the metallic nanoparticle; these charges will distribute depending on the geometry of the particle, giving rise to different modes. For small nanoparticles, usually are only excited dipolar LSP resonances, but quadrupoles, octupoles and higher order modes can also be excited.

When metallic nanoparticles are closely placed and excited at their LSP resonance, it is possible to couple them via near field interaction, leading to higher order LSP modes. To understand this coupling mechanism, let us take a look to **Figure 8**, where a dimmer of spherical metallic nanoparticles oriented along the x axis(dimmer axis), is excited with electric field oscillating in z and x directions.

When the electric field is oriented along the z axis, perpendicular to the dimer axis, the dipolar resonances of the nanoparticles are oriented also in the z direction. If the dipoles are in phase (**Figure 8a**), the dimmer also presents a dipolar reso-nance. If the dipoles are out-of-phase (**Figure 8b**), the dimmer presents a quadrupolar resonance. Since the distribution of the charges is perpendicular to the dimmer axis, both modes are referred as dipolar and quadrupolar transverse modes, respectively.

If the electric field oscillates along the x axis, the dipoles of the nanoparticles will be oriented along the dimmer axis, thus, the coupled modes are called dipolar longitudinal modes. If the dipoles are in phase (**Figure 8c**), the resonance wavelength of the longitudinal mode will be shorter than the resonance wavelength of the out-of-phase dipoles (**Figure 8d**).

In the same way, a periodic array of metallic nanoparticles can be coupled, allowing light propagation. Thus, when properly excited, a periodic array of metallic nanoparticles can be regarded as a plasmonic waveguide. These resonances are named plasmonic chain modes, and their waveguiding properties will depend on the shape and period of the nanoparticles, as well as the orientation of the incident electromagnetic field. Besides energy transportation capabilities, these modes have been widely studied because they allow a strong enhancement of the electromagnetic field in a localized nanometric region.

If we consider that the nanoparticles in **Figure 8** are invariant in the out-of-plane y direction, the same coupling mechanism is preserved. These structures are named metallic nanowires (MNW): metallic nanostructures with nanometric cross section and micrometric lengths. As the length of the nanowires is times longer than the incident wavelength, the absorption of light prevents from the formation of cavity modes in this direction. This means that plasmonic resonances in metallic nanowires can be excited only if the electric field is perpendicular to the invariant

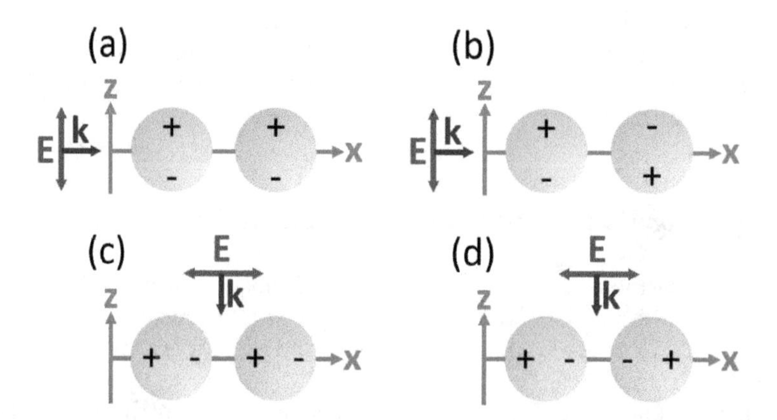

Figure 8.
LSP coupled modes for spherical nanoparticle dimmers. When the electric field is perpendicular to the main axis of the dimmer are excited (a) dipolar and (b) quadrupolar transverse modes. When the electric field is parallel to the main axis of the dimmer, dipolar longitudinal modes of (c) shorter and (d) longer resonant wavelengths are obtained.

y axis, i. e. with TM polarized light. Thus, the plasmonic chain modes in metallic nanowires will mainly depend on the geometry of their cross section [26–28].

3.1 Plasmonic chain modes in MNW with rectangular cross section

Let us consider an infinite periodic array of gold nanowires of width $w = 80$ nm, gap between them $g = 50$ nm (period $\Lambda = 130$ nm) and thickness $t = 20$ nm, embedded in a homogeneous dielectric medium of refractive index $n_d = 1.5$ (glass), as depicted in **Figure 9a**. The structure in invariant along the out-of-plane direction.

To perform a modal analysis, it is required to compute the dispersion curves of these system. Different numerical methods can be employed, like effective index method [29–31], source model technique [32], rigorous coupled wave analysis (RCWA) [33–35], or Fourier modal method (FMM) [36–39], among others.

In our case, we will make use of the FMM to compute the dispersion curves of the periodic structure. This rigorous method computes the Maxwell equations in the frequency domain. To solve them, a unit cell of the periodic structure, as well as the dielectric function and electromagnetic field are expanded in Fourier series. This formulation leads to an eigenvalue matrix formulation that can be used to obtain the modes of the nanowires in a multilayered media. Also, by adding perfectly matched layers (PML), it is possible to compute the beam propagation in a finite periodic structure. It is not our intention to show how to implement this numerical method, but to analyze plasmonic chain modes in periodic MNW. For a better comprehension about this method, we invite the reader to look at references [37, 38].

The plot in **Figure 9b** corresponds to the dispersion curves of the system under study. As we are dealing with a periodic structure, it is useful to represent this plot in terms of the propagation constant (along the periodicity direction), normalized to the Bragg condition ($\beta = \pi/\Lambda$), which defines the first Brillouin zone (vertical black line). The red dotted curve represents the modes supported by the MNW, as they are confined below the glass light-line. To understand the behavior of this

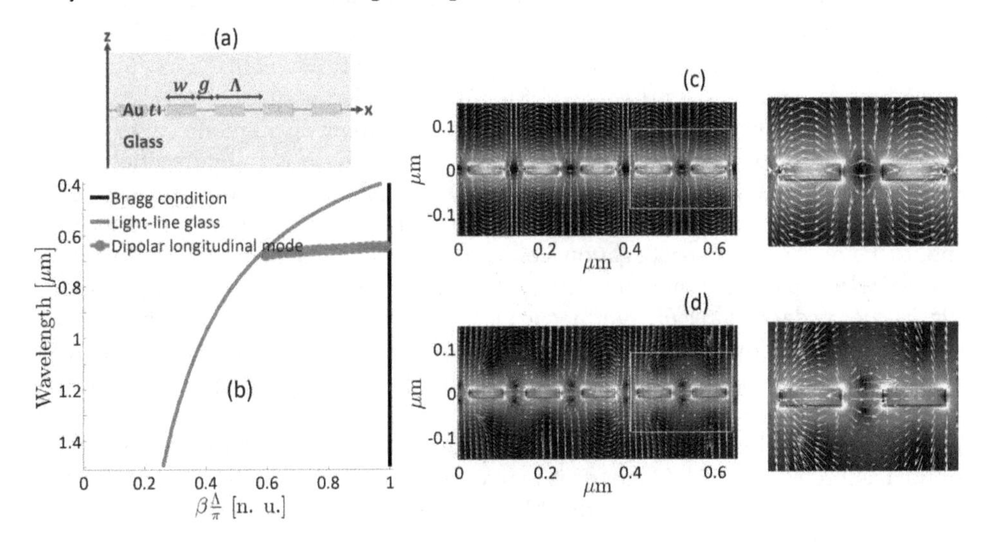

Figure 9.
(a) Schematic representation of a periodic infinite array of gold nanowires of width $w = 80$ nm, gap $g = 50$ nm, period $\Lambda = 130$ nm and thickness $t = 20$ nm in a homogeneous dielectric medium. (b) the confined modes below the glass light-line (green curve) correspond to a dipolar longitudinal mode (red dotted curve). (c) the energy density map and electric field lines computed at the Bragg condition ($\lambda = 645$ nm), reveals the dipolar longitudinal coupling between MNW. (d) out of the Bragg condition, at $\lambda = 667$ nm, we still observe the dipolar coupling.

plasmonic mode, it is necessary to observe the distribution of the charges in the MNW.

In **Figure 9c** are shown the energy density map and electric field lines distribution of the plasmonic mode at $\lambda = 645$ nm. At this wavelength, the red curve crosses the Bragg condition, defining a stationary mode. As we can observe, MNW are coupled, with a phase shift of π rad between them. Thus, it is a plasmonic chain mode. Out of the Bragg condition, for instance, at $\lambda = 667$ nm ($\beta = 0.641\pi/\Lambda$), the chain mode becomes propagative and the electric field lines remain almost longitudinally oriented inside the MNW (**Figure 9d**). In view of the phase shift and the orientation of the electric field lines, the plasmonic chain mode results from coupled dipolar resonances oriented along the x axis. Thus, the red dotted curve corresponds to the dispersion relation of a dipolar longitudinal plasmonic chain mode.

As defined for a SPP, the propagation length of this plasmonic chain mode can be computed through the relationship

$$L_p = \frac{1}{2\beta''}, \tag{12}$$

where β'' is the imaginary part of the propagation constant. In our structure, the propagation length varies from $L_p = 200$ nm (for wavevectors close to the glass light-line) up to $L_p = 1.14$ μm (for wavevectors near the Bragg condition).

3.2 Hybrid plasmonic chain modes in MNW integrated to a dielectric waveguide

Now, let us study a hybrid photonic-plasmonic system consisting of a dielectric waveguide of thickness $d = 200$ and refractive index $n_{wg} = 2.0$, on top of which is placed, at a distance $h = 30$ nm, a periodic array of gold nanowires with the same parameters than those of the previous subsection, as depicted in **Figure 10a**.

The dispersion curves in the first Brillouin zone, obtained with the FMM, are shown in **Figure 10b**. The green and magenta curves represent the glass and waveguide light lines, respectively, while the black vertical line represents the Bragg condition. The blue triangles curve corresponds to the fundamental TM_0 photonic guided mode without the presence of the MNW, while the green asterisk curve is the dispersion curve of the dipolar longitudinal plasmonic chain mode (without the presence of the waveguide). As can be observed, there is an intersection point between the TM_0 and dipolar longitudinal modes in the guided region around a wavelength value $\lambda = 664$ nm (**Figure 10c**). This crossing point means that both modes have the same propagation constant ($\beta = 0.672(\pi/\Lambda)$), hence, they will couple when placing them near to each other, leading to a hybrid photonic-plasmonic mode. As a result, symmetric and anti-symmetric modes will arise. This hybridization is corroborated when computing the dispersion curves of the integrated system, being observed two curves: a lower branch passing from guided to confined region (red circles) and an upper branch in the guided region. In analogy with a two coupled waveguide system, this situation means that, for the lower branch, energy from the TM_0 mode is converted into plasmonic chain mode; for the upper branch, the energy from the plasmonic chain mode is converted into TM_0 mode. This anti-crossing phenomenon is the characteristic signature of strong coupling between guided modes.

To corroborate the symmetry of these modes, in **Figure 11** we plot the energy density maps and electric field lines distribution for both modes at the Bragg condition. **Figure 11a** corresponds to the energy density map computed for the upper branch at $\lambda = 473$ nm, where we can observe that the energy is distributed in

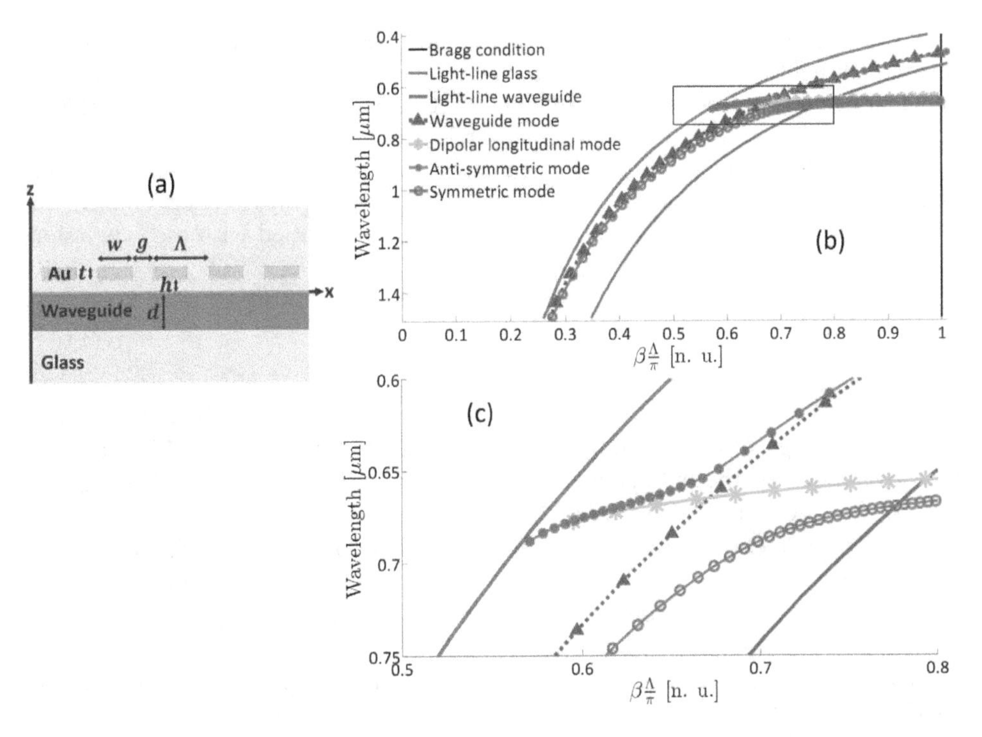

Figure 10.
(a) Schematic representation of the integrated structure consisting of an infinite periodic arrays of gold nanowires ($w = 80$ nm, $g = 50$ nm, $\Lambda = 130$ nm, $t = 20$ nm) surrounded by glass ($n_s = 1.5$), placed on top of a dielectric waveguide ($d = 200$ nm, $n_{wg} = 2.0$) at a distance $h = 30$ nm. (b) Dispersion curves of the integrated system. (c) Inset at the anti-crossing region.

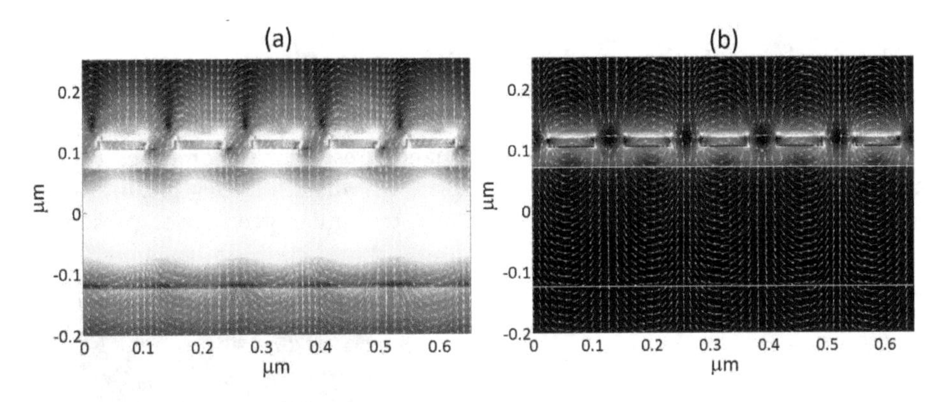

Figure 11.
Energy density maps and electric field lines distribution for (a) the antisymmetric mode at $\lambda = 473$ nm, and (b) for the symmetric mode at $\lambda = 660$ nm.

the dielectric waveguide and the periodic array of MNW. When looking at the electric field lines, it is observed a phase difference between these vectors above and below the MNW, defining an antisymmetric quadrupolar chain mode. For the case of the lower branch, computed at $\lambda = 660$ nm (**Figure 11b**), the energy is confined to the MNW and not in the waveguide, as expected from the dispersion curves. For this case, the electric field lines above and below the MNW are in phase, defining a symmetric dipolar chain mode.

Finally, by making use of the aperiodic Fourier modal method [27, 39], and considering a finite number of 27 MNW, we computed the transmission (red curve), reflection (blue dashed curve), and absorption (black dotted curve) spectra of light propagating through the integrated system, normalized to the incident beam (**Figure 12a**). The reflection spectrum (blue dashed curve) exhibits a maximum peak around $\lambda = 474$ nm, while the transmission spectrum shows a broad depth around $\lambda = 520$ nm and a narrower depth around $\lambda = 678$ nm. The first depth (close to the maximum in reflection) is due to a Bragg reflection induced by the periodic array of MNW. This condition can be verified with the expression

$$\lambda_{Bragg} = \frac{2n_{eff}\Lambda}{m}, \tag{13}$$

where n_{eff} is the effective index of the mode, Λ the period of the MNW array, and m the Bragg order. According to the dispersion curves, at $\lambda = 520$ nm, the effective index of the antisymmetric mode is $n_{eff} = 1.801$, and considering the period $\Lambda = 130$ nm, the first order Bragg reflection will occur at $\lambda_{Bragg} = 468$ nm, which is a value close to the maximum in the reflection spectrum. The near field map in **Figure 12b**, shows the amplitude of the H_y field component, where are observed periodic lobes inside the waveguide core due to the Bragg reflection.

The second depth on the transmission spectrum corresponds to the excitation of the dipolar longitudinal plasmonic chain mode. In the dispersion curves, this wavelength value corresponds to the anti-crossing point between symmetric and antisymmetric modes, around $\beta = 0.672(\pi/\Lambda)$. As expected, at this wavelength the plasmonic chain mode is efficiently coupled to the photonic TM_0 mode of the waveguide, leading to an anergy exchange between the photonic waveguide and the periodic array of MNW, as illustrated in the near field map of **Figure 12c**.

3.3 Hybrid plasmonic chain modes in MNW of triangular cross section integrated to a dielectric waveguide

Among the large variety of shapes in nanowires, sharp geometries such as nanotips stimulate a great interest in applications where a strong localization of the electromagnetic fi eld is required. These triangular geometries present an

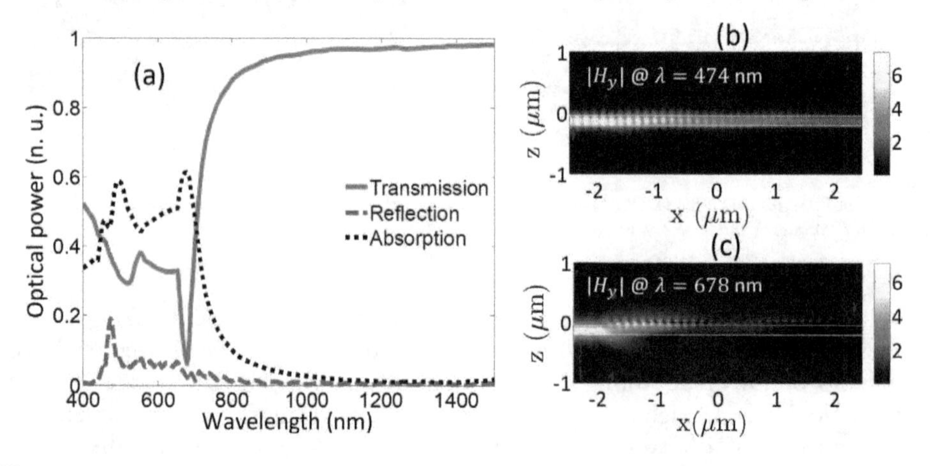

Figure 12.
(a) Normalized transmission (red), reflection (blue dashed) and absorption (black dotted) spectra of light propagating through the waveguide in the integrated device. (b) At $\lambda = 474$ nm, the H_y field component exhibits a Bragg reflection, while at (c) $\lambda = 678$ nm, the it is observed an energy exchange between the waveguide and periodic array of MNW.

excitation of LSP resonances polarized along their tip axis. In this section, we will study the excitation of plasmonic chain modes in periodic arrays of gold nanowires with triangular cross section through the photonic mode of a dielectric waveguide.

Firstly, we will study the plasmonic chain modes of the MNW placed on top of a glass substrate and the photonic modes of the dielectric waveguide that will bea used to excite them. The MNW consist of an infinite periodic array of gold nanowires with triangular cross section of height $t = 144$ nm, width $w = 72$ nm, and period $\Lambda = 200$ nm, placed on top of a glass substrate ($n_{sub} = 1.5$). The superstrate is air ($n_{sup} = 1.0$), the tip radius of the nanowires is $r = 5$ nm, and the system is invariant in the out-of-plane y direction, as depicted in **Figure 13a**. Then dielectric waveguide consist of a core of thickness $d = 200$ nm end refractive index $n_{wg} = 2.0$, buried a distance $h = 30$ nm from the glass/air interface (**Figure 13b**).

The dispersion curves in the first Brillouin zone are plotted in **Figure 13c**, where can be observed the light lines of air superstrate (white curve), glass substrate (black curve) and core of the waveguide (green curve). The colored regions correspond to the normalized absorption spectra obtained when illuminating the structure from the substrate with a plane wave and mapped into $\beta = k_0 n_{sub} \sin\theta_{inc}\Lambda/\pi$ (see reference [28] for a detailed explanation). The blue triangles curve in the guided region correspond to the fundamental TM_0 photonic mode of the isolated waveguide. The array of MNW supports plasmonic chain modes. The first one is a

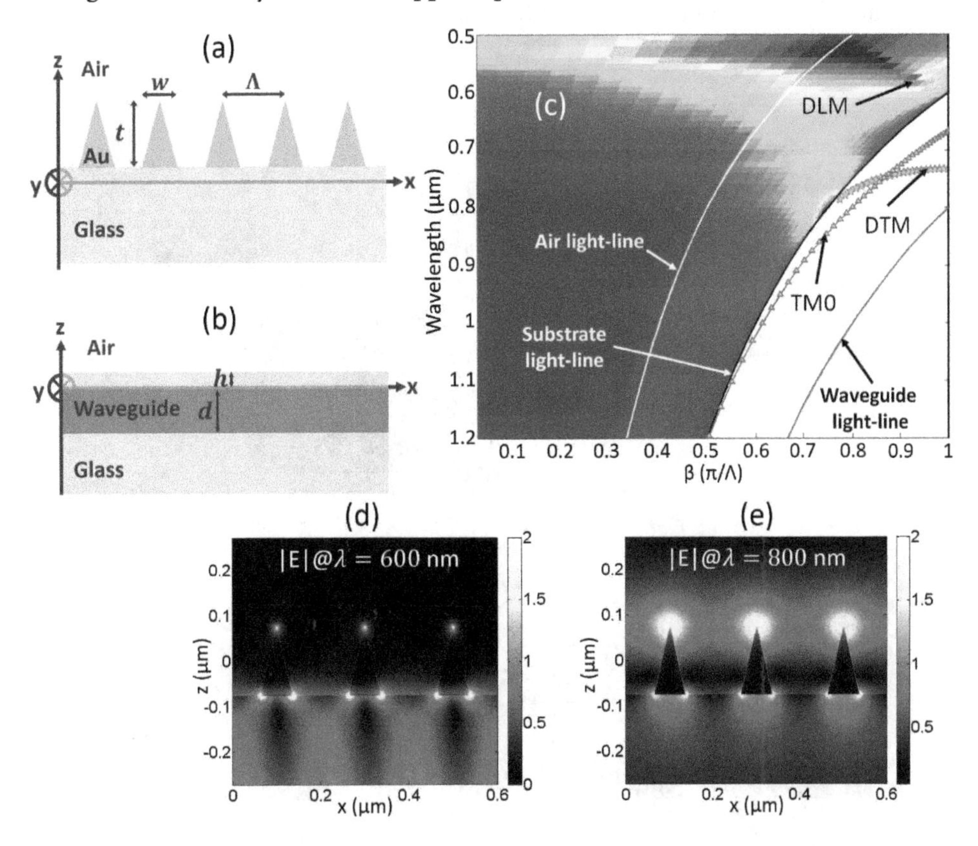

Figure 13.
Schematic representation of (a) periodic array of gold nanowires with triangular cross section ($w = 72$ nm, $t = 144$ nm, $\Lambda = 200$ nm) on top of a glass substrate surrounded by air, and (b) a dielectric waveguide of thickness $d = 200$ nm and refractive index $n_{wg} = 2.0$ buried a depth $h_1 = 30$ nm in a glass substrate. (c) Dispersion curves of the TM_0 mode supported by the waveguide (blue triangles), dipolar longitudinal mode (radiated to the substrate) and dipolar longitudinal mode (green stars). Energy density maps of (d) DLM at $\lambda = 600$ nm and (e) DTM at $\lambda = 800$ nm.

dipolar longitudinal mode (DLM) radiating into the substrate (which is barely excited at the Bragg condition). This mode is characterized by charges coupling and electromagnetic field enhancement at the base (bottom apexes) of the nanowires, as can be observed in the energy density map in **Figure 13d**. The second plasmonic chain mode (green stars) is a dipolar transverse mode (DTM) resulting from coupling of dipoles oriented along the z axis. This guided mode is characterized by a strong electromagnetic field enhancement at the upper apexes of the nanowires, as can be observed in **Figure 13e**. We can notice in the dispersion curves that at $\lambda = 747$ nm ($\beta = 0.86\pi/\Lambda$), the DTM crosses the TM_0 mode of the waveguide, situation that will lead to a strong coupling between these modes when integrating both waveguiding systems.

When integrating the MNW on top of the dielectric waveguide (**Figure 14a**), the dispersion curves shows three modes in the guided region (**Figure 14b**), one corresponding to the DLM (magenta circles) and two other branches corresponding to the anti-symmetric (blue circles) and symmetric (red dots) dipolar transverse chain modes. The mode splitting between these two last resonances arises from strong coupling between dipolar transverse and TM_0 modes. The near field maps in **Figure 14c** present the amplitude and phase of the H_y field for the DLM, where is observed a field enhancement below each nanowire and in the waveguide region. For the antisymmetric DTM (**Figure 14d**) the field is mainly enhanced in the waveguide region, with a phase difference of π rad between the MNW and the waveguide; for the symmetric DTM (**Figure 14e**), the field is enhanced between the

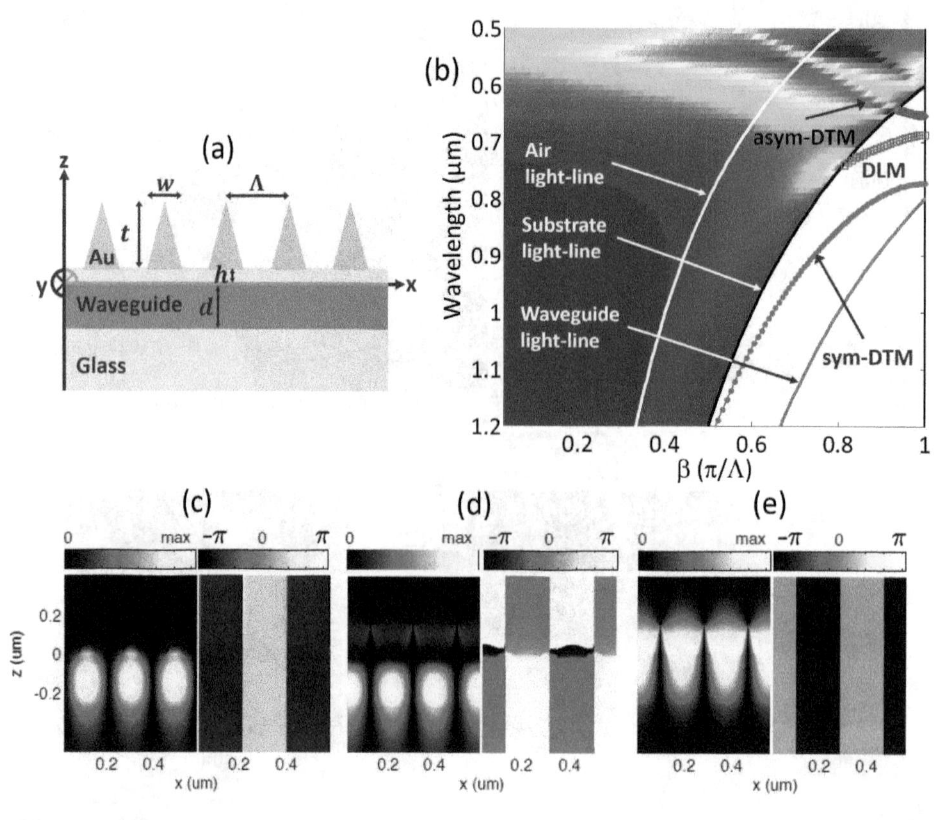

Figure 14.
(a) Schematic representation of an infinite periodic array of MNW integrated on top of a dielectric waveguide buried in a glass substrate. The superstrate is air. (b) In the dispersion curves are observed three chain modes in the guided region: DLM (magenta circles), asymmetric DTM (blue circles) and symmetric DTM (red dots). Distribution of the amplitude and phase of magnetic field in the out of plane direction of the (c) DLM, (d) asymmetric DTM and (e) symmetric DTM.

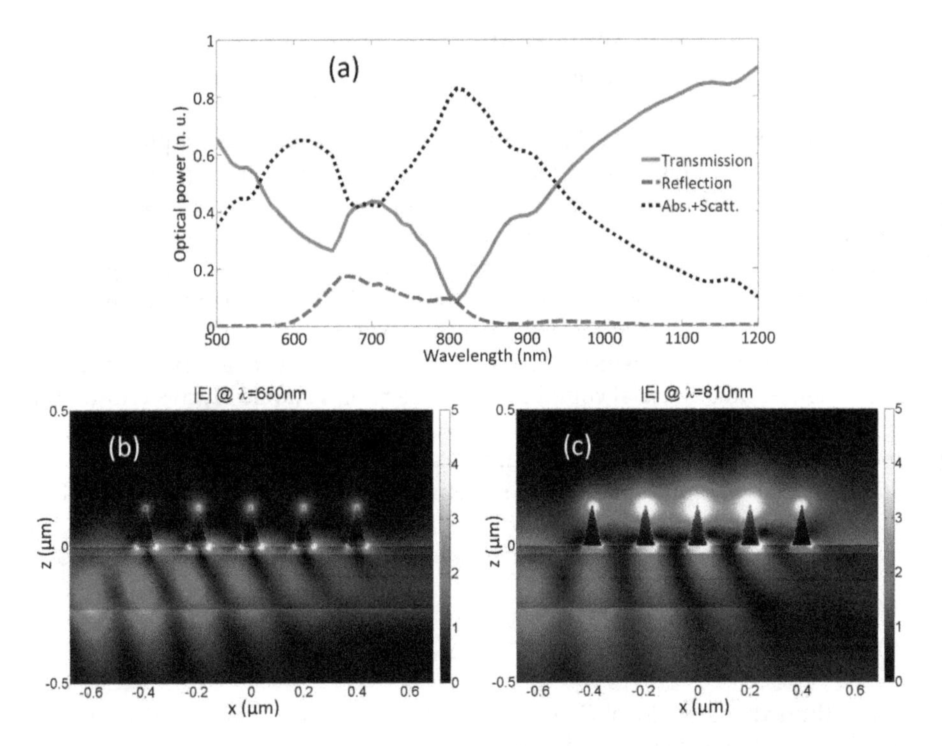

Figure 15.
(a) Normalized transmission (red), reflection (blue dashed) and absorption plus scattering (black dotted) curves of a periodic array of 5 gold nanowires integrated on top of the dielectric waveguide. (b) At $\lambda = 650$ nm, the energy density map shows a field enhancement at the lower apexes of the MNW. (c) At $\lambda = 810$ nm, the energy density map shows a strong enhancement of the field at the upper apexes of the MNW.

MNW with no phase difference between waveguide and MNW, being defined the symmetry of the modes.

To corroborate the coupling between the photonic and plasmonic chain modes, we simulated the beam propagation along the integrated structure. For this simulation, we used a finite number of 5 MNW and excited the waveguide with its fundamental TM_0 mode. The transmission (red curve), reflection (blue dashed curve) and absorption plus scattering (black dotted curve) where normalized to the incident electromagnetic field and the results are shown on **Figure 15a**.

The transmission shows a first minimum value around $\lambda = 650$ nm, corresponding to the excitation of DLM and antisymmetric DTM. This excitation if corroborated in the energy density map of **Figure 15b**, where is observed a field enhancement at the bottom apexes of the nanowires. The second minimum observed in the transmission spectrum around $\lambda = 810$ nm, correspond to the efficient excitation of the symmetric DTM, being characterized by a strong enhancement of the electromagnetic field at the upper apexes of the nanowires, as can be observed in the energy density map of **Figure 15c**. According to the numer-ical calculations, at this wavelength the amplitude of the electromagnetic field measured 10 nm above the apex of the nanowires, is 8.7 times stronger in compar-ison to the amplitude of the incident electromagnetic field. This field enhancement is referred as tip-localized surface plasmon resonance.

4. Conclusions

As we have studied in this chapter, mode hybridization between plasmonic and photonic guided modes offers the possibility to design integrated devices for light confinement in nanometric volumes. Also, as light propagates in dielectric structures, these integrated devices reduce intrinsic propagation losses in conventional plasmonic waveguides.

When properly excited, we have studied how localized surface plasmons can couple in periodic arrays of metallic nanowires, leading to light propagation. In other words, periodic arrays of MNW can behave as plasmonic waveguides. Depending on the geometry of their cross section (shape and aspect ratio), the electromagnetic field can be strongly confined and localized, desired property for number of applications, like excitation of quantum dots or single photon emitters, surface enhanced Raman spectroscopy, and biosensing.

The examples and explanations brought in this chapter can also be expanded to other geometries and material combinations. The reader should always remind that mode hybridization is the core of the physics behind the design of integrated photonic-plasmonic devices for light guiding applications.

These hybrid photonic-plasmonic systems offers the capability of matching diffraction limited guided optics, with nanometric materials, opening new perspectives for the development of a new generation of integrated optical devices.

Acknowledgements

The authors thank National Council of Science and Technology, CONACYT, for partial financial support (Basic Scientific Research, Grant No. A1-S-21527).

Conflict of interest

The authors declare no conflicts of interest.

Author details

Ricardo Téllez-Limón[1]* and Rafael Salas-Montiel[2]

1 CONACYT – Center for Scientific Research and Higher Education at Ensenada (CICESE), Unit Monterrey, Nuevo Leon, Mexico

2 L2n – Laboratory Light, Nanomaterials and Nanotechnologies, CNRS ERL 7004 and University of Technology at Troyes, Troyes, France

*Address all correspondence to: rtellez@conacyt.mx; rafael.salas@utt.fr

References

[1] Novotny L, Hecht B. Principles of Nano-Optics. 2nd ed. Cambridge: Cambridge University Press; 2012. 564 p. DOI: 10.1017/CBO9780511794193

[2] Lévêque G, Martin, O. J. F. Optical interactions in a plasmonic particle coupled to a metallic film, Optics Express, 2006; 14:9971–9981. DOI: 10.1364/OE.14.009971

[3] Dong, J-W, Deng, Z-L. Direct eigenmode analysis of plasmonic modes in metal nanoparticle chain with layered medium, Optics Letters, 2013; 38:2244–2246. DOI: 10.1364/OL. 38.002244

[4] Quinten M, Leitner A, Krenn J. R, Aussenegg F. R. Electromagnetic energy transport via linear chains of silver nanoparticles, Optics Letters, 19 98; 23: 1331–1333. DOI: 10.1364/OL.23. 001331

[5] Brongersma M. L, Hartman J. W, Atwater H. A. Electromagnetic energy transfer and switching in nanoparticle chain arrays below the diffraction limit, Physical Review B, 2000; 62: R16356-R16359. DOI: 10.1103/PhysRe vB.62. R16356

[6] Fung K. H, Chan C. T. Plasmonic modes in periodic metal nanoparticle chains: a direct dynamic eigenmode analysis, Optics Letters, 2007; 32:973–975. DOI: 10.1364/OL.32.000973

[7] Compaijen P. J, Malyshev V. A, Knoester J. Elliptically polarized modes for the unidirectional excitation of surface plasmon polaritons, Optics Express, 2016; 24:3858–3872. DOI: 10.1364/ OE.24.003858

[8] Février M, Gogol P, Aassime A, Mégy R, Delacour C, Chelnokov A, Apuzzo A, Blaize S, Lourtioz, J-M, Dagens B. Giant Coupling Effect between Metal Nanoparticle Chain and Optical Waveguide, Nano Letters, 2012; 12: 1032–1037. DOI: 10.1021/ nl204265f

[9] Saha S, Dutta A, Kinsey N, Kildishev A. V, Shalaev V. M, Boltasseva A. On-Chip Hybrid Photonic-Plasmonic Waveguides with Ultrathin Titanium Nitride Films, ACS Photonics, 2018; 5:4423–4431. DOI: 10.1021/acsphoton ics.8b00885

[10] Saleh B. E. A, Teich M. C. Fundamentals of Photonics. 2nd ed. New York: John Wiley & Sons, Inc; 2007. 947 p. DOI: 10.1002/0471213748

[11] Maier S. A, Plasmonics: Fundamentals and Applications. New York: Springer US; 2007. 224 p. DOI: 10.1007/0–387–37825-1

[12] Jackson J. D. Classical electrodynamics. 3rd ed. New York: Wiley; 1998. 832 p. ISBN:9780471309321

[13] Beltran Madrigal J, Tellez-Limon R, Gardillou F, Barbier D, Geng W, Couteau C, Salas-Montiel R, Blaize S. Hybrid integrated optical waveguides in glass for enhanced visible photoluminescence of nanoemitters, Applied Optics, 2016; 55:10263–10268. DOI: 10.1364/ AO.55.010263

[14] Tellez-Limon R, Blaize S, Gardillou F, Coello V, Salas-Montiel R. Excitation of surface plasmon polaritons in a gold nanoslab on ion-exchanged waveguide technology, Applied Optics, 2020; 59: 572–578. DOI: 10.1364/AO.381915

[15] Inclán Ladino A, Mendoza-Hernández J, Arroyo-Carrasco M. L, Salas-Montiel R, García-Méndez M, Coello V, Tellez-Limon R. Large depth of focus plasmonic metalenses based on Fresnel biprism, AIP Advances, 2020; 10: 045025. DOI: 10.1063/5.0004208

[16] Katsidis C. C, Siapkas D. I. General transfer-matrix method for optical multilayer systems with coherent,

partially coherent, and incoherent interference, Applied Optics, 2002; 41: 3978–3987. DOI: 10.1364/AO.41.003978

[17] Anemogiannis E, Glytsis E. N. Multilayer waveguides: efficient numerical analysis of general structures, Journal of Lightwave Technology, 19 92; 10:1344–1351. DOI: 10.1109/50.166 774

[18] Anemogiannis E, Glytsis E. N, Gaylord T. K. Determination of guided and leaky modes in lossless and lossy planar multilayer optical waveguides: reflection pole method and wavevector density method, Journal of Lightwave Technology, 1999; 17:92 9–941. DOI: 10.1109/50.762914

[19] Kocabaş Ş . E, Veronis G, Miller D. A. B, Fan S. Modal analysis and coupling in metal-insulator-metal waveguides, Physical Review B, 2009; 79: 035120. DOI: 10.1103/PhysRevB. 79.035120

[20] Davis T. J. Surface plasmon modes in multi-layer thin-films, Optics Communications, 2009; 282:1 35–140. DOI: 10.1016/j.optcom.2008. 09.043

[21] Nesterov M. L, Kats A. V, Turitsyn S. K. Extremely short-length surface plasmon resonance devices, Optics Express, 2008; 16: 20227–20240. DOI: 10.1364/OE.16.020227

[22] Kekatpure R. D, Hryciw A. C, Barnard E. S, Brongersma M. L. Solving dielectric and plasmonic waveguide dispersion relations on a pocket calculator, Optics Express, 2009; 17: 24112–24129.DOI:10.1364/OE.17.024112

[23] Haeseler F. V, Peitgen H. O. Newton's method and complex dynamical systems, Acta Applicandae Mathematica, 1988; 13:3–58. DOI: 10.10 07/BF00047501

[24] Garrido Alzar C. L, Martinez M. A. G, Nussenzveig P. Classical analog of electromagnetically induced transparency, American Journal of Physics; 2001; 70:37–41. DOI: 10.1119/ 1.1412644

[25] Novotny L. Strong coupling, energy splitting, and level crossings: A classical perspective, American Journal of Physics, 2010; 78:1199–1202. DOI: 10.1119/ 1.3471177

[26] Tellez-Limon R, Fevrier M, Apuzzo A, Salas-Montiel R, Blaize S. Modal analysis of LSP propagation in an integrated chain of gold nanowires. In: Proceedings SPIE Nanophotonic Materials X, 13 September 2013; San Diego. California: Proc. SPIE; 2013. p. 88070J

[27] Tellez-Limon R, Fevrier M, Apuzzo A, Salas-Montiel R, Blaize S. Theoretical analysis of Bloch mode propagation in an integrated chain of gold nanowires, Photonics Research, 2014; 2:24–30. DOI: 10.1364/PRJ.2.000024

[28] Tellez-Limon R, Février M, Apuzzo A, Salas-Montiel R, Blaize S. Journal of the Optical Society of America B, 2017; 34: 2147–2154. DOI: 10.1364/JOSAB.34. 002147

[29] Hocker G. B, Burns W. K. Mode dispersion in diffused channel waveguides by the effective index method, Applied Optics, 1977; 16:113–118. DOI: 10.1364/AO.16.000113

[30] Payne F. P. A new theory of rectangular optical waveguides, Optical and Quantum Electronics, 1982; 14:525–537. DOI: 10.1007/BF00610308

[31] Patchett S, Khorasaninejad M, Nixon O, Saini S. S. Effective index approximation for ordered silicon nanowire arrays, Journal of the Optical Society of America B, 2013; 30:306–313. DOI: 10.1364/JOSAB.30.000306

[32] Szafranek D, Leviatan Y. A Source-Model Technique for analysis of wave guiding along chains of metallic

nanowires in layered media, Optics Express, 2011; 19:25397–25411. DOI: 10.1364/OE.19.025397

[33] Moharam M. G, Gaylord, T. K. Rigorous coupled-wave analysis of planar-grating diffraction, Journal of the Optical Society of America, 1981; 71: 811–818. DOI: 10.1364/JOSA.71.000811

[34] Moharam M. G, Grann E. B, Pommet D. A, Gaylord T. K. Formulation for stable and efficient implementation of the rigorous coupled-wave analysis of binary gratings, Journal of the Optical Society of America A, 1995; 12: 1068–1076. DOI: 10.1364/JOSAA.12.001068

[35] Liu H, Lalanne P. Comprehensive microscopic model of the extraordinary optical transmission, Journal of the Optical Society of America A, 2010; 27: 2542–2550. DOI: 10.1364/JOSAA.27.002542

[36] Noponen E, Turunen J. Eigenmode method for electromagnetic synthesis of diffractive elements with three-dimensional profiles, Journal of the Optical Society of America A, 1994; 11:2494–2502. DOI:10.1364/JOSAA.11.002494

[37] Li L. New formulation of the Fourier modal method for crossed surface-relief gratings, Journal of the Optical Society of America A, 1997; 14: 2758–2767. DOI: 10.1364/JOSAA.14.002758

[38] Kim H, Park J, Lee B. Fourier modal method and its applications in computational nanophotonics. 1st ed. Boca Raton: CRC Press; 2012. 326 p. DOI: 10.1201/b11710

[39] Bykov D. A, Bezus E. A, Doskolovich L. L. Use of aperiodic Fourier modal method for calculating complex-frequency eigenmodes of long-period photonic crystal slabs, Optics Express, 2017; 25: 27298–27309. DOI: 10.1364/OE.25.027298

12

Solar Thermal Conversion of Plasmonic Nanofluids: Fundamentals and Applications

Meijie Chen, Xingyu Chen and Dongling Wu

Abstract

Plasmonic nanofluids show great interests for light-matter applications due to the tunable optical properties. By tuning the nanoparticle (NP) parameters (material, shape, and size) or base fluid, plasmonic nanofluids can either absorb or transmit the specific solar spectrum and thus making nanofluids ideal candidates for various solar applications, such as: full spectrum absorption in direct solar absorption collectors, selective absorption or transmittance in solar photovoltaic/thermal (PV/T) systems, and local heating in the solar evaporation or nanobubble generation. In this chapter, we first summarized the preparation methods of plasmonic nanofluids, including the NP preparation based on the top-down and bottom-up, and the nanofluid preparation based on one-step and two-step. And then solar absorption performance of plasmonic nanofluids based on the theoretical and experimental design were discussed to broaden the absorption spectrum of plasmonic nanofluids. At last, solar thermal applications and challenges, including the applications of direct solar absorption collectors, solar PT/V systems, solar distillation, were introduced to promote the development of plasmon nanofluids.

Keywords: solar thermal, plasmonic, nanofluid, absorption, nanoparticle

1. Introduction

Nowadays, with the development of society and the improvement of people's living standards, environment issues, such as: greenhouse effect, acid rain, and haze, has become more serious. Developing green energy technology has attracted more researchers' attention, especially for solar energy, which is universal, harmless, huge, and sustainable. Solar utilizations can be divided into two main categories: solar-electric and solar-thermal. And both of them are needed to enhance the solar absorption performance of working media at its first step of solar conversion applications.

Solar thermal conversion is one of the most simple and direct ways of solar utilizations by heating the working mediums directly for follow-up usages, which can be widely used in the solar thermal collectors [1], solar distillation [2, 3] and so on [4]. Therefore, it's critical to improve the solar absorption performance of working mediums for the solar thermal conversion applications. For example, based on the surface absorber, various nanostructure coatings (e.g., grating, porous structure, and so on [5, 6]) were designed to achieve the selective absorption ability, which serves as solar selective absorbers by heating the surface and transferring the heat to the working fluid for the follow-up applications. The heat loss from the absorbed surface due to the high temperature and heat transferred resistance between the absorbed surface and working should be considered during the design processes, which also limits its large-scale practical application at the high or middle temperature solar thermal conversion applications.

Instead of absorbing solar energy by a surface, work fluid can be used to absorb solar energy, which serves as both the solar absorber and heat transfer medium and can avoid the local high temperature area and reduce the heat transfer resistance. However, the common working fluids such as: water, oil, and alcohol usually have the limited solar absorption ability [7]. It was found that adding nanoparticles (NPs) to these working fluid (i.e., nanofluid) can greatly improve the solar collector efficiency [8, 9]. Nanofluid is a suspension of NPs (1–100 nm) in a conventional base fluid, which was first used by Choi in 1995 [10]. Nanofluids show unique characteristics in many aspects, including the heat transfer [11, 12] and the solar absorption ability due to the interaction between the light and NPs at nanoscale [9, 13]. For example, carbon nanotube, graphite and the other black carbon NPs were added into the base fluid to achieve the great solar absorption performance [14].

Plasmonic nanofluids show great interests to improve the absorption ability by dispersing plasmonic NPs in the base fluid stability. Due to the surface plasmon resonance (SPR) around the NP surface [15], the incident electric coupled with the free electron oscillation around the NP surface at the resonance frequency can strongly enhance the absorption performance of NPs [16] in **Figure 1**. The optical absorption performance of nanofluids can be enhanced by tuning the NP shape, size, or base fluid. Using plasmonic nanofluids as the absorber and heat transfer medium in the solar thermal applications shows great potential due to the excellent optical and thermal characteristics. To choose a proper nanofluids for specific solar thermal applications (such as: solar collectors, solar PV/T systems), many researchers investigated the optical and thermal properties of various nanofluids. For example, for the direct absorption solar collectors (DASCs), nanofluids as the absorber need to absorb the solar radiation in the full solar spectrum (0.3–2.5 um). While the nanofluid only serves as a beam splitter (i.e., selective absorber) in solar PV/T systems, which absorbs the useless spectrum for the PV cell and avoids heating the PV cells to improve the overall PV/T efficiency [4]. Hence, the optical absorption performance of plasmonic nanofluids should be considered in different solar thermal applications.

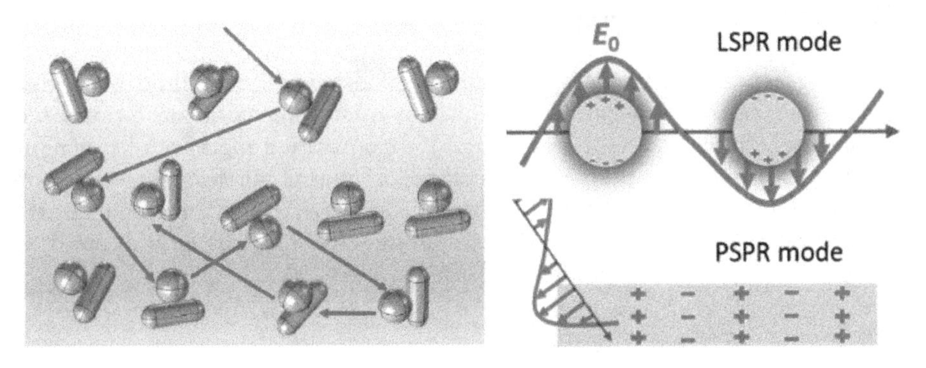

Figure 1.
Light propagation in the nanofluid [17] and the surface plasmon resonance (SPR) around the NP surface, dividing into localized and propagating surface plasmon resonance (LSPR and PSPR).

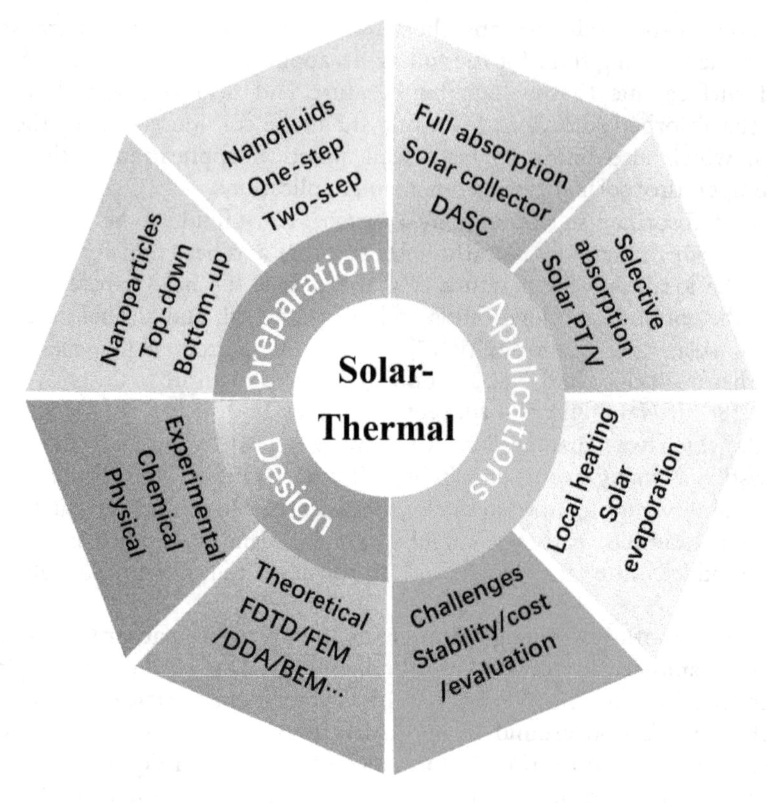

Figure 2.
The main parts of this chapter for solar thermal conversion of plasmonic nanofluids.

In this chapter, we focus on the solar thermal conversion of plasmonic nanofluids in **Figure 2**, which consists of the following three parts: 1) plasmonic nanofluid preparation including NPs and nanofluids; 2) solar absorption of plasmonic nanofluids based on the theoretical and experimental design; 3) solar thermal applications and challenges, including direct solar absorption collectors, solar PT/V systems, solar evaporation, other applications and challenges. To increase the understanding of previous studies, related analyses and calculation techniques are illustrated. This chapter is expected to provide researchers with deep insight into the solar thermal conversion of plasmonic nanofluids and facilitate future studies in this field.

2. Plasmonic nanofluid preparation

As discussed above, the NP parameters and dispersed base fluid have great effect on the optical absorption and solar thermal conversion performance of plasmonic nanofluids. We will first discuss the preparation methods of plasmonic nanofluids, and then the preparation methods of plasmonic NPs (**Figure 3**) are summarized due to the great interaction of NP parameters with the light. In this section, some common methods to prepare plasmonic NPs or nanofluids are listed and their advantages or limitation would also be discussed.

2.1 Plasmonic nanofluid preparation

The preparation method of nanofluids can be classified into two main categories in **Figure 3a**: one-step method and two-step method [18].

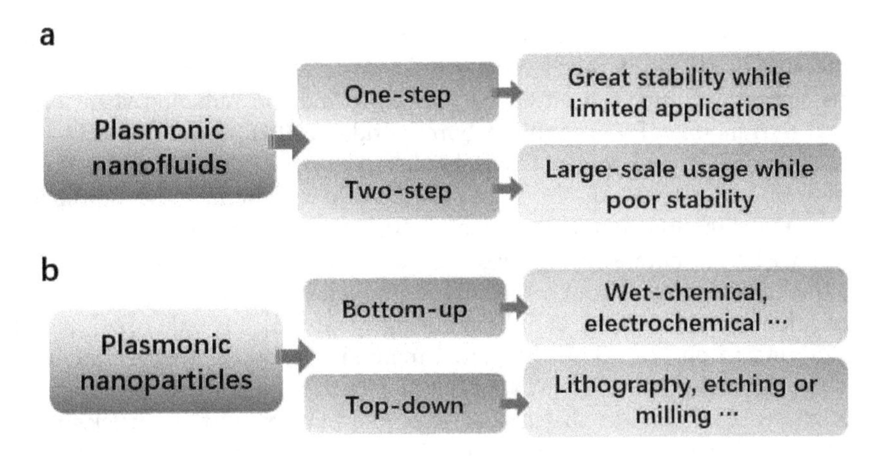

Figure 3.
Preparation methods of (a) plasmonic nanofluids and (b) nanoparticles.

The one-step method is to disperse NPs in the NP synthesis process by using the physical method or wet-chemical method. The prepared nanofluid is relatively stable by avoiding the NP separation process (e.g., centrifuge or drying) and redisperse process (e.g., stirring or ultrasonic oscillation). Hence, the one-step method can reduce NP agglomeration or sedimentation in the nanofluids. For example, an ultrasound-assisted one-step method was used to prepare spherical and plate-shaped Au NPs with the NP size of $10 \sim 300$ nm [19]. But the dispersed base fluid has limitations, which usually consists of the residual chemical reagent in the NP synthesize process and the nanofluid could not be applied in a large-scale range.

The two-step method separates the NP synthesis process from the dispersion process, which is widely used in the large-scale applications. In this method, various NPs, such as: nanospheres, nanorods, nanotubes, and so on, are in the state of dry powders and then these NPs can be dispersed into different base fluids by stirring or ultrasonic oscillation for different applications. Due to the redispersion process of NPs, the stability is worse than that of the one-step method, but the two-step method has a simple preparation process, a high NP controllability, and wide application ranges. It can be seen from the two-step method that the NP parameters, such as the size, morphology, and dielectric environment of the base fluid determine their unique SPRs. Therefore, for solar thermal conversion applications of plasmonic nanofluids, preparing plasmonic NPs with the controllable morphol-ogy and size is the prerequisite and basis for their applications. And the fabrication of these plasmonic NPs will be discussed in the next section.

2.2 Plasmonic nanoparticle fabrication

The fraction processes of plasmonic NPs can be divided into two main categories in **Figure 3b**: the top-down based on the lithography, etching or milling, and the bottom-up, including the seed-mediated growth, chemical reduction, electrochem-ical method, and so on. Given that the NP shape can significantly affect the way it interacts with light and its SPRs, researchers have made great efforts to develop preparation methods for plasmonic NPs with the reproducible control of the size and shape. Nowadays, it's possible to fabricate high-quality plasmonic NPs (e.g., Au or Ag) with the target SPR wavelengths or near-field enhancement by enabling a systematic study of SPR dependencies on the size, shape, and structure of plasmonic NPs.

The wet-chemical method is one of the most common bottom-up methods to tune the size or shape of metal NPs. For Au NPs, chloroauric acid ($HAuCl_4$)in aqueous solution is usually used as the precursor and the reducing agent can be polyvinyl pyrrolidone (PVP), Sodium borohydride ($NaBH_4$), ascorbic acid (AA), sodium citrate (SC), and so on. The reduction reaction is also determined by the temperature. And various Au NP shapes can be achieved by introducing the additive, e.g., etyltrimethylammonium bromide (CTAB) for nanorods [20], sodium citrate or trisodium citrate for spheres [21], silver iron (Ag^+) for thorns [22]. Au NPs with the size ranged from 5 nm to 150 nm can be obtained by changing the concentration of the precursor and the reducing agent [23], PH value [24], temperature [25] and so on [26]. Seed-medicated method is another effective method to control the size and shape of Au NPs, which divides the reaction into nucleation and growth stages separately, so that the size and morphology of the particles can be controlled in a larger range [27]. By controlling the growth rate of different crystal planes, the final synthesized NPs can deviate from the initial seed crystal structure by the seed growth method. For example, cube Au NPs were prepared by the amount of ascorbic acid (AA) in the growth solution and the size can be controlled by the amount of seed solution or the growth number [28, 29]. Ag NPs also can be prepared by the similar methods using $AgNO_3$ or other silver salts. To scalable and green prepare metal NPs, a rotating electrodeposition and separation (REDS) technique developed, which entails electrochemically depositing NPs onto a continuously rotating metal foil and subsequently harvesting them through mechanical delamination. A wide array of elemental nanoparticles (e.g., Ag, Au, Ni, Cu), alloys nanoparticles (e.g., FeCoNi and FeCoNiW), and metal oxide nanomaterials (e.g., CO_3O_4) were synthesized by REDS [30].

Besides the bottom-up method, top-down method is another to prepare NPs, which mainly including some physical method, such as: electron beam lithography (EBL), milling, annealing, laser-melting, and so on. The fabrication process of EBL is similar to classical photolithography, an electron beam is used to mark the pattern in the resist instead of light. [31]. Despite the low fabrication throughput and the fact that very small structures may be at the physical limit in terms of electronic function, the technique can find applications in preparation of reproducible large-scale arrays of plasmonic NP with arbitrary two-dimensional shapes. Among the disadvantages are the technological requirements such as high vacuum and a scanning electron microscope system, much longer time to write a pattern than photolithography. Using conventional lithography techniques many shapes of nanoparticles on surfaces can be achieved. However, large scale fabrication using reproducible patterns, inverse replication or transfer of NPs between substrates, and three-dimensional nanostructures including deep etching has been increasingly demanded in plasmonic nanostructures [32, 33].

After obtaining different plasmonic NPs, it should be dispersed into the working fluid stability to form various nanofluids. The fabrication of NPs based on the top-down method is expensive and consumed on the materials, which could not meet the scale requirement although the high-quality NPs can be achieved. Wet-chemical method can be an efficient method to achieve plasmonic NPs in the solar thermal conversion applications.

2.3 Nanofluid stability

Nanofluid is defined as dispersing NPs stability into the base fluid, which is not simply mixing solid NP phase and liquid phase. It's a complex colloid by dispersing specific functional NPs in the base fluid (e.g., water, oil and so on). The main challenge for nanofluid applications is how to produce well-dispersion nanofluids.

Owing to the interaction among different NPs at the nanoscale and gravity at Earth, NPs are usually agglomerated due to Van der Waals force and then trend to be sediment at the bottom [34]. As a result, the agglomeration and sedimentation of NPs in the base fluid would affect the optical absorption and heat transfer performance, weakening the system efficiency. In addition, recent studies showed that the agglomeration or sedimentation can be worse under harsh operating conditions, such as: high temperature and pressure [35, 36]. Many methods were used to evaluate the stability of nanofluids, the simplest and direct method is the sedimentation method [18]. Interface electromotive force analysis is another common method to observe the stability of nanofluids, but this method is limited by the viscosity and concentration of the fluid [37]. Wang et al. [38] used an ultraviolet–visible spectrophotometer to study the stability of nanofluids. The NP concentration can be obtained by measuring the change in the light absorption rate of the system with the sedimentation time because the NP concentration is a linear relationship with the absorbance of nanofluid at the low concentration.

The stability of plasmonic nanofluids is also one of the major issues limiting the applications of nanofluids. Many researchers have made much efforts to improve the stability of the plasmonic nanofluids from the aspect of long-time and high-temperature dispersion [39]. For example, Au@SiO_2 and Ag@SiO_2 core-shell NPs were synthesized using a low-temperature two-step solution process. Results showed that the synthesized metal@SiO_2 nanofluids exhibited excellent dispersion stability of 93.7% for Au@SiO_2 and 100% for Ag@SiO_2 in 6 months without using any surfac-tants, and they also showed a good thermal stability after thermal exposure at 150° C for an hour [40]. An ultrastable nanofluids with the broadband photothermal absorp-tion was achieved using citrate and polyethylene glycol-coated Au NPs, circumventing the need for free surfactants. Electrostatic stabilization provided superior colloidal stability and more consistent optical properties; chemical and colloidal stability was verified for 16 months, the longest demonstration of stable nanofluids under ambient storage in the solar literature [41]. Besides the base fluid water used above, the base fluid oil was also studied to improve the stability. A facile and effective strategy, including controlled high-temperature synthesis of nanoparticles, surface modification of particles, and post-modification particle size partition, was designed to prepare stably dispersed silicone-oil-based nanofluids that enable high-temperature operation [42]. A low cost, and scalable method was reported to synthesize solar selective nanofluids from 'used engine oil' with the excellent long-term stability and photo-thermal conversion efficiency. Results showed that their stability and functional char-acteristics can retain even after extended periods (72hours) of high temperature(300°C) heating, ultra violet light exposure and thermal cyclic loading [43].

3. Solar absorption of plasmonic nanofluids

The excellent optical absorption performance of plasmonic NPs make it to be a great candidate in the solar thermal conversion applications, which is critical for the solar thermal conversion applications. And the optical properties of nanofluids can be controlled by the NP size, shape, concentration and base fluid. In this section, we will discuss the optical properties of NPs or nanofluids from two aspects: theoretical design and experimental design.

3.1 Theoretical design

To achieve the optical properties of nanofluids, including transmittance, reflectance, and absorptance, the optical performance of single NP is usually determined firstly. And the dielectric function of materials are required for the optical simulation, which is taken from the experimental data of bulk materials (e.g., Johnson and Christy [44],) or a model approximating experimental results (e.g., Drude method). The Drude model is the simplest of all, but disregards radiation damping. Even today, mainly because of the simplicity, the Drude model is still used to describe the dielectric functions in many calculations. In some problems, the classi-cal models of dielectric functions are unsatisfactory but, at the same time, full quantum theories involve a very complex treatment including non-local effects[45], polarizabilities including non-linear terms [46], electron densities calculation using mean-field theories [47] and temperature dependent effects [48]. The need for quantum treatment of the optical properties of small particles has been evidenced in recent experimental studies [49]. In large particles the resonances are influenced by retardation effects and are strongly dependent on the size of particles, but the dielectric function can be assumed as that of bulk. Based on the dielectric function, Au, Ag and Al are the three most used materials in plasmonics. Their SPR wavelengths are at visible or UV spectral bands and, therefore, of great potential in solar thermal applications.

Mie theory: Mie theory is a simple and theoretical method to calculate the optical properties of sphere NPs in a homogenous medium, which uses a series of coefficients a_n and b_n for the scattered fields and c_n and d_n for the internal fields to determine the scattering fields. The scattering and extinction cross sections can be calculated as: [50].

$$C_{scat} = \frac{2\pi}{k^2}\sum_{n=1}^{\infty}(2n+1)\left(|a_n|^2 + |b_n|^2\right) \quad (1)$$

$$C_{ext} = \frac{2\pi}{k^2}\sum_{n=1}^{\infty}(2n+1)\,\mathrm{Re}\,(a_n + b_n) \quad (2)$$

The absorption cross section can be obtained as: $C_{abs} = C_{ext} - C_{scat}$. Despite to the less computation load, it is possible to obtain cross-sections for many wavelengths in a few seconds, using a common PC. However, a large number of terms is required for accurate cross-section calculations of spheres with very large size parameter [51]. The Mie theory has been extended to permit calculations for ellipsoidal shape, multilayer or several spheres [52].

DDA: To calculate the light scattering of an arbitrary shape NP, discrete dipole approximation (DDA) was first presented by Purcell and Pennypacker [53] by using a grid of dipoles. To occupy by the scattering target, DDA method discretizes the volume by an array of N dipoles using Clausius–Mossotti polarizability α_j for each dipole, which interacts with the incident field and the neighbors. The polarization of dipole j located at r_j can be determined by $P_j = \alpha_j E_j$, and the field can be calculated as:

$$E_j = E_j^{inc} - \sum_{k \neq j} A_{jk} P_k \quad (3)$$

where $E_j^{inc} = E_0 e^{ikr - i\omega t}$. A_{jk} is the matric of dipole interaction and retardation effect.

To achieve accurate and reproduce the calculation results, two validity conditions should be verified in DDA: (a) the dipole lattice spacing d should be small enough, i.e., $|m|kd \leq 1$, where m is the complex refractive index of the scattering target. (b) d must be small enough to refabricate accurately the NP shape. For small plasmonic NPs, or small inter-particle separations, d must be smaller than 1 nm.

BEM: Boundary element method (BEM) is another method to calculate the optical properties of plasmonic nanostructures, which was introduced by García de Abajo and Howie [54] using the following equations:

$$\phi(r) = \phi_j^{ext}(r) + \int_{S_j} G_j(|r-s|)\sigma_j(s)ds \quad (4)$$

$$A(r) = A_j^{ext}(r) + \int_{S_j} G_j(|r-s|)h_j(s)ds \quad (5)$$

where $\phi(r)$ is the electric potential, $A(r)$ is the vector potential, $\sigma_j(r')$ is the surface charge density and $h_j(r')$ is the surface current density and G_j is the Green's function of Helmholtz equation inside each homogeneous medium of dielectric function. S_j is the boundary of the medium j. s is the point of the boundary between medias. r is the point inside the medium. $\phi^{ext}(r)$ and $A^{ext}(r)$ are the potentials at the interface caused by external sources and the full space is filled by a homogeneous medium j. Therefore, a much smaller number of elements is required to evaluate the fields than volume integral based methods, but involving a complex parameteriza-tion of the boundary elements.

FDTD: Finite-difference time-domain (FDTD) is one of the most popular optical calculation methods in plasmonic nanostructures, which is first developed by Yee in 1966 [55]. The basis of the model is from Maxwell equations in electrodynamics. The second-order precision central difference is used to approximate the discretization of the differential form of Maxwell equations, thereby a set of time-domain propulsion formulas can be used to deal with electromagnetic wave propagation problems. Since the FDTD method directly discretizes the time-domain wave equation, it will not limit its application range due to mathematical models, and can effectively simulate various complex structures.

The popularity of this method has strongly increased in the last two decades, mainly due to the simplicity of implementation, support of arbitrary NP shape, allowing to investigate linear and non-linear properties of NPs, using Maxwell's equations directly without approximations. There are, however, some undesired effects, like the staircase of fields in non-rectangular boundaries, mainly in code implementations without adaptive meshing. To avoid this, very fine discretization or sub-pixel smoothing of the dielectric function must be applied [56]. The dielectric function of the materials requires analytical expressions (e.g., Drude–Lorentz) [57].

FEM: Finite element method (FEM) was developed to solve differential equations of boundary-value problems [58]. Physical problems described by differential equations over a domain, like for example the Helmholtz equation in real three-dimensional space. Hence, electromagnetic (EM) field propagation around the sin-gle NP can be described by the Helmholtz Equation [59]:

$$\nabla \times \left(\mu_r^{-1} \nabla \times \boldsymbol{E}\right) - k_0^2 \varepsilon_r \boldsymbol{E} = 0 \quad (6)$$

where $\boldsymbol{E}$ is the electric field of the medium, j is the current density, k_0 is the wavenumber, ε_r is the dielectric function, which is calculated as $\varepsilon_r = (n - ik)^2$, n and k are the complex refractive indices. Within each element $\boldsymbol{E}$ is approximated using a basis function expansion $\boldsymbol{E} = \sum_{j=1}^{n} \boldsymbol{N}_j \xi_j$, where the sum is over n interpolation point. $\boldsymbol{N}_j$ is chosen basis function and ξ_j is the unknown coefficient. A solution can be obtained by using the variational principle to determine ξ_j. To obtain a meaningful solution, $\boldsymbol{N}_j$ is required to satisfy Gauss's law and appropriate boundary conditions on the surface of all elements. During the last decade, an increasing number of publications on plasmonic nanostructures done with COMSOL Multiphysics has appeared in the literature.

A short comparison of these above calculation methods for NPs are listed in **Table 1**. The choice of the calculation method depends on many factors, such as: NP size, shape and dielectric environment. But the general method, such as: FEM, and FDTD, can be used in most situations by applying the periodic boundary conditions at the sides of the unit cell.

The optical properties of nanofluids also can be calculated based on the above method, such as: FDTD or FEM, which are the direct way to achieve the absorption performance of nanofluids without the strict assumptions. But the computation load is large for nanofluids since the geometry size (~ mm) is much larger than NP size or mesh size (~ nm). Therefore, the optical properties of nanofluids can be obtained from the optical properties of single NP due to the low NP concentration of plasmonic nanofluids in the solar thermal conversion applications and the independent scattering can be applied in the calculation of nanofluids.

One method is to avoid the scattering effect of NPs due to the small size, resulting in the negligible scattering effect in the nanofluids. Therefore, the absorp-tion efficiency of the nanofluids can be obtained by the independent scattering approximation, which can be described as [9]:

$$k_{a\lambda,nf} = k_{a\lambda,bf} + k_{a\lambda,np} = \frac{4\pi\kappa}{\lambda} + \frac{f_v C_{abs}}{V_{np}} \tag{7}$$

where $k_{a\lambda,nf}$, $k_{a\lambda,bf}$, and $k_{a\lambda,np}$ are the absorption coefficients of the nanofluid, the base fluid water and the NPs respectively. κ is the absorption index of water. f_v is the NP volume fraction. V_{np} is the single NP volume. C_{abs} is the absorption cross section of the NP. Based on the Beer–Lambert law [61], the radiation intensity decays exponentially along the transmission direction. Therefore, the solar absorption efficiency η_{abs} can be calculated as:

$$\eta_{abs} = \frac{\int_{0.3\mu m}^{2.5\mu m} I_{abs}(\lambda)d\lambda}{\int_{0.3\mu m}^{2.5\mu m} I_s(\lambda)d\lambda} = \frac{\int_{0.3\mu m}^{2.5\mu m} I_s(\lambda)\left(1 - e^{-Hk_{a\lambda,nf}}\right)d\lambda}{\int_{0.3\mu m}^{2.5\mu m} I_s(\lambda)d\lambda} \tag{8}$$

where $I_s(\lambda)$ is the solar spectra at AM 1.5. $I_{abs}(\lambda)$ is the absorbed spectra. H is the depth.

Method	Description	Advantages	Limitations
Mie theory	Theoretical result	Accurate solution	Valid for simple shapes (such as: sphere)
DDA	Approximate result based on discrete dipoles	Arbitrary shape	Time-consuming; large memory space for large NPs
BEM	Numeric solution with surface discretization	Fast calculation than volume methods	Complex parameterization of boundary elements
FEM	Differential equations solved over a domain	General application	Time-consuming; large memory space for complex structures
FDTD	Discretization of Maxwell equation with Yee cell	General application	Computational stability depends on dielectric functions; Time-consuming for spectral calculations

Table 1.
Comparison of calculation methods for plasmonic NPs [60].

The optical properties of nanofluids also can be solved by the Monte Carlo (MC) method to obtain the solar absorption performance of nanofluids. MC technique is a flexible method for simulating light propagation in the medium. The simulation is based on the random walks that photons make as they travel, which are chosen by statistically sampling the probability distributions for step size and angular deflec-tion per scattering event. After propagating many photons, the net distribution of all the photon paths yields an accurate approximation to reality. In this method,the scattering effect is considered by the scattering efficiency and scattering phase function. The absorptance, transmittance, and reflectance of nanofluids can be calculated by counting the fate of photons.

For the plasmonic nanofluids applied in the solar thermal applications, the absorption spectral distribution is one of the most important parameters, which is proportional to the NP parameters (concentration, shape and size), Qin et al. theoretically optimized the spectral absorption coefficient of an ideal plasmonic nanofluid for a DASC to maximize the thermal efficiency while maintaining the magnitude of the average absorption coefficient at a certain value [62]. However, considering that the SPR frequency of metallic NPs, such as Au, Ag, and Al, is usually located in the ultraviolet to visible range. The actual plasmonic nanofluids usually have the narrow absorption band due to the SPRs. Two strategies can be adopted to overcome this shortage in the NP theoretical design process.

One is to blend NPs with different absorption peaks to form hybrid plasmonic nanofluids for full utilization of solar energy in a broad spectrum. For example, an ideal distribution of spherical metal NPs, including nanospheres and nanoshells, were designed to match the AM 1.5 solar spectrum with an determination of absorbing and scattering distributions [63]. Based on MC method and FEM, four type of Au nanoshells were blended in the base fluid to enhance the solar absorption performance of plasmonic nanofluids with an extremely low particle concentration (e.g., approximately 70% for a 0.05% particle volume fraction) [64]. By applying the customized genetic algorithm, an optimal combination for a blended nanofluid (metal nanosphere, metal@SiO_2 core-shell, and metal nanorod) was designed with the desired spectral distribution of the absorption coefficient [65]. Besides the core-shell NPs, other NP shapes were also designed to expand the absorbance over the entire solar spectrum [66, 67]. Although different blended NPs were designed to broad the absorption spectrum, the comparison or enhancement is usually done based on the single-element nanofluid, which is not enough to compare with the other blend styles.

The other is to design complex NP structures with multiple absorption peaks at different wavelengths or coupled with the great intrinsic absorption materials. For example, core-shell NPs (Al@CdS [68], Ag@SiO_2@CdS [69], Au@C [70], Ag@TiO_2 [71], and gallium-doped zinc oxide@Cu [72, 73]) were the direct way to enhance the solar absorption performance due to the enhancement and tunable SPRs of shell and intrinsic absorption of core by optical simulations, thus broadening the absorption spectrum and improving the solar absorption performance of plasmonic nanofluids. Results also found that Ag NPs with sharp edges can induce multiple absorption peaks due to both LSPR and lightning rod effect to broad the absorption spectrum [74]. In addition, a plasmonic dimer nanofluid, consisting of the rod and sphere, was proposed to enhance the solar absorption performance by LSPR, PSPR, and gap resonance between the rod and sphere at different wavelengths [17], which was also similar as the thorny NPs [75].

3.2 Experimental design

Although various NP structures were designed to enhance the solar absorption performance of plasmonic nanofluids theoretically, the synthesizes of these complex structures are still difficult and more efforts still are needed to precious control the NP size parameters (e.g., size or shape) experimentally.

Compared with the other common nanofluids (e.g., SiO_2, TiO_2, Al), plasmonic Au nanofluids with the small NP size were prepared experimentally to obtain the great solar thermal conversion efficiency [21, 76]. However, the absorption peaks of these common metals usually locate in the visible part especially for the metal sphere. Multi-element NPs (such as: alloy NP [77]) can further enhance solar absorption ability compared with the single-element NPs by tuning the LSPR peak. Various core-shell NPs were also prepared to enhance the solar absorption perfor-mance of plasmonic nanofluids [73, 78]. For example: Ag@CdS core-shell NPs were synthesized by a facile method and the optical absorption performance of Ag@CdS nanofluids was enhanced in a wide range of visible light compared with bare Ag and CdS NPs [79]. Sn@SiO_2@Ag core-shell NPs were prepared with good abilities of both optical absorption and thermal energy storage [80]. Ag shell can improve light absorption due to LSPR effect, which was also can be found experimentally for CuO@Ag [81] and TiO_2@Ag plasmonic nanofluids compared with CuO, TiO_2, and Ag nanofluids [82].

Another simple way is to blend sphere NPs with different materials [83–85]. Various NPs have been blended experimentally to enhance the solar absorption performance of plasmonic nanofluids. For example, hybrid nanofluids containing reduced graphene oxides decorated with Ag NPs [86], multi-wall carbon nanotubes and SiO_2@Ag NPs [87], Fe_3O_4, Cu and Au NPs [88], and Au and TiN NPs [89] showed great solar absorption performance by tuning the ratios of different com-ponents to broaden the absorption spectrum. LSPR effect around plasmonic NPs and intrinsic absorption of semiconductor NPs make the hybrid nanofluids possess superior optical absorption to bare NPs at the same concentration. Besides the blended nanofluids with different NP materials discussed above, the other route is to blend the NPs with different shapes. For example, by mixing Au NPs(such as: nanorods [90]) with different shapes in water, a blended plasmonic nanofluid was prepared and absorption spectrum can be broadened due to the various LSPR peaks of different NP shapes [84]. The blended nanofluids based on Ag triangular nanosheets and Au nanorods, were proposed and a high efficiency of 76.9% is achieved experimentally with a very low volume concentration(0.0001%) [91].

As discussed above, blending different NPs is a simple way to achieve multi absorption peaks. However, compared with the single component NPs, the interac-tion between the different NPs in the blended nanofluids is limited due to indepen-dent scattering at the low NP fraction, leading that the solar thermal conversion efficiency of blended NPs was almost equal to the arithmetic sum of the efficiency of each component NPs without enhanced coupled effect between different NPs with the incident light [83]. Designing complex structures with multi-resonance peaks experimentally can be an efficient way to enhance the solar absorption per-formance of plasmonic nanofluids. For example, Au thorn [92] and Au dimer [93] were designed experimentally enhance the light absorption performance of plasmonic nanofluids. In addition, Janus NPs also showed great optical absorption performance due to the complex structure experimentally [94].

4. Applications

Nanofluids can either absorb or transmit specific solar spectrum and thus mak-ing assorted nanofluids ideal candidates for various solar applications [95]. Based on the tunable optical absorption performance of plasmonic nanofluids, several applications, including full spectrum absorption in direct solar absorption collector, selective absorption in solar PT/V systems, and local heating in solar evaporation or steam generation, are discussed below in **Figure 4**.

Some efforts have been made to investigate the solar thermal conversion performance of stationary plasmonic nanofluids based on the direct solar absorption collectors (DASCs). A one-dimensional transient heat transfer analysis was carried out to analyze the effects of NP volume fraction, collector height, irradiation time, solar flux, and NP material on the collector efficiency. Results showed that the plasmonic nanofluids (e.g., Au and Ag) achieved the better collector efficiency in the stationary state [98]. Solar thermal conversion performance of Au nanofluids in a cylindrical tube under natural solar irradiation conditions was studied and a efficiency of 76.0% at a concentration of 5.8 ppm can be achieved [99]. Although Au nanofluids have high solar absorption performance, their expensive cost limits their practical use [100]. The solar thermal conversion performance of six (Ag, Cu Zn Fe, Si and Al_2O_3) common NPs in direct absorption solar collectors (DASC) was investigated under a focused simulated solar flux. Ag nanofluid turned out to be the best among all due its strong plasmonic resonance nature [101]. Stable silver nanofluids were prepared through a high-pressure homogenizer and the outdoor experiments were conducted under sunlight on a rooftop continuously for $\sim$10 h and the excellent photothermal conversion capability even under very low concen-trations can be achieved [102].

Recently, the direct-absorption parabolic-trough solar collector (DAPTSC) using the flow nanofluids has been proposed, and its thermal efficiency has been reported to be 5–10% higher than the conventional surface-based parabolic-trough solar collector. In order to reduce the cost of a collector and avoid NP agglomeration when using plasmonic nanofluids, the configuration with the lowest possible absorption coefficient but with the reasonably high temperature gain as well as efficiency was explored [103]. For the collector design, an extra glass tube inside was inserted so the nanofluid was separated into two concentric segmentations (i.e., an inner section and an outer section), and a nanofluid of lower concentration was applied in the outer section while a nanofluid of a higher concentration in the inner section. Results showed that at the same NP concentration parameter, the DAPTSCs with two concentric segmentations of nanofluids outperform those with one uniform nanofluid for all considered configurations [104]. Furthermore, the transparent DAPTSC was improved by applying a reflective coating on the upper half of the inner glass tube outer surface such that the optical path length was doubled compared to that of the transparent DAPTSC, allowing a reduction in the absorption coefficient of the nanofluid [105]. In addition, by replacing the semi-cylindrical reflective coating with a semi-cylindrical absorbing coating for exploiting both volumetric and surface absorption of the solar radiation. The DAPTSC with a

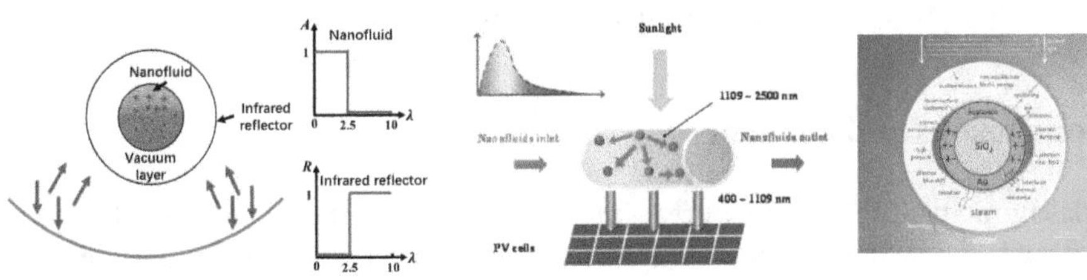

Figure 4.
Solar thermal applications of plasmonic nanofluids, including nanofluid based direct solar absorption collector, solar spectral beam splitter in solar PV/T systems [96], solar steam or nanobubble generation in solar evaporation [97].

hybrid of volumetric and surface absorption can achieve a significantly higher thermal efficiency than the previous design of a DAPTSC with a reflective coating [8]. An innovative nanofluid enabled pump-free DASC concept was presented by combining the advantages of volumetric solar harvesting and oscillating heat pipes to enhance the solar harvesting and spontaneously transfer the heat into targeted areas, providing a novel approach for efficient solar energy utilization [106].

Nanofluid-based spectral beam splitters have become dramatically popular for PV/T applications due to it can achieve tunable optical properties inexpensively [107]. For example, $CoSO_4$-based Ag nanofluid was developed to be utilized as fluid optical filter for hybrid PV/T system with silicon concentrator solar cell [108]. Furthermore, Ag NPs suspended in hybrid $CoSO_4$ and propylene glycol base fluids were prepared for both silicon and GaAs cells. Ag/$CoSO_4$-PG nanofluid filters exhibited broad absorption outside solar wavelengths and showed high transmittance in wavelength range used by the two types of cells efficiently [109]. More review about the application of nanofluids in solar PT/V systems can be found in [96, 110].

Steam generation by nanofluid under solar radiation has attracted intensive attention recently. Due to strong absorption of solar energy, NP-based solar vapor generation could have wide applications in many areas including desalination, sterilization and power generation. Steam generation of Au nanofluids under focused sunlight of 5 sun and 10 sun were performed. Results showed that localized energy trapping at the surface of nanofluid was responsible for the fast vapor generation [111]. The total efficiency reached 65% using a plasmonic Au nanofluid (178 ppm) under 10 sun, achieving a $\sim$ 300% enhancement in efficiency compared with the pure water [112]. Optimizing the range of nanofluid concentration and optical depth can be used for future solar vapor generator design. To further increase the sunlight intensity to 220 sun, experiment results coupled with the simulation model indicated that the initial stage of steam generation is mainly caused by localized boiling and vaporization in the superheated region due to highly non-uniform temperature and radiation energy distribution, albeit the bulk fluid is still subcooled [113]. A similar experiment under a sunlight intensity of 280 sun was also conducted to investigate the steam production phenomenon using Au nanofluids [114]. To further improve the solar evaporation, bubbles were also introduced into dilute plasmonic nanofluids to enhance solar water evaporation, which acted as light scattering centers to extend the incident light pathway and provided large gas–liquid interfaces for moisture capture as well as kinetic energy from bubble bursting to improve vapor diffusion [115]. Well-controlled experiments were performed to clarify the mechanism of the solar evaporation process using plasmonic Au nanofluid, carbon black nanofluid, and micro-sized porous medium. The results showed that Au nanofluids are not feasible for solar evaporation applications due to the high cost and low absorptance. High nanofluid concentration is needed to trap the solar energy in a thin layer at the liquid-gaseous interface, resulting in a local higher temperature and a higher evaporation rate [116, 117].

5. Conclusions and challenges

Plasmonic nanofluids show great interest to improve the absorption ability due to the surface plasmon resonance (SPR) around the NP surface. By designing the NP parameters (material, shape, and size) or base fluid, plasmonic nanofluids can either absorb or transmit specific solar spectrum and thus making nanofluids ideal candidates for various solar applications in full spectrum absorption in direct solar absorption collectors, selective absorption in solar PT/V systems, and local heating in solar evaporation. As discussed above, some efforts have been made to improve the solar thermal conversion applications of plasmonic nanofluids. Some challenges are still needed to overcome for the further development of plasmonic nanofluid applications. The first one is the stability of nanofluids including the long-time and high-temperature. Currently, many works were conducted in the lab and a great solar thermal conversion performance can be achieved in a short period. The performance of nanofluids in the actual applications should be considered. The second one is that the performance evaluation standard of nanofluids should be unified. Many experiments were conducted and compared in the unique experimental conditions by oneself and a more extensive evolution method is needed for different researchers to compare the performance of different nanofluids. The last one is the cost of plasmonic nanofluids. The common plasmonic metal NPs, such as: Au, Ag, Cu, are expensive in the actual applications. More cheap NPs, including the low cost of preparation processes and materials, should be developed in the further.

Acknowledgements

This work was financially supported by the Central South University and the National Natural Science Foundation of China (Grant No. 52006246).

Conflict of interest

The authors declare that they have no known competing financial interests or personal relationships that could have appeared to influence the work reported in this paper.

Author details

Meijie Chen*, Xingyu Chen and Dongling Wu
School of Energy Science and Engineering, Central South University, Changsha, China

*Address all correspondence to: chenmeijie@csu.edu.cn

References

[1] Chen M, He Y, Zhu J, Shuai Y, Jiang B, Huang Y. An experimental investigation on sunlight absorption characteristics of silver nanofluids. Sol Energy 2015;115:85–94. https://doi.org/10.1016/j.solener.2015.01.031.

[2] Wang X, He Y, Chen M, Hu Y. ZnO-Au composite hierarchical particles dispersed oil-based nanofluids for direct absorption solar collectors. Sol Energy Mater Sol Cells 2018;179:185–93. https://doi.org/10.1016/j.solmat.2017.11.012.

[3] Huang J, He Y, Chen M, Wang X. Separating photo-thermal conversion and steam generation process for evaporation enhancement using a solar absorber. Appl Energy 2019:244–52. https://doi.org/10.1016/j.apenergy.2018.11.090.

[4] Li H, He Y, Wang C, Wang X, Hu Y. Tunable thermal and electricity generation enabled by spectrally selective absorption nanoparticles for photovoltaic/thermal applications. Appl Energy 2019;236:117–26. https://doi.org/10.1016/j.apenergy.2018.11.085.

[5] Ye Q, Chen M, Cai W. Numerically investigating a wide-angle polarization-independent ultra-broadband solar selective absorber for high-efficiency solar thermal energy conversion. Sol Energy 2019;184:489–96. https://doi.org/10.1016/j.solener.2019.04.037.

[6] Chen M, He Y. Plasmonic nanostructures for broadband solar absorption based on the intrinsic absorption of metals. Sol Energy Mater Sol Cells 2018;188:156–63. https://doi. org/10.1016/j.solmat.2018.09.003.

[7] Otanicar TP, Phelan PE, Golden JS. Optical properties of liquids for direct absorption solar thermal energy systems. Sol Energy 2009;83:969–77. https://doi.org/10.1016/j.solener.2008.12.009.

[8] Qin C, Lee J, Lee BJ. A hybrid direct-absorption parabolic-trough solar collector combining both volumetric and surface absorption. Appl Therm Eng 2021;185:116333. https://doi.org/10.1016/j.applthermaleng.2020.116333.

[9] Chen M, He Y, Zhu J, Wen D. Investigating the collector efficiency of silver nanofluids based direct absorption solar collectors. Appl Energy 2016;181:65–74. https://doi.org/10.1016/j. apenergy.2016.08.054.

[10] Choi SUS. Enhancing thermal conductivity of fluids with nanoparticles. Am. Soc. Mech. Eng. Fluids Eng. Div. FED, vol. 231, 1995, p. 99–105.

[11] Qi C, Hu J, Liu M, Guo L, Rao Z. Experimental study on thermo-hydraulic performances of CPU cooled by nanofluids. Energy Convers Manag 2017;153:557–65. https://doi.org/10.1016/j.enconman.2017.10.041.

[12] Hu Y, He Y, Gao H, Zhang Z. Forced convective heat transfer characteristics of solar salt-based SiO2 nanofluids in solar energy applications. Appl Therm Eng 2019;155:650–9. https://doi.org/10.1016/j.applthermaleng.2019.04.109.

[13] Hazra SK, Ghosh S, Nandi TK. Photo-thermal conversion characteristics of carbon black-ethylene glycol nanofluids for applications in direct absorption solar collectors. Appl Therm Eng 2019;163:114402. https://doi.org/10.1016/j.applthermaleng.2019.114402.

[14] Guo C, Liu C, Jiao S, Wang R, Rao Z. Introducing optical fi ber as internal light source into direct absorption solar collector for enhancing photo-thermal conversion performance of MWCNT-H 2 O nano fl uids. Appl Therm Eng 2020; 173:115207. https://doi.org/10.1016/j.a pplthermaleng.2020.115207.

[15] Zhang H, Chen HJ, Du X, Wen D. Photothermal conversion characteristics of gold nanoparticle dispersions. Sol Energy 2014;100:141–7. https:// doi.org/ 10.1016/j.solener.2013.12.004.

[16] Prashant K. Jain, Xiaohua Huang IHE-S and MAE-S. Noble Metals on the Nanoscale: Optical and Photothermal Properties and Some Applications in Imaging, Sensing, Biology, and Medicine. Acc Chem Res 2008;41:1578–1586. https://doi.org/10.1021/a r7002804.

[17] Chen Z, Chen M, Yan H, Zhou P, Chen XY. Enhanced solar thermal conversion performance of plasmonic gold dimer nanofluids. Appl Therm Eng 2020;178:115561. https://doi.org/ 10.1016/j.applthermaleng.2020.115561.

[18] Li Y, Zhou J, Tung S, Schneider E, Xi S. A review on development of nanofluid preparation and characterization. Powder Technol 2009; 196:89–101. https://doi.org/10.1016/j. powt ec.2009.07.025.

[19] Chen HJ, Wen D. Ultrasonic-aided fabrication of gold nanofluids. Nanoscale Res Lett 2011;6:1–8. https:// doi.org/10.1186/1556-276X-6-198.

[20] Daniel MC, Astruc D. Gold Nanoparticles: Assembly, Supramolecular Chemistry, Quantum-Size-Related Properties, and Applications Toward Biology, Catalysis, and Nanotechnology. Chem Rev 2004; 104:293–346. https://doi.org/10.1021/cr030698+.

[21] Chen M, He Y, Huang J, Zhu J. Investigation into Au nanofluids for solar photothermal conversion. Int J Heat Mass Transf 2017;108:1894–900. https:// doi. org/ 10.1016/j.ijheatmasstra nsfer.2017.01.005.

[22] He Y, Chen M, Wang X, Hu Y. Plasmonic multi-thorny Gold nanostructures for enhanced solar thermal conversion. Sol Energy 2018; 171:73–82. https://doi.org/10.1016/j.sole ner.2018.06.071.

[23] FRENS G. Controlled Nucleation for the Regulation of the Particle Size in Monodisperse Gold Suspensions. Nat Phys Sci 1973;241:20–2. https://doi.org/ 10.1038/physci241020a0.

[24] Ji X, Song X, Li J, Bai Y, Yang W, Peng X. Size control of gold nanocrystals in citrate reduction: The third role of citrate. J Am Chem Soc 2007;129:13939–48. https://doi.org/10.1021/ja074447k.

[25] Bastús NG, Comenge J, Puntes V. Kinetically controlled seeded growth synthesis of citrate-stabilized gold nanoparticles of up to 200 nm: Size focusing versus ostwald ripening. Langmuir 2011;27:11098–105. https:// doi.org/10.1021/la201938u.

[26] Ojea-Jiménez I, Romero FM, Bastús NG, Puntes V. Small gold nanoparticles synthesized with sodium citrate and heavy water: Insights into the reaction mechanism. J Phys Chem C 2010;114:18 00–4.https://doi.org/10.1021/jp9091305.

[27] Xia Y, Xiong Y, Lim B, Skrabalak SE. Shape-controlled synthesis of metal nanocrystals: Simple chemistry meets complex physics? Angew Chemie - Int Ed 2009;48:60–103. https://doi.org/ 10.1002/anie.200802248.

[28] Wu HL, Kuo CH, Huang MH. Seed-mediated synthesis of gold nanocrystals with systematic shape evolution from cubic to trisoctahedral and rhombic dodecahedral structures. Langmuir 20 10;26:12307–13. https://doi.org/ 10.10 21/la1015065.

[29] Jana NR, Gearheart L, Murphy CJ. Evidence for seed-mediated nucleation in the chemical reduction of gold salts to gold nanoparticles. Chem Mater 2001; 13:2313–22. https://doi.org/10.10 21/cm000662n.

[30] Huang Y, Yang C, Lang J, Zhang S, Feng S, Schaefer LA, et al. Metal Nanoparticle Harvesting by Continuous Rotating Electrodeposition and Separation. Matter 2020;3:1294–307.https://doi.org/10.1016/j.matt. 2020.08.019.

[31] Rechberger W, Hohenau A, Leitner A, Krenn JR, Lamprecht B, Aussenegg FR. Optical properties of two interacting gold nanoparticles. Opt Commun 2003;220:137–41. https://doi. org/10.1016/S0030-4018(03)01357-9.

[32] Atwater HA, Polman A. Plasmonics for improved photovoltaic devices. Nat Mater 2010;9:205–13. https://doi.org/ 10.1038/nmat2629.

[33] Gonçalves MR, Makaryan T, Enderle F, Wiedemann S, Plettl A, Marti O, et al. Plasmonic nanostructures fabricated using nanosphere-lithography, soft-lithography and plasma etching. Beilstein J Nanotechnol 2011;2:448–58. https://doi.org/10.3762/bjnano.2.49.

[34] Yu F, Chen Y, Liang X, Xu J, Lee C, Liang Q, et al. Dispersion stability of thermal nanofluids. Prog Nat Sci Mater Int 2017;27:531–42. https://doi.org/10.1016/j.pnsc.2017.08.010.

[35] Taylor RA, Phelan PE, Adrian RJ, Gunawan A, Otanicar TP. Characterization of light-induced, volumetric steam generation in nanofluids. Int J Therm Sci 2012;56:1–11. https://doi.org/10.1016/j.ijthermalsc i.2012.01.012.

[36] Mahian O, Kianifar A, Kalogirou SA, Pop I, Wongwises S. A review of the applications of nanofluids in solar energy. Int J Heat Mass Transf 2013;57: 582–94. https://doi.org/10.1016/j.ijheatmasstransfer.2012.10.037.

[37] Yu W, France DM, Routbort JL, Choi SUS. Review and comparison of nanofluid thermal conductivity and heat transfer enhancements. Heat Transf Eng 2008;29:432–60. https://doi.org/10.10 80/01457630701850851.

[38] Wu D, Zhu H, Wang L, Liu L. Critical Issues in Nanofluids Preparation, Characterization and Thermal Conductivity. Curr Nanosci 2009;5:103–12. https://doi.org/10.2174/157341309787314548.

[39] Walshe J, Amarandei G, Ahmed H, McCormack S, Doran J. Development of poly-vinyl alcohol stabilized silver nanofluids for solar thermal applications. Sol Energy Mater Sol Cells 2019;201:110085. https://doi.org/ 10.1016/j.solmat.2019.110085.

[40] Lee R, Kim JB, Qin C, Lee H, Lee BJ, Jung GY. Synthesis of Therminol-based plasmonic nanofluids with core/shell nanoparticles and characterization of their absorption/scattering coefficients. Sol Energy Mater Sol Cells 2020; 209: 11044. https://doi.org/10.1016/j.solmat.2020.110442.

[41] Sharaf OZ, Rizk N, Joshi CP, Abi Jaoudé M, Al-Khateeb AN, Kyritsis DC, et al. Ultrastable plasmonic nanofluids in optimized direct absorption solar collectors. Energy Convers Manag 2019; 199:112010. https://doi.org/10.1016/j.enconman.2019.112010.

[42] Chen Y, Quan X, Wang Z, Lee C, Wang Z, Tao P, et al. Stably dispersed high-temperature Fe3O4/silicone-oil nanofluids for direct solar thermal energy harvesting. J Mater Chem A 2016;4: 17503–11. https://doi.org/ 10.1039/c6ta07773k.

[43] Singh N, Khullar V. Efficient Volumetric Absorption Solar Thermal Platforms Employing Thermally Stable - Solar Selective Nanofluids Engineered from Used Engine Oil. Sci Rep 2019;9:1–12. https: // doi. org/10. 1038/s41598-019-47126-3.

[44] Johnson PB, Christy RW. Optical constants of the noble metals. Phys Rev

B 1972;6:4370–9. https://doi.org/ 10.11 03/PhysRevB.6.4370.

[45] David C, García De Abajo FJ. Spatial nonlocality in the optical response of metal nanoparticles. J Phys Chem C 2011;115:19470–5. https:// doi.org/ 10.1021/jp204261u.

[46] Renger J, Quidant R, Van Hulst N, Novotny L. Surface-enhanced nonlinear four-wave mixing. Phys Rev Lett 2010; 104:046803. https://doi.org/10.1103/Ph ysRevLett.104.046803.

[47] Barrera RG, Fuchs R. Theory of electron energy loss in a random system of spheres. Phys Rev B 1995;52:3256–73. https://doi.org/10.1103/PhysRe vB.52.32 56.

[48] Link S, El-Sayed MA. Size and temperature dependence of the plasmon absorption of colloidal gold nanoparticles. J Phys Chem B 1999;103: 4212–7. https://doi.org/10.1021/jp9847 96o.

[49] Scholl JA, Koh AL, Dionne JA. Quantum plasmon resonances of individual metallic nanoparticles. Nature 2012;483:421–7. https://doi.org/ 10.1038/nature10904.

[50] Bohren CF. Absorption and scattering of light by small particles. 1983. https://doi.org/10.1088/ 0031-91 12/35/3/025.

[51] Wiscombe WJ. Improved Mie scattering algorithms. Appl Opt 1980;19: 1505. https://doi.org/10.1364/ao. 19.00 1505.

[52] Xu Y lin, Gustafson BÅS. A generalized multiparticle Mie-solution: Further experimental verification. J Quant Spectrosc Radiat Transf 2001;70: 395–419. https://doi.org/10. 1016/S0022-4073(01)00019-X.

[53] Purcell EM, Pennypacker CR. Scattering and Absorption of Light by Nonspherical Dielectric Grains. Astrophys J 1973;186:705. https://doi. org/10.1086/152538.

[54] García De Abajo FJ. Optical excitations in electron microscopy. Rev Mod Phys 2010;82:209–75. https://doi. org/10.1103/RevModPhys.82.209.

[55] Yee KS. Numerical Solution of Initial Boundary Value Problems Involving Maxwell's Equations in Isotropic Media. IEEE Trans Antennas Propag 1966;14: 302–7. https:// doi.org/ 10.1109/TAP.1966.1138693.

[56] Farjadpour A, Roundy D, Rodriguez A, Ibanescu M, Bermel P, Joannopoulos JD, et al. Improving accuracy by subpixel smoothing in the finite-difference time domain. Opt Lett 2006;31:2972. https://doi.org/10.1364/ ol.31.002972.

[57] Hao F, Nordlander P. Efficient dielectric function for FDTD simulation of the optical properties of silver and gold nanoparticles. Chem Phys Lett 2007;446:115–8. https://doi.org/ 10.10 16/j.cplett.2007.08.027.

[58] Salazar-Palma M, García-Castillo LE, Sarkar TK. The finite element method in electromagnetics. Eur. Congr. Comput. Methods Appl. Sci. Eng. ECCOMAS 2000, 2000.

[59] Zhao J, Pinchuk AO, Mcmahon JM, Li S, Ausman LK, Atkinson AL, et al. Methods for Describing the Electromagnetic Properties of Silver and Gold Nanoparticles. Chem Soc Rev 2008; 41:1710–20.

[60] Gonçalves MR. Plasmonic nanoparticles: Fabrication, simulation and experiments. J Phys D Appl Phys 2014;47:213001. https://doi.org/ 10.10 88/0022-3727/47/21/213001.

[61] Swinehart DF. The Beer-Lambert law. J Chem Educ 1962;39:333–5. https:// doi.org/10.1021/ed039p333.

[62] Qin C, Kang K, Lee I, Lee BJ. Optimization of the spectral absorption coefficient of a plasmonic nanofluid for a direct absorption solar collector. Sol Energy 2018;169:231–6. https://doi.org/10.1016/j.solener.2018.04.056.

[63] Cole JR, Halas NJ. Optimized plasmonic nanoparticle distributions for solar spectrum harvesting. Appl Phys Lett 2006;89:153120. https://doi.org/ 10.1063/1.2360918.

[64] Lee BJ, Park K, Walsh T, Xu L. Radiative heat transfer analysis in plasmonic nanofluids for direct solar thermal absorption. J Sol Energy Eng Trans ASME 2012;134. https://doi.org/10.1115/1.4005756.

[65] Seo J, Qin C, Lee J, Lee BJ. Tailoring the Spectral Absorption Coefficient of aBlended Plasmonic Nanofluid Using a CustomizedGenetic Algorithm. Sci Rep 2020;10:1–10. https://doi.org/10.1038/s41598-020-65811-6.

[66] Du M, Tang GH. Plasmonic nanofluids based on gold nanorods/nanoellipsoids/nanosheets for solar energy harvesting. Sol Energy 2016;137: 393–400. https://doi.org/10.1016/j.solener.2016.08.029.

[67] Mallah AR, Kazi SN, Zubir MNM, Badarudin A. Blended morphologies of plasmonic nanofluids for direct absorption applications. Appl Energy 2018; 229:505–21. https://doi.org/ 10.1016/j.apenergy.2018.07.113.

[68] Duan H, Xuan Y. Enhanced optical absorption of the plasmonic nanoshell suspension based on the solar photocatalytic hydrogen production system. Appl Energy 2014;114:22–9. https://doi.org/10.1016/j.apenergy.2013.09.035.

[69] Duan H, Xuan Y. Enhancement of light absorption of cadmium sulfide nanoparticle at specific wave band by plasmon resonance shifts. Phys E Low-Dimensional Syst Nanostructures 2011; 43:1475–80. https://doi.org/10.1016/j.physe.2011.04.010.

[70] Wang Z, Quan X, Zhang Z, Cheng P. Optical absorption of carbon-gold core-shell nanoparticles. J Quant Spectrosc Radiat Transf 2018;205:291–8. https://doi.org/10.1016/j. jqsrt.2017.08.001.

[71] Liu X, Xuan Y. Defects-assisted solar absorption of plasmonic nanoshell-based nanofluids. Sol Energy 2017;146: 503–10. https://doi.org/10.1016/j.solener.2017.03.024.

[72] Yu X, Xuan Y. Solar absorption properties of embellished GZO/Cu Janus nanoparticles. Energy Procedia 2019; 158:345–50. https://doi.org/10.1016/j.egypro.2019.01.100.

[73] Liu X, Xuan Y. Full-spectrum volumetric solar thermal conversion: Via photonic nanofluids. Nanoscale 2017;9: 14854–60. https://doi.org/10.1039/c7nr03912c.

[74] Qin C, Kim JB, Gonome H, Lee BJ. Absorption characteristics of nanoparticles with sharp edges for a direct-absorption solar collector. Renew Energy 2020;145:21–8. https://doi.org/10.1016/j.renene.2019.05.133.

[75] Chen M, Wang X, Hu Y, He Y. Coupled plasmon resonances of Au thorn nanoparticles to enhance solar absorption performance. J Quant Spectrosc Radiat Transf 2020;250: 107029. https://doi.org/10.1016/j. jqsrt.2020.107029.

[76] Chen M, He Y, Zhu J, Kim DR. Enhancement of photo-thermal conversion using gold nanofluids with different particle sizes. Energy Convers Manag 2016;112:21–30. https://doi.org/10.1016/j.enconman.2016.01.009.

[77] Chen M, He Y, Zhu J. Preparation of Au – Ag bimetallic nanoparticles for

enhanced solar photothermal conversion. Int J Heat Mass Transf 2017; 114:1098–104. https://doi.org/10.1016/j. ijheatmasstransfer.2017.07.005.

[78] Xuan Y, Duan H, Li Q. Enhancement of solar energy absorption using a plasmonic nanofluid based on TiO $_2$ / Ag composite nanoparticles. RSC Adv 2014;4:16206–13. https://doi.org/ 10.1039/C4RA00630E.

[79] Duan H, Xuan Y. Synthesis and optical absorption of Ag/CdS core/shell plasmonic nanostructure. Sol Energy Mater Sol Cells 2014;121:8–13. https://doi.org/10.1016/j.solmat.2013.10.011.

[80] Zeng J, Xuan Y, Duan H. Tin-silica-silver composite nanoparticles for medium-to-high temperature volumetric absorption solar collectors. Sol Energy Mater Sol Cells 2016;157: 930–6. https://doi.org/10.1016/j.solmat.2016.08.012.

[81] Yu X, Xuan Y. Investigation on thermo-optical properties of CuO/Ag plasmonic nanofluids. Sol Energy 2018; 160:200–7. https://doi.org/10.1016/j.sole ner.2017.12.007.

[82] Xuan Y, Duan H, Li Q. Enhancement of solar energy absorption using a plasmonic nanofluid based on TiO_2/Ag composite nanoparticles. RSC Adv 2014;4:16206–13. https://doi.org/ 10.1039/c4ra00630e.

[83] Chen M, He Y, Huang J, Zhu J. Synthesis and solar photo-thermal conversion of Au, Ag, and Au-Ag blended plasmonic nanoparticles. Energy Convers Manag 2016;127:293–300. https://doi. org / 10.1016 / j.enconman.2016.09.015.

[84] Duan H, Zheng Y, Xu C, Shang Y, Ding F. Experimental investigation on the plasmonic blended nanofluid for efficient solar absorption. Appl Therm Eng 2019;161:114192. https://doi.org/10.1016/j.applthermaleng.2019.114192.

[85] Qu J, Zhang R, Wang Z, Wang Q. Photo-thermal conversion properties of hybrid CuO-MWCNT/H_2O nanofluids for direct solar thermal energy harvest. Appl Therm Eng 2019;147:390–8. https: // doi. org / 10.1016/j.applthermaleng.2018.10.094.

[86] Mehrali M, Ghatkesar MK, Pecnik R. Full-spectrum volumetric solar thermal conversion via graphene/silver hybrid plasmonic nanofluids. Appl Energy 2018;224:103–15. https://doi.org/10.1016/j.apenergy.2018.04.065.

[87] Zeng J, Xuan Y. Enhanced solar thermal conversion and thermal conduction of MWCNT-SiO_2/Ag binary nanofluids. Appl Energy 2018;212:809–19. https: // doi. org / 10.1016/j.apenergy.2017.12.083.

[88] Jin X, Lin G, Zeiny A, Jin H, Bai L, Wen D. Solar photothermal conversion characteristics of hybrid nanofluids: An experimental and numerical study. Renew Energy 2019;141:937–49. https://doi.org/10.1016/j.renene.2019.04.016.

[89] Wang L, Zhu G, Wang M, Yu W, Zeng J, Yu X, et al. Dual plasmonic Au/TiN nanofluids for efficient solar photothermal conversion. Sol Energy 2019;184:240–8. https://doi.org/ 10.1016/j.solener.2019.04.013.

[90] Jeon J, Park S, Lee BJ. Analysis on the performance of a flat-plate volumetric solar collector using blended plasmonic nanofluid. Sol Energy 2016; 132:247–56. https://doi.org/10.1016/j.solener.2016.03.022.

[91] Wang TM, Tang GH, Du M. Photothermal conversion enhancement of triangular nanosheets for solar energy harvest. Appl Therm Eng 2020;173: 115182. https://doi.org/10.1016/j.applthe rmaleng.2020.115182.

[92] Chen M, He Y, Ye Q, Wang X, Hu Y. Shape-dependent solar thermal conversion properties of plasmonic Au

nanoparticles under di ff erent light fi lter conditions. Sol Energy 2019;182: 340_7. https: //doi. org/10.1016/j.solene r.2019.02.070.

[93] Huang J, Liu C, Zhu Y, Masala S, Alarousu E, Han Y, et al. Harnessing structural darkness in the visible and infrared wavelengths for a new source of light. Nat Nanotechnol 2016;11:60–6. https://doi.org/10.1038/nnano. 2015. 228.

[94] Zeng J, Xuan Y. Tunable full-spectrum photo-thermal conversion features of magnetic-plasmonic Fe_3O_4/TiN nanofluid. Nano Energy 2018;51: 754–63. https://doi.org/10.1016/j.nanoe n.2018.07.034.

[95] Sajid MU, Bicer Y. Nanofluids as solar spectrum splitters: A critical review. Sol Energy 2020;207:974–1001. https://doi.org/10.1016/j.solener. 2020. 07.009.

[96] Liang H, Wang F, Yang L, Cheng Z, Shuai Y. Progress in full spectrum solar energy utilization by spectral beam splitting hybrid PV / T system. Renew Sustain Energy Rev 2021;141:110785. https://doi.org/10.1016/j.rser.2021.110785.

[97] Polman A. Solar steam nanobubbles. ACS Nano 2013; 7(1): 15–18. https://doi. org/10.1021/nn305869y.

[98] Chen M, He Y, Zhu J, Wen D. Investigating the collector efficiency of silver nanofluids based direct absorption solar collectors. Appl Energy 2016;181: 65–74. https://doi.org/10.1016/j.apenerg y.2016.08.054.

[99] Jin H, Lin G, Bai L, Amjad M, Bandarra Filho EP, Wen D. Photothermal conversion efficiency of nanofluids: An experimental and numerical study. Sol Energy 2016;139: 278–89. https://doi.org/10.1016/j.solene r.2016.09.021.

[100] Zeiny A, Jin H, Bai L, Lin G, Wen D. A comparative study of direct absorption nanofluids for solar thermal applications. Sol Energy 2018;161:74–82. https://doi. org/10.1016/j.solener.2017.12.037.

[101] Amjad M, Jin H, Du X, Wen D. Experimental photothermal performance of nanofluids under concentrated solar flux. Sol Energy Mater Sol Cells 2018;182:255–62. https://doi.org/10.101 6/j.solmat.2018.03.044.

[102] Bandarra Filho EP, Mendoza OSH, Beicker CLL, Menezes A, Wen D. Experimental investigation of a silver nanoparticle-based direct absorption solar thermal system. Energy Convers Manag 2014;84:261–7. https://doi.org/ 10.1016/j.enconman.2014.04.009.

[103] Qin C, Kang K, Lee I, Lee BJ. Optimization of a direct absorption solar collector with blended plasmonic nanofluids. Sol Energy 2017;150:512–20. https : // doi. org / 10. 1016 /j.solene r.2017.05.007.

[104] Qin C, Kim JB, Lee J, Lee BJ. Comparative analysis of direct-absorption parabolic-trough solar collectors considering concentric nanofluid segmentation. Int J Energy Res 2020;44: 4015–4025. https://doi.org/ 10.1002/ er.5165.

[105] Qin C, Kim JB, Lee BJ. Performance analysis of a direct-absorption parabolic-trough solar collector using plasmonic nanofluids. Renew Energy 2019;143:24–33. https://doi. org/10.1016/ j.renene.20 19.04.146.

[106] Jin H, Lin G, Guo Y, Bai L, Wen D. Nanoparticles enabled pump-free direct absorption solar collectors. Renew Energy 2020;145:2337–44. https:// doi. org/10.1016/j.renene.2019.07.108.

[107] Du M, Tang GH, Wang TM. Exergy analysis of a hybrid PV/T system based on plasmonic nanofluids and silica aerogel glazing. Sol Energy 2019;183: 501–11. https: // doi.org /10.1016/j.solene r.2019.03.057.

[108] Han X, Chen X, Wang Q, Alelyani SM, Qu J. Investigation of $CoSO_4$-based Ag nanofluids as spectral beam splitters for hybrid PV/T applications. Sol Energy 2019;177:387–94. https://doi.org/ 10.1016/j.solener. 2018.11.037.

[109] Han X, Chen X, Sun Y, Qu J. Performance improvement of a PV/T system utilizing Ag/$CoSO_4$-propylene glycol nanofluid optical filter. Energy 2020;192:116611. https://doi.org/ 10.10 16/j.energy.2019.116611.

[110] Abdelrazik AS, Al-Sulaiman FA, Saidur R, Ben-Mansour R. A review on recent development for the design and packaging of hybrid photovoltaic/ thermal (PV/T) solar systems. Renew Sustain Energy Rev 2018;95:110_29. https://doi.org/10.1016/j.rser. 2018. 07. 013.

[111] Jin H, Lin G, Zeiny A, Bai L, Wen D. Nanoparticle-based solar vapor generation: An experimental and numerical study. Energy 2019;178:447–59. https://doi.org/10.1016/j.energy. 2019. 04. 085.

[112] Wang X, He Y, Liu X, Shi L, Zhu J. Investigation of photothermal heating enabled by plasmonic nanofluids for direct solar steam generation. Sol Energy 2017;157:35–46. https://doi.org/ 10.1016/j.solener.2017.08.015.

[113] Jin H, Lin G, Bai L, Zeiny A, Wen D. Steam generation in a nanoparticle-based solar receiver. Nano Energy 20 16;28:397–406. https://doi. org/10.10 16/j.nanoen.2016.08.011.

[114] Amjad M, Raza G, Xin Y, Pervaiz S, Xu J, Du X, et al. Volumetric solar heating and steam generation via gold nanofluids. Appl Energy 2017;206:393–400. https://doi.org/10.1016/j.apenergy. 2017.08.144.

[115] Yao G, Xu J, Liu G. Solar steam generation enabled by bubbly flow nanofluids. Sol Energy Mater Sol Cells 2020;206:110292. https://doi.org/ 10.10 16/j.solmat.2019.110292.

[116] Zeiny A, Wen D. Nanofluids-based and porous media-based solar evaporation: A comparative study. AIP Conf. Proc., vol. 2123, 2019, p. 020088. https://doi.org/10.1063/1.5117015.

[117] Zeiny A, Jin H, Lin G, Song P, Wen D. Solar evaporation via nanofluids: A comparative study. Renew Energy 2018;122:443–54. https://doi. org/10.1016/j.renene.2018.01.043.

Permissions

All chapters in this book were first published by InTech Open; hereby published with permission under the Creative Commons Attribution License or equivalent. Every chapter published in this book has been scrutinized by our experts. Their significance has been extensively debated. The topics covered herein carry significant findings which will fuel the growth of the discipline. They may even be implemented as practical applications or may be referred to as a beginning point for another development.

The contributors of this book come from diverse backgrounds, making this book a truly international effort. This book will bring forth new frontiers with its revolutionizing research information and detailed analysis of the nascent developments around the world.

We would like to thank all the contributing authors for lending their expertise to make the book truly unique. They have played a crucial role in the development of this book. Without their invaluable contributions this book wouldn't have been possible. They have made vital efforts to compile up to date information on the varied aspects of this subject to make this book a valuable addition to the collection of many professionals and students.

This book was conceptualized with the vision of imparting up-to-date information and advanced data in this field. To ensure the same, a matchless editorial board was set up. Every individual on the board went through rigorous rounds of assessment to prove their worth. After which they invested a large part of their time researching and compiling the most relevant data for our readers.

The editorial board has been involved in producing this book since its inception. They have spent rigorous hours researching and exploring the diverse topics which have resulted in the successful publishing of this book. They have passed on their knowledge of decades through this book. To expedite this challenging task, the publisher supported the team at every step. A small team of assistant editors was also appointed to further simplify the editing procedure and attain best results for the readers.

Apart from the editorial board, the designing team has also invested a significant amount of their time in understanding the subject and creating the most relevant covers. They scrutinized every image to scout for the most suitable representation of the subject and create an appropriate cover for the book.

The publishing team has been an ardent support to the editorial, designing and production team. Their endless efforts to recruit the best for this project, has resulted in the accomplishment of this book. They are a veteran in the field of academics and their pool of knowledge is as vast as their experience in printing. Their expertise and guidance has proved useful at every step. Their uncompromising quality standards have made this book an exceptional effort. Their encouragement from time to time has been an inspiration for everyone.

The publisher and the editorial board hope that this book will prove to be a valuable piece of knowledge for researchers, students, practitioners and scholars across the globe.

List of Contributors

Hamed Ghodsi and Hassan Kaatuzian
Photonics Research Laboratory (PRL), Electrical Engineering Department, Amirkabir University of Technology (Tehran Polytechnique), Tehran, Iran

Boris I. Lembrikov, David Ianetz and Yossef Ben Ezra
Department of Electrical Engineering and Electronics, Holon Institute of Technology (HIT), Holon, Israel

Chuchuan Hong and Justus Chukwunonso Ndukaife
Department of Electrical Engineering and Computer Science, Vanderbilt University, Nashville, TN, USA
Vanderbilt Institute of Nanoscale Science and Engineering, Vanderbilt University, Nashville, TN, USA

Sen Yang
Vanderbilt Institute of Nanoscale Science and Engineering, Vanderbilt University, Nashville, TN, USA
Interdisciplinary Materials Science, Vanderbilt University, TN, USA

Shiva Hayati Raad and Zahra Atlasbaf
Department of Electrical and Computer Engineering, Tarbiat Modares University, Tehran, Iran

Mauro Cuevas
Faculty of Engineering and Information Technology, University of Belgrano, Buenos Aires, Argentina

Karlo Queiroz da Costa, Gleida Tayanna Conde de Sousa, Paulo Rodrigues Amaral, Tiago Dos Santos Garcia and Pitther Negrão dos Santos
Department of Electrical Engineering, Federal University of Para, Belém-PA, Brazil

Janilson Leão Souza
Federal Institute of Education, Science and Technology of Para, Tucuruí-PA, Brazil

Rishabh Rastogi, Matteo Beggiato and Sivashankar Krishnamoorthy
Materials Research and Technology Department, Luxembourg Institute of Science and Technology, Belvaux, Luxembourg

Pierre Michel Adam
Institute Charles Delaunay CNRS, Physics, Mechanics, Materials and Nanotechnology, Department (PM2N), Nanotechnologies, Light, Nanomaterials and Nanotechnology team (L2N), University of Technology of Troyes (UTT), Troyes Cedex, France

Saulius Juodkazis
Faculty of Science, Engineering and Technology, Swinburne University of Technology, Hawthorn, VIC, Australia

Rajkumar Devasenathipathy, De-Yin Wu and Zhong-Qun Tian
State Key Laboratory of Physical Chemistry of Solid Surfaces, Department of Chemistry, College of Chemistry and Chemical Engineering, Xiamen University, Xiamen, China

Tao Dong, Yue Xu and Jingwen He
State Key Laboratory of Space-Ground Integrated Information Technology, Beijing, China
Beijing Institute of Satellite Information Engineering, Beijing, China

Andrii Iurov
Department of Physics and Computer Science, Medgar Evers College of the City University of New York, Brooklyn, NY, USA

Godfrey Gumbs
Department of Physics and Astronomy, Hunter College of the City University of New York, NY, USA
Donostia International Physics Center (DIPC), San Sebastian, Basque Country, Spain

Danhong Huang
Air Force Research Laboratory, Space Vehicles Directorate, Kirtland Air Force Base, NM, USA

Yu Liu, Wenjie Chen, Linyuan Zhao, Weijun Chen, Jun Xu and Renrong Liang
Institute of Microelectronics, Tsinghua National Laboratory for Information Science and Technology (TNList), Tsinghua University, Beijing, China

Junxiong Guo
State Key Laboratory of Electronic Thin Films and Integrated Devices, School of Electronic Science and Engineering (National Exemplary School of Microelectronics), University of Electronic Science and Technology of China, Chengdu, China

Ricardo Téllez-Limón
CONACYT - Center for Scientific Research and Higher Education at Ensenada (CICESE), Unit Monterrey, Nuevo Leon, Mexico

Rafael Salas-Montiel
L2n - Laboratory Light, Nanomaterials and Nanotechnologies, CNRS ERL 7004 and University of Technology at Troyes, Troyes, France

Jianfeng Jiang
School of Microelectronics, Shandong University, Jinan, China

Meijie Chen, Xingyu Chen and Dongling Wu
School of Energy Science and Engineering, Central South University, Changsha, China

Index